Progress in the Chemistry of Organic Natural Products

The volumes of this classic series, now referred to simply as "Zechmeister" after its founder, Laszlo Zechmeister, have appeared under the Springer Imprint ever since the series' inauguration in 1938. It is therefore not really surprising to find out that the list of contributing authors, who were awarded a Nobel Prize, is quite long: Kurt Alder, Derek H.R. Barton, George Wells Beadle, Dorothy Crowfoot-Hodgkin, Otto Diels, Hans von Euler-Chelpin, Paul Karrer, Luis Federico Leloir, Linus Pauling, Vladimir Prelog, with Walter Norman Haworth and Adolf F.J. Butenandt serving as members of the editorial board.

The volumes contain contributions on various topics related to the origin, distribution, chemistry, synthesis, biochemistry, function or use of various classes of naturally-occurring substances ranging from small molecules to biopolymers.

Each contribution is written by a recognized authority in the field and provides a comprehensive and up-to-date review of the topic in question. Addressed to biologists, technologists, and chemists alike, the series can be used by the expert as a source of information and literature citations and by the non-expert as a means of orientation in a rapidly developing discipline.

All contributions are listed in PubMed.

Progress in the Chemistry of Organic Natural Products

Volume 122

A. Douglas Kinghorn · Heinz Falk ·
Simon Gibbons · Yoshinori Asakawa · Ji-Kai Liu ·
Verena M. Dirsch
Editors

Progress in the Chemistry of Organic Natural Products 122

Botanical Dietary Supplements and Herbal Medicines

By Eric D. Salinas-Arellano, Ines Y. Castro-Dionicio, Jonathan G. Jeyaraj, Nathan P. Mirtallo Ezzone, Esperanza J. Carcache de Blanco, Ahmed Osman, Amar G. Chittiboyina, Bharathi Avula, Zulfiqar Ali, Sebastian J. Adams, Ikhlas A. Khan, Roy Upton, Natascha Techen and Iffat Parveen

Springer

Editors
A. Douglas Kinghorn
College of Pharmacy
Ohio State University
Columbus, OH, USA

Heinz Falk
Institute of Organic Chemistry
Johannes Kepler University of Linz
Linz, Austria

Simon Gibbons
Centre for Natural Products Discovery
Liverpool John Moores University
Liverpool, UK

Yoshinori Asakawa
Faculty of Pharmaceutical Sciences
Tokushima Bunri University
Tokushima, Japan

Ji-Kai Liu
School of Pharmaceutical Sciences
South Central University for Nationalities
Wuhan, China

Verena M. Dirsch
Department of Pharmaceutical Sciences
University of Vienna
Vienna, Austria

ISSN 2191-7043 ISSN 2192-4309 (electronic)
Progress in the Chemistry of Organic Natural Products
ISBN 978-3-031-26767-3 ISBN 978-3-031-26768-0 (eBook)
https://doi.org/10.1007/978-3-031-26768-0

This Springer imprint is published by the registered company Springer Nature Switzerland AG
The registered company address is: Gewerbestrasse 11, 6330 Cham, Switzerland

Preface

In 1994, the Dietary Supplement Health and Education Act was passed by the United States Congress. The main purposes of this legislation were to increase consumer access to dietary supplements taken orally, such as vitamins, minerals, amino acids, and herbs and other botanicals, and also to enhance the regulation and labeling requirements for these products. The Center of Food Safety and Applied Nutrition of the United States Food and Drug Administration oversees both dietary supplement products themselves and their ingredients. Dietary supplements are not required to undergo any formal Food and Drug Administration approval process before marketing, although they can be withdrawn for sale, particularly if a lack of safety is demonstrated [1].

Botanical dietary supplements, which are referred to also as "herbals" and "botanicals", represent about 20% of the total dietary supplement industry sales in the United States [1]. Many of the plants used as dietary supplements are the same as those used in Europe and elsewhere as regulated phytomedicines [1, 2]. From recently published data, the total value of the botanical segment of the market in the United States was $12,350 billion in 2021, with the continued herbal sales growth focused on their purported value in maintaining immune and digestive health, in addition to their use in mood support and energy-enhancing products [3]. In the United States, scientific research on botanical products is spearheaded by the Office of Dietary Supplements [2] and the National Center for Complementary and Integrative Health (NCCIH) [4], both of the National Institutes of Health. Monographs to guide the standardization and quality control of botanical products are developed annually by personnel of the "United States Pharmacopeia" [5].

However, certain problems are evident concerning the current use of botanical dietary supplements. Thus, the increased consumption of these herbal products has led to reports of their toxicity when ingested in excessive amounts [1]. Also, products that are sold to the public may be mislabeled [6] or adulterated [7], while others have been found to contain illicit synthetic drugs in their formulations [8].

In this volume, four chapters are included from experienced investigators working in the botanical dietary supplement field. In the first chapter "Phytochemical Profiles and Biological Studies of Selected Botanical Dietary Supplements Used in the United States", Eric D. Salinas-Arellano and co-authors have provided detailed information on the biologically active secondary metabolites of some 30 botanicals that have a wide current use in the United States, with concomitant descriptions of recent in vitro and in vivo biological activities of these same constituents. The structural range of the compounds covered in this chapter is extensive, and the range of technical journals cited where relevant literature reports have appeared is extremely broad. In the second chapter "Quality Consistency of Herbal Products: Chemical Evaluation", contributed by Ahmed Osman and colleagues, an updated survey of the different chromatographic, spectroscopic, and metabolomics techniques that can be used for the quality control of botanical products is provided. The third chapter "Nomenclature: Herbal Taxonomy in the Global Commerce of Botanicals" of this volume, by Roy Upton, delves into the importance of plant nomenclature in the correct taxonomic identification of source plants used in botanical preparations. In the final chapter "Deoxyribonucleic Acid Barcoding for the Identification of Botanicals", Natascha Techen and her associates discuss the use of a newer approach to species identification for plants used as dietary supplements, using deoxyribonucleic molecular barcoding techniques. It is intended that this volume will provide a timely account of contemporary methods that are being utilized for the scientific study of botanical dietary supplements and herbal medicines.

Mexico City, Mexico — Rachel Mata
Columbus, OH, USA — A. Douglas Kinghorn
Linz, Austria — Heinz Falk
Liverpool, UK — Simon Gibbons
Tokushima, Japan — Yoshinori Asakawa
Wuhan, China — Ji-Kai Liu
Vienna, Austria — Verena M. Dirsch

References

1. Carcache de Blanco EJ, Kinghorn AD (2021) Botanical dietary products. In: Adejare A. (ed), Remington: The science and practice of pharmacy, 23rd edn. Academic Press, San Diego, p 45
2. Durazzo A, Sorkin BC, Lucarini M, Gusev PA, Kuszak AJ, Crawford C, Boyd C, Deuster PA, Saldanha LG, Gurley BJ, Pehrsson PR, Harnly JM, Turrini A, Andrews KW, Lindsey AT, Heinrich M, Dwyer JT (2022) Analytical challenges and metrological approaches to ensuring dietary supplement quality: international perspectives. Front Pharmacol 12:714434
3. Smith T, Resetar H, Morton C (2022) US sales of herbal supplements increase by 9.7% in 2021. HerbalGram, issue 136:42
4. Still P, Chen W, Weber W, Hopp DC (2022) NCCIH priorities for natural products research. Planta Med 88:698
5. Lu Z, Handy SM, Zhang N, Zheng Q, Xu Q, Ambrose M, Giancaspro G, Sarma ND. (2022) Development and validation of a species-specific PCR method for the identification of ginseng species using orthogonal approaches. Planta Med 88:1004

6. Crawford C, Avula B, Lindsey AT, Walter MS, Katragunta K, Khan I A, Duester PA. (2022). Analysis of select dietary supplement products marketed to support or boost the immune system. JAMA Netw Open 5:e2226040
7. Cohen PA, Avula B, Khan I (2022) The unapproved drug centrophenoxine (meclofenoxate) in cognitive enhancement dietary supplements. Clin Toxicol 60:1156
8. Gafner S, Blumenthal M, Foster S, Cardellina II JH, Khan IA, Upton R. Botanical ingredient forensics: Detection of attempts to deceive commonly used analytical methods for authenticating herbal dietary and food ingredients and supplements (2023) J Nat Prod 86:460

Contents

Contributors

Sebastian J. Adams School of Pharmacy, National Center for Natural Products Research, The University of Mississippi, University, MS, USA

Zulfiqar Ali School of Pharmacy, National Center for Natural Products Research, The University of Mississippi, University, MS, USA

Bharathi Avula School of Pharmacy, National Center for Natural Products Research, The University of Mississippi, University, MS, USA

Esperanza J. Carcache de Blanco Medicinal Chemistry and Pharmacognosy, College of Pharmacy, The Ohio State University, Columbus, OH, USA

Ines Y. Castro-Dionicio Medicinal Chemistry and Pharmacognosy, College of Pharmacy, The Ohio State University, Columbus, OH, USA

Amar G. Chittiboyina School of Pharmacy, National Center for Natural Products Research, The University of Mississippi, University, MS, USA

Jonathan G. Jeyaraj Medicinal Chemistry and Pharmacognosy, College of Pharmacy, The Ohio State University, Columbus, OH, USA

Ikhlas A. Khan School of Pharmacy, National Center for Natural Products Research, The University of Mississippi, University, MS, USA

Nathan P. Mirtallo Ezzone Medicinal Chemistry and Pharmacognosy, College of Pharmacy, The Ohio State University, Columbus, OH, USA

Ahmed Osman School of Pharmacy, National Center for Natural Products Research, The University of Mississippi, University, MS, USA

Iffat Parveen School of Pharmacy, National Center for Natural Product Research, The University of Mississippi, University, MS, USA

Eric D. Salinas-Arellano Medicinal Chemistry and Pharmacognosy, College of Pharmacy, The Ohio State University, Columbus, OH, USA

Natascha Techen School of Pharmacy, National Center for Natural Product Research, The University of Mississippi, University, MS, USA

Roy Upton American Herbal Pharmacopoeia, Scotts Valley, CA, USA

Phytochemical Profiles and Biological Studies of Selected Botanical Dietary Supplements Used in the United States

Eric D. Salinas-Arellano, Ines Y. Castro-Dionicio, Jonathan G. Jeyaraj, Nathan P. Mirtallo Ezzone, and Esperanza J. Carcache de Blanco

Contents

E. D. Salinas-Arellano · I. Y. Castro-Dionicio · J. G. Jeyaraj · N. P. Mirtallo Ezzone ·
E. J. Carcache de Blanco (✉)
Medicinal Chemistry and Pharmacognosy, College of Pharmacy, The Ohio State University, 500 West 12th Avenue, Columbus, OH 43210, USA
e-mail: carcachedeblanco.1@osu.edu

E. D. Salinas-Arellano
e-mail: salinas.80@osu.edu

I. Y. Castro-Dionicio
e-mail: castro.206@osu.edu

J. G. Jeyaraj
e-mail: jeyaraj.5@buckeyemail.osu.edu

N. P. Mirtallo Ezzone
e-mail: ezzone.4@buckeyemail.osu.edu

A. D. Kinghorn, H. Falk, S. Gibbons, Y. Asakawa, J.-K. Liu, V. M. Dirsch (eds.), *Progress in the Chemistry of Organic Natural Products 122*, Progress in the Chemistry of Organic Natural Products 122, https://doi.org/10.1007/978-3-031-26768-0_1

1 Introduction

Nature has been the source of many current drug treatments. Several World Health Organization (WHO) reports indicate that a large part of the global population chooses traditional herbal preparations as a first option in the event of illness [1–3]. In addition, the consumption of botanical dietary supplements, as complementary and alternative therapies, continues to expand worldwide [2, 4, 5]. For example, WHO recognizes the importance of herbal preparations and has provided guidance on several occasions to improve the regulatory aspects of medicinal plants. In 1988, the WHO published a document titled "Quality Control Methods for Medicinal Plants Materials", to ensure the quality of crude drugs and herbal preparations. This publication established recommendations and procedures for quality assurance. In 2013, the WHO also published a further document, "WHO Strategy on Traditional Medicine 2014–2023". The main objectives of this document were (a) to support member countries in harnessing traditional medicine to achieve wellness and primary care for their health problems; and (b) promote the safe and effective use of traditional medicines by regulating products, practices and professionals. Thus, the focus of this document was to strengthen the safety, quality, and efficacy of herbal products through regulation [1].

Many countries regulate the medicinal and food uses of herbal food supplements, and some others have specific requirements for this type of product since typically they are not classified as drugs. The European Union regulates the medicinal uses of botanical dietary supplements through the directive on Traditional Herbal Medicinal Products and Food Supplements under the Food Supplements Directive 2002/46/EC [4, 6, 7]. In the United States, the Dietary Supplement and Health Education Act of 1994, as regulated by the Food and Drug Administration (FDA), dietary supplements are regulated as foods, rather than as drugs. Due to increases in sales of dietary supplements, the FDA established the Office of Dietary Supplement Programs (ODSP) to increase surveillance. Product manufacturers must now submit any new dietary ingredient notification to the ODSP for review. Furthermore, botanical dietary supplements must be prepared using good manufacturing practices (GMPs) and must be certified for safety [4, 6, 7].

Botanical dietary supplements typically are composed of complex mixtures of different parts of the same species, so the composition and concentration of their chemical constituents may vary depending on exogenous (climate, altitude, and soil nutrients) and endogenous (genetic and epigenetic) factors. Therefore, quality control and quality assurance are essential to ensure the efficacy and safety of herbal preparations. The quality control of herbal drugs and derivatives aims to guarantee their identity, purity, and content in active principles and/or markers and to be able to obtain reproducibility of the safety and efficacy parameters of herbal products [6, 8]. Quality assurance of botanical dietary supplements requires a good integrated system of GMPs, storage, distribution, and quality control to ensure consumer protection. The safety and quality of herbal food supplements are closely related to the state of knowledge on the pharmacological activity of the phytochemical components of a plant extract [6, 8].

Efficacy and safety of herbal dietary supplements are important parameters that require studies such as dose standardization, homogenization of herbal products, identification of marker compounds, quantification of active compounds, pharmacokinetic parameters of the molecules found in the herbal preparation, and identification of toxic metabolites and/or adverse effects [5, 7]. Botanical, chemical and genetic tests allow for the identification of crude drugs. Botanical taxonomic identification is performed at the macroscopic level by making a reliable comparison with voucher specimens deposited in a herbarium. This analysis takes into account the shape, size, color, odor, surface characteristics, texture and appearance of the cut surface of the plant. Microscopic inspection is indispensable for the identification of plant material and is an initial test for impurities [1, 8, 9]. In turn, purity tests seek to ensure that the plant is free of contaminants, such as metals, aflatoxins, radioactivity, pesticides, foreign plant material, mold, insects and other chemical impurities [8–10].

Chemical identity assays include characteristic reactions of plant extracts using chromatographic techniques that could be hyphenated with mass spectrometry. The evaluation of chemical identity is mainly focused on the detection of marker compounds and obtaining chromatographic profiles that serve as a "fingerprint" and provide a comprehensive representation of the composition, distribution and concentration of the chemical compounds present in the plant preparations [6, 8]. Marker compounds are defined as chemical constituents, useful for analytical analysis, where they may or may not exhibit biological activity. Currently, the types of markers that can be identified in crude drugs or herbal preparations are specified in monographs and pharmacopoeias. Chromatographic profiles are generated by several chromatographic techniques that include thin-layer chromatography (TLC), high-performance liquid chromatography (HPLC), and gas chromatography (GC). Other available techniques to obtain the chemical profile are mass spectrometry, infrared spectroscopy, and metabolomic profiling and quantitative determination based on nuclear magnetic resonance spectra [3, 6]. (See Chapter "Quality Consistency of Herbal Products: Chemical Evaluation", in this volume.)

In summary, the worldwide use of medicinal plants or food supplements suggests they are highly useful in health preservation. Therefore, it is essential to source appropriately medicinal plants, authenticate plant material, characterize isolated bioactive compounds, determine the chemical composition, and perform clinical, and toxicological studies of the components in traditional preparations, which are indispensable to ensure the efficacy and safety of herbal products.

2 Overview of Botanical Dietary Supplements Used in the United States

In the United States, the Dietary Supplement and Health Education Act of 1994 requires all new products to be registered. However, many products with a long history of use are available to the U.S. population. Since the mid-1990s there has been an upward trend in the use of botanical dietary supplements (BDS) by the general public, for research, and in clinical studies [11]. This was reflected by rising annual sales of BDS from \$2.9 billion in 1995 to \$11.3 billion in 2020 in the United States as reported by the "Nutrition Business Journal" [11, 12]. In 2020, there was an increase in sales by 17.3% from 2019, according to this same publication. The mass market channel increased by 25.1% from 2019 and totaled \$2.131 billion in 2020, which showed the strongest growth in 2020. This surpassed twice the sales growth of 9.4% in this channel from 2018 to 2019. Direct sales of herbal supplements increased by 23.7% in 2020, which included online sales and was more than double the percentage growth of 11.5% seen in 2019. Sales in the so-called natural, health food, and specialty channel totaled \$2.950 billion in 2020, an increase of 1.6%. Despite the moderate sales growth in this latter category, total sales in this channel have been higher than in the mass market channel since at least 2005 [12].

Top ten selling botanical dietary supplements in 2020 in the U.S. Mainstream Multi-Outlet channel include elderberry (*Sambucus nigra* and *S. canadensis*), horehound (*Marrubium vulgare*), cranberry (*Vaccinium macrocarpon*), turmeric (*Curcuma longa*), apple cider vinegar (*Malus* spp.), ginger (*Zingiber officinale*), Echinacea (*Echinacea* spp.), garlic (*Allium sativum*), fenugreek (*Trigonella foenum-graecum*), and wheatgrass/barley grass (*Triticum aestivum/Hordeum vulgare*) [12]. According to the same source, the top ten selling botanicals in the U.S. Natural Channel included cannabidiol/CBD (*Cannabis sativa*), elderberry (*Sambucus nigra* and *S. canadensis*), turmeric (*Curcuma longa*), wheatgrass/barley grass (*Triticum aestivum/Hordeum vulgare*), mushrooms, aloe (*Aloe vera*), Ashwagandha (*Withania somnifera*), oregano (*Origanum vulgare*), Echinacea (*Echinacea* spp.), and flax seed/flax oil (*Linum usitatissimum*) [12]. Some of these species not only have a long history of use, but several studies have also validated their utilization and have provided clinical evidence of efficacy. However, others lack enough scientific and clinical evidence to properly support their use. The increase in consumption and popularity of these botanical products has elicited a greater need to assess and ensure their safety and overall quality to protect consumers, particularly for botanicals with prior limited studies [11].

An additional fifteen botanicals that were among the most widely sold in 2020 by the U.S. Mainstream Multi-Outlet channel include saw palmetto (*Serenoa repens*), Ashwagandha (*Withania somnifera*), green tea (*Camellia sinensis*), ivy leaf (*Hedera helix*), ginkgo (*Ginkgo biloba*), cannabidiol/CBD (*Cannabis sativa*), black cohosh (*Actaea racemosa*), beta-sitosterol, red yeast rice (*Oryza sativa*), aloe (*Aloe vera*), St John's wort (*Hypericum perforatum*), flax seed/flax oil (*Linum usitatisimum*), milk thistle (*Silybum marianum*), yohimbe (*Pausinystalia johimbe*, syn. *Corynanthe johimbe*), and goji berry (*Lycium* spp.) According to the same source, the additional top fifteen selling botanicals in the U.S. Natural Channel added to the list of other botanicals included an Echinacea-goldenseal combination (*Echinacea* spp./*Hydrastis canadensis*), cranberry (*Vaccinium macrocarpon*), garlic (*Allium sativum*), maca (*Lepidium meyenii*), valerian (*Valeriana officinalis*), nigella (*Nigella sativa*), quercetin, chlorophyll/chlorella (*Chlorella vulgaris*), horsetail (*Equisetum* spp.), reishi mushrooms (*Ganoderma lucidum*), beet root (*Beta vulgaris*), and apple cider vinegar (*Malus* spp.) [12]. Similarly, as in the top ten products, some of these products have supportive studies for their use while others lack enough scientific and clinical support. Moreover, *Cannabis sativa* and its bioactive constituents are listed in both groups above a dietary supplement. However, the present position of the Food and Drug Administration (FDA) states: "Based on available evidence, FDA has concluded that THC and CBD products are excluded from the dietary supplement definition under section 201(ff)(3)(B) of the FD&C Act (21 U.S.C. § 321(ff)(3)(B))". Under that provision, if a substance (such as THC or CBD) is an active ingredient in a drug product that has been approved under section 505 of the FD&C Act (21 U.S.C. § 355), or has been authorized for investigation as a new drug for which substantial clinical investigations have been instituted and for which the existence of such investigations has been made public, then products containing that substance are excluded from the definition of a dietary supplement. The FDA considers a substance to be "authorized for investigation as a new drug" if it is the subject of an Investigational New Drug application (IND) that has gone into effect. Under FDA's regulations (21 CFR 312.2), unless a clinical investigation meets the limited criteria in that regulation, an IND is required for all clinical investigations of products that are subject to section 505 of the FD&C Act" [13]. The above statement is supported by the following argument also released by the FDA as follows: "When a substance is excluded from the dietary supplement definition under section 201(ff)(3)(B) of the FD&C Act, the exclusion applies unless FDA, in the agency's discretion, has issued a regulation, after notice and comment, finding that the article would be lawful under the FD&C Act. To date, no such regulation has been issued for any substance" [13]. Due to the present FDA position on the status of *Cannabis sativa* as a dietary supplement and on the drug status of some of the constituents isolated from *C. sativa* and that have been investigated in depth, Sects. 4 and 5 of this book chapter will not include studies on the bioactive constituents of this botanical [14, 15]. Thus, this chapter summarizes the main studies that validate the use of the most consumed botanical products that have a long history of use by the U.S. population, both with and without adequate clinical evidence.

3 Chemotaxonomy of Botanicals Used as Dietary Supplements

Chemotaxonomy is a field of study for which the practical application began in the early twentieth century, with the aim of classifying a group of organisms according to profiles of their chemical components [16]. In this chapter, the most widely used botanicals in the United States, according to the American Botanical Council and from the 2020 Report from the Nutrition Business Journal, are listed in Table 1 [12]. As depicted in Table 1, botanical dietary supplements come from different sources widespread in the plant kingdom and therefore their overall chemical composition is diverse in Nature. The different classes of secondary metabolites that have been identified in the different species listed in Table 1 include alkaloids, fatty acids, phenolics (e.g., flavonoids and tannins), and terpenoids. The phytochemical studies as well as the biological activities of the secondary metabolites that could be responsible for validating the use of the botanicals with the most wide use will be discussed in Sects. 4 and 5 of this review chapter.

The advent of modern molecular methods together with the traditional morphological-based plant taxonomy allowed the construction of a modern plant classification system. In Scheme 1, the taxonomic classification of selected botanicals used as dietary supplements in the United States has been compiled with the aim of

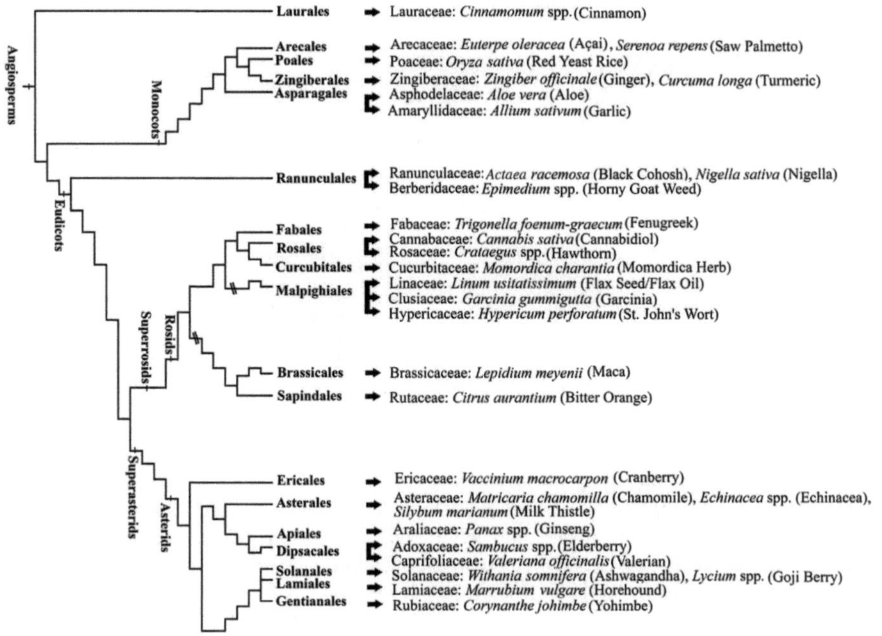

Scheme 1 Taxonomic relationship between selected plants used as dietary supplements, according to the APG IV (2016) classification (orders) of flowering plants

emphasizing the taxonomic diversity in the plant kingdom. As depicted in the scheme, botanical dietary supplements are diverse mono- and of dicotyledonous species and are classified mostly as part of the angiosperm vascular plants. *Ginkgo biloba* is not part of the scheme, and it is classified as part of the gymnosperm seed-producing plants. In this context, chemotaxonomy is nowadays considered as a complementary approach aimed at describing the set of specialized metabolites in a given defined taxon [17]. For instance, sulfur-containing glucosinolates are characteristic of the family Brassicaceae, while tetracyclic cucurbitane triterpenoids are isolated from plants of the family Cucurbitaceae. Other examples of compounds predominantly found in certain plant families include Amaryllidaceae alkaloids (Amaryllidaceae) and tropane alkaloids (Solanaceae).

Rapid advancements of the analytical techniques used for the characterization of specialized plant metabolites have provided more information about the identities and quantities in which these specialized products or compounds (Scheme 2) are likely to be present in some commercially available botanicals [18]. This contributes to a more accurate authentication of the plant material and extracts. For example,

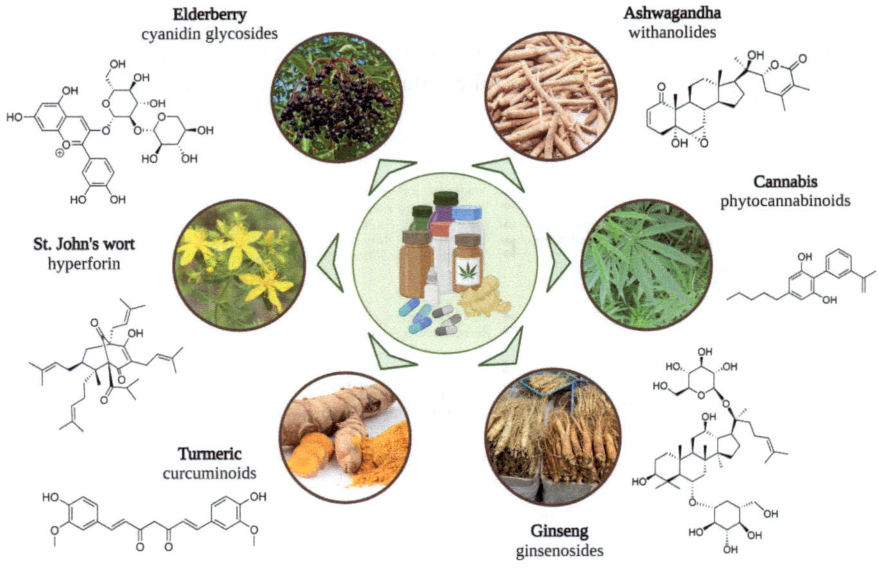

Scheme 2 Chemical quality markers that are usually analyzed in selected botanical dietary supplements. Sources "*Cannabis sativa* by James St. John CC BY 2.0 (https://plants.ces.ncsu.edu/plants/cannabis-sativa/)", "*Hypericum perforatum* by C.T. Johansson CC BY-SA 3.0 (https://commons.wikimedia.org/wiki/File:Hypericum_perforatum-IMG_4353.jpg)", "Leaves and black fruits of *Sambucus nigra* by H. Zell CC BY-SA 3.0 (https://plants.ces.ncsu.edu/plants/sambucus-nigra/)", "Turmeric (*Curcuma longa*): fresh rhizome and powder by Simon A. Eugster CC BY 3.0 (https://commons.wikimedia.org/wiki/File:Curcuma_longa_roots.jpg)", "White and Red Ginseng by Eugene Kim CC BY 2.0 (https://www.flickr.com/photos/eekim/5152649004)", "Withania: Graded, dried *Withania somnifera* roots by Piyush Kothari CC BY-SA 4.0 (https://commons.wikimedia.org/wiki/File:Ashwagandha_Roots.jpg)"

the curcuminoid content has been evaluated in turmeric (*Curcuma longa*) dietary supplements sold in the United States [19].

In addition, other modern methods such as chemical fingerprinting and metabolomic studies are being applied to enable a more comprehensive profile of the metabolites present in a given botanical sample and to characterize the relationships between its metabolome and corresponding genotype, origin, quality, or other attributes [18]. As an example, a comprehensive characterization of the composition of phytocannabinoids present in *Cannabis sativa* indicates that there are new subgroups within the traditional classification of chemotypes [20], although at the present time, as mentioned above, this botanical is not classified as a botanical dietary supplement, according to the FDA. Thus, chemotaxonomy goes hand-in-hand with the overall study and development of these botanicals since taxonomic classification can help predict the potential chemical composition of a botanical and hence the associated biological activities.

Table 1 Botanicals with the most widely reported use in the United States [12]

Latin binomial	Family	Latin binomial	Family
Actaea racemosa	Ranunculaceae	*Marrubium vulgare*	Lamiaceae
Allium sativum	Amaryllidaceae	*Matricaria chamomilla*	Asteraceae
Aloe vera	Asphodelaceae	*Momordica charantia*	Cucurbitaceae
Cannabis sativa	Cannabaceae	*Nigella sativa*	Ranunculaceae
Cinnamomum verum	Lauraceae	*Oryza sativa*	Poaceae
Citrus aurantium	Rutaceae	*Panax* spp.	Araliaceae
Crataegus spp.	Rosaceae	*Pausinystalia johimbe*	Rubiaceae
Curcuma longa	Zingiberaceae	*Sambucus nigra*	Adoxaceae
Echinacea spp.	Asteraceae	*Sambucus canadensis*	Adoxaceae
Epimedium spp.	Berberidaceae	*Serenoa repens*	Arecaceae
Euterpe oleracea	Arecaceae	*Silybum marianum*	Asteraceae
Garcinia gummi-gutta	Clusiaceae	*Trigonella foenum-graecum*	Fabaceae
Ginkgo biloba	Ginkgoaceae	*Vaccinium macrocarpon*	Ericaceae
Hypericum perforatum	Hypericaceae	*Valeriana officinalis*	Valerianaceae
Lepidium meyenni	Brassicaceae	*Withania somnifera*	Solanaceae
Linum usitatissimum	Linaceae	*Zingiber officinale*	Zingiberaceae
Lycium spp.	Solanaceae		

4 Phytochemical Studies on Selected Botanicals

Botanicals used as dietary supplements in the United States contain secondary metabolites that have been classified among different chemical classes of compounds including alkaloids, fatty acids, flavonoids, lignans, saponins, sterols, and terpenes. Below, the chemical composition of a selected number of botanicals is presented based on their natural source of origin.

4.1 Actaea racemosa

Actaea racemosa L. (syn: *Cimicifuga racemosa* (L.) Nutt.) (Plate 1), known as Black Cohosh, is a species distributed in North America and its dried roots and rhizomes have long been used by native Americans to alleviate menopausal symptoms. The

Plate 1 Partial representation of *Actaea racemosa* (family Ranunculaceae). Online source "*Actaea racemosa*". Photograph courtesy W. John Hayden (https://vnps.org/wildflower-of-the-year-2017-actaea-racemosa-black-cohosh/). Accessed Oct. 19, 2022

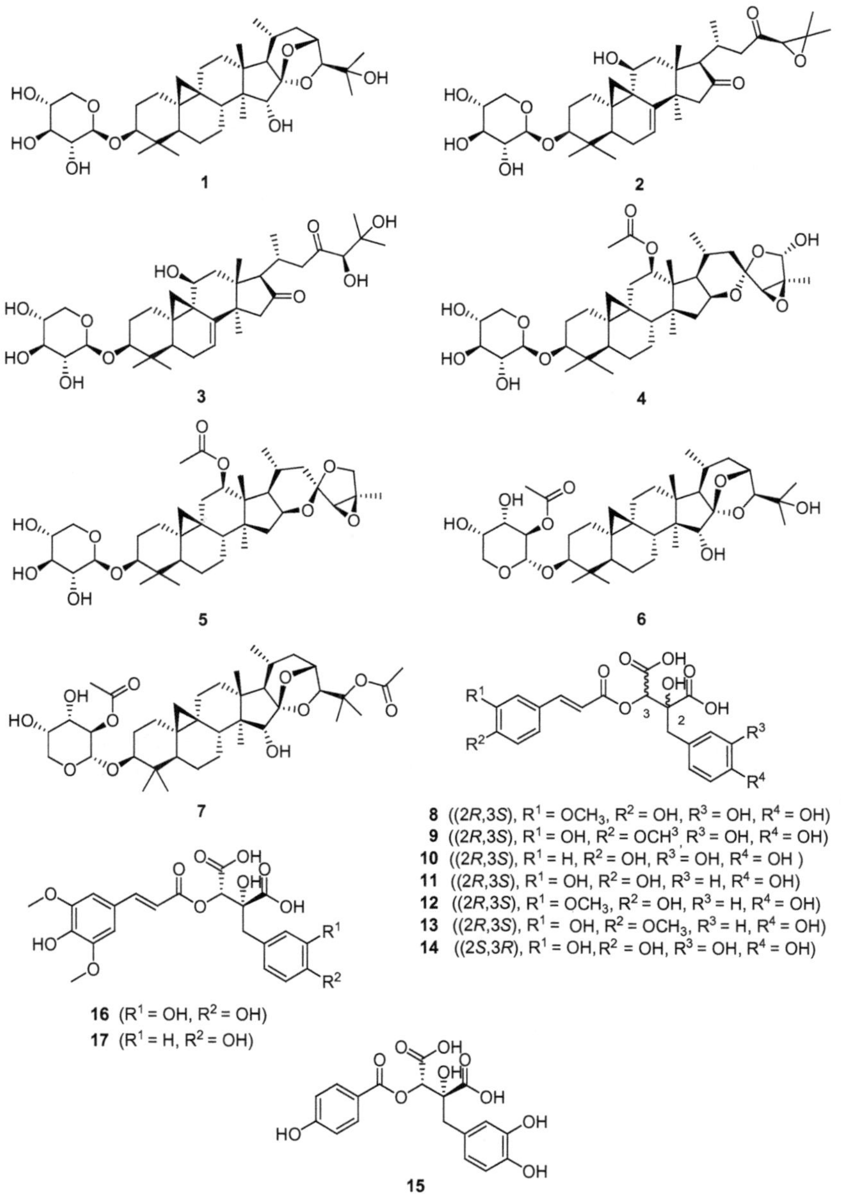

Fig. 1 Structures of cimigenol-3-*O*-*β*-D-xyloside (**1**), cimicifugoside H-1 (**2**), cimicifugoside H-2 (**3**), actein (**4**), 23-*epi*-26-deoxyactein (**5**), cimicifoetisides A (**6**) and B (**7**) and cimicifugic acids A–J (**8–17**) isolated from *Actaea racemosa*

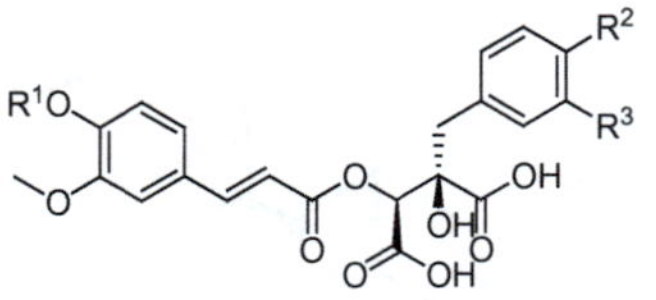

18 (R^1 = *β*-D-Gal, R^2 = OH, R^3 = OH)
19 (R^1 = *β*-D-Gal, R^2 = OH, R^3 = H)
20 (R^1 = *β*-D-Glc, R^2 = OH, R^3 = H)

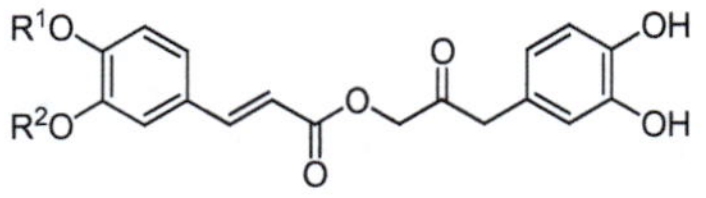

21 (R^1 = CH_3, R_2 = H)
22 (R^1 = H, R^2 = CH_3)

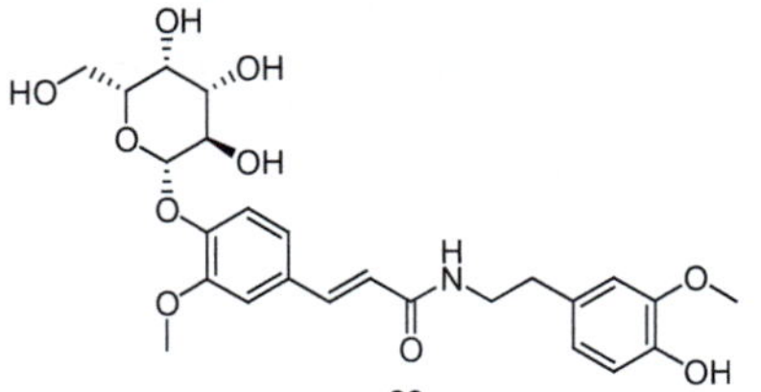

23

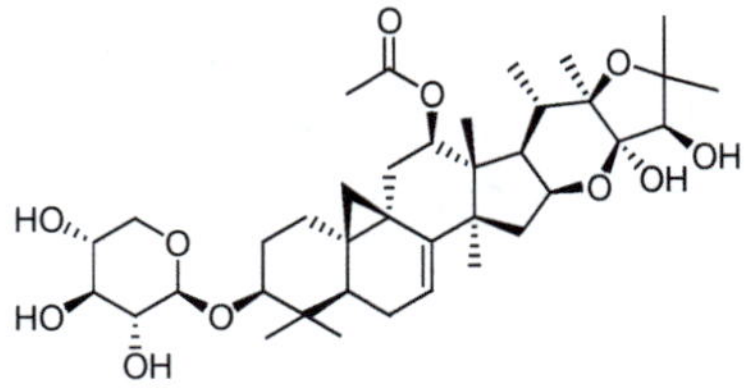

24

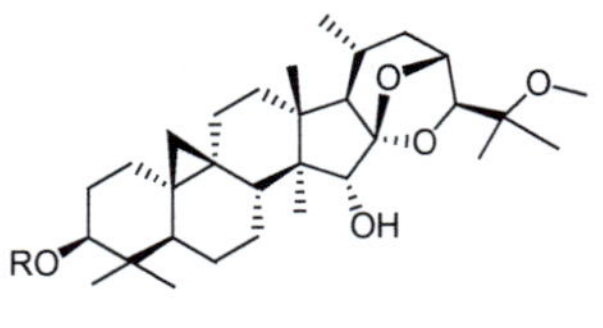

25 (R = *α*-L-Ara)
26 (R = *β*-D-Xyl)

Fig. 2 Structures of shomasides A–C (**18**–**20**), cymiracemates A (**21**) and B (**22**), cimocifugamide (**23**), cimiracemoside A (**24**), 25-*O*-methylcimigenol-3-*O*-*α*-L-arabinopyranoside (**25**) and 25-*O*-methylcimigenol-3-*O*-*β*-D-xylopyranoside (**26**) isolated from *Actaea racemosa*

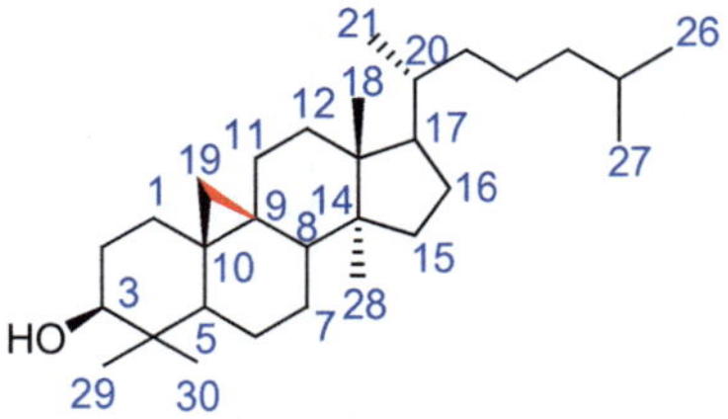

Position	Typical functional group and/or sugar moieties
C-1	Hydroxy
C-7, C-8	Double bond
C-12, C-15	Acetoxy
C-3	*O*-ß-D-xylopyranosyl or *O*-*α*-L-arabinopyranosyl

Scheme 3 Core substitution patterns for the cycloartane triterpenes of *Actaea racemosa*

genus *Actaea* (family Ranunculaceae) encompasses about 28 species of perennial and herbaceous plants distributed widely in temperate regions of the northern hemisphere especially in East Asia, Europe, and North America [21].

Three species native to Asia, *Actaea cimicifuga* L. (syn: *Cimicifuga foetida* L.), *Actaea dahurica* (Turcz. ex Fisch. & C.A. Mey) Franch. (syn: *Cimicifuga dahurica* (Turcz.) Maxim) and *Actaea heracleifolia* (Kom.) J. Compton (syn: *Cimicifuga heracleifolia* Kom.), are commonly known as "Shengma" and reported to possess anti-inflammatory, antipyretic, and analgesic activities. These and other species are reported as common adulterants of *A. racemosa* [22].

Actaea species are sources of 9,19-cycloartane-type triterpenoid saponins, which are tetracyclic triterpenoids with an extra cyclopropane ring system, formed by a C-atom bridge between C-9 and C-19 [21]. The structures of the A–D rings do not vary greatly among this diverse family of cycloartane-type saponins, except for the seven-membered B-ring derivatives (e.g. podocarpasides A–G) that occur in *Actaea podocarpa* [23]. However, the lateral chain varies substantially due to differences in oxygenation and the mode of ring formation. Based on these variations, several subtypes of 9,19-cycloartane-type triterpenoid saponins have been identified as being present in *Actaea* species. Most of these are of the cimigenol type, such as cimigenol-3-*O*-*β*-D-xyloside (**1**), the shengmanol type, as in cimicifugosides H-1 (**2**) and H-2 (**3**), and the actein type, such as actein (**4**) and 23-*epi*-26-deoxyactein (**5**) [21, 24]. The C-3 hydroxy group is frequently glycosylated, usually with xylose or arabinose. Acetylated monosaccharides are rare, as exemplified by cimicifoetisides A (**6**) and B (**7**) obtained from *A. cimicifuga* (Fig. 1) [25, 26].

In addition to common hydroxycinnamic acids such as caffeic acid, ferulic acid and isoferulic acid, other derivatives have been isolated from *Actaea* species, including cimicifugic acids A-J (**8–17**), shomasides A-C (**18–20**) and cymiracemates A (**21**) and B (**22**) [21] (Fig. 2). Hydroxycinnamic acid amide (HCAA) glycosides are conjugated compounds of hydroxycinnamic acids and tyramine, dopamine or methyldopamine via an amide bond. Only six HCCA glycosides, such as cimocifugamide (**23**), have been isolated from *Actaea* species and these have recently received attention for their use in the authentication of black cohosh [27].

A phytochemical study of the rhizomes of *A. racemosa* led to the isolation of glycosylated triterpenes, phenolic compounds, chromones and nitrogen-containing compounds [21, 28]. From an ethanol extract (a commercial preparation), triterpenoids containing the 9,19-cycloartane group have been identified, which are considered as the active principles of the plant (Scheme 3) [24, 29, 30]. Specifically, the structural diversity of *Actaea* triterpenes is attributed to substitution in the core with functional groups such as a double bond at between C-7 and C-8, hydroxy groups at C-1, C-12 and/or C-15, sugar moieties (e.g. 3-*O*-*β*-D-xylopyranosyl or 3-*O*-*α*-L-arabinopyranosyl) at C-3, and acetoxy groups at C-12 and/or C-15 (Scheme 3) [24, 28–30]. The cimigenol and actein types are the major cycloartane triterpene constituents, representing characteristic biosynthetic scaffolds of the triterpenoids of *A. racemosa* (Fig. 1) [24, 30]. These compounds contain 3-, 5-, and 6-membered oxygen heterocycles, which are side-chain variants and are attached to C-16 and C-17.

The presence of cimicigenol-3-*O*-β-D-xyloside (**1**), actein (**4**), 23-*epi*-26-deoxyactin (**5**) (Fig. 1) and cimiracemoside A (**24**) (Fig. 2) have been reported from the EtOAc fraction of the roots and rhizomes of *A. racemosa.* [24]. Other compounds reported from *Actaea racemosa* include 25-*O*-methylcimigenol-3-*O*-α-L-arabinopyranoside (**25**), and 25-*O*-methylcimigenol-3-*O*-β-D-xylopyranoside (**26**) (Fig. 2) [24]. Biological studies performed on the phytochemicals found in this species are discussed in Sect. 5.1.

4.2 Allium sativum

The genus *Allium* (Amaryllidaceae) is comprised of about 1000 species of perennial and bulbous plants endemic to the dry and temperate regions of the northern hemisphere and currently distributed worldwide [31]. This genus includes important representative edible species such as garlic (*Allium sativum* L.) (Plate 2), onion (*Allium cepa* L.), leek (*Allium ampeloprasum* L.), and chive (*Allium schoenoprasum* L.). These species are characterized by their content of organosulfur compounds, flavonoids and steroidal saponins. *Allium sativum* or garlic, is a strongly aromatic

Plate 2 Flowers of *Allium sativum* (Family Amaryllidaceae). Online source "*Allium sativum*". Photograph courtesy Aleksandrs Balodis (https://commons.wikimedia.org/wiki/File:Allium_sativum_02.JPG). Accessed Oct. 27, 2022

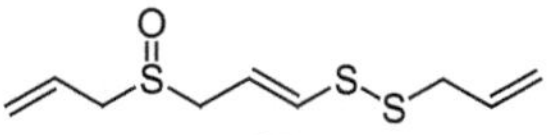
27
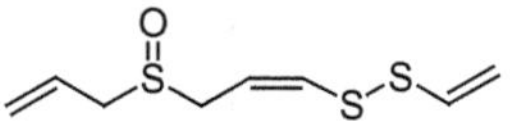
28
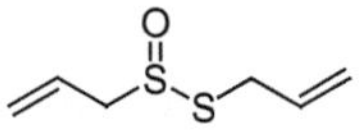
29
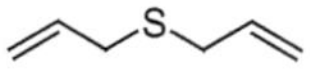
30
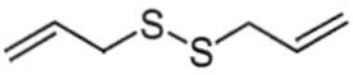
31
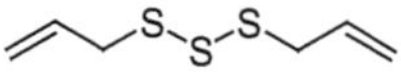
32
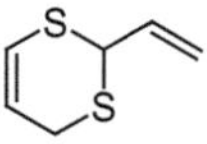
33
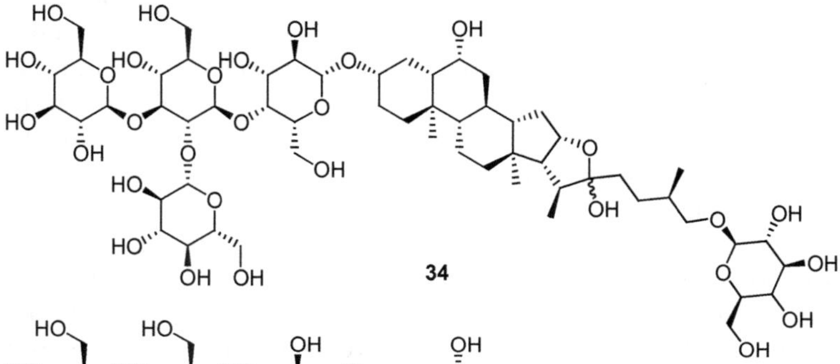
34
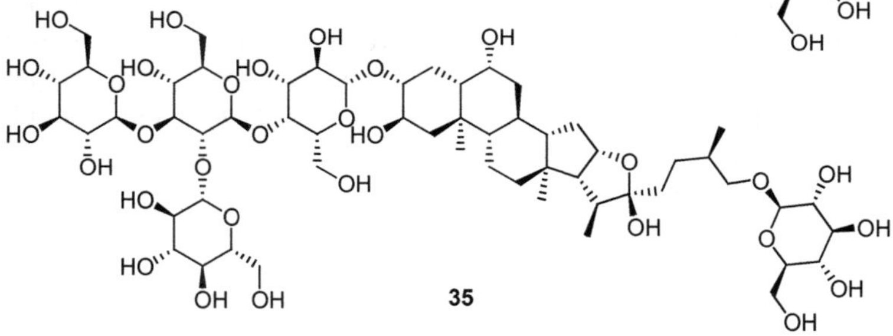
35
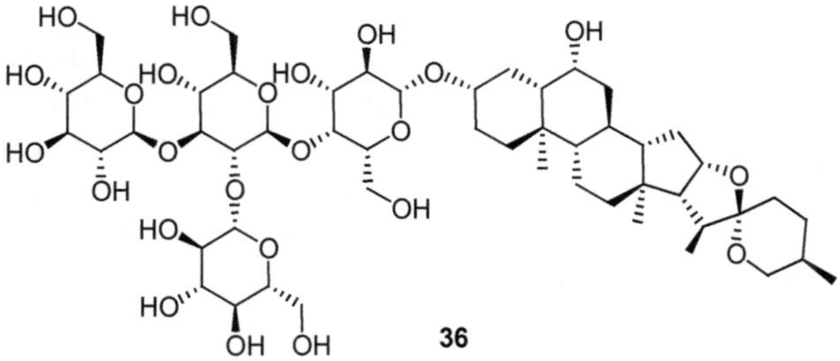
36
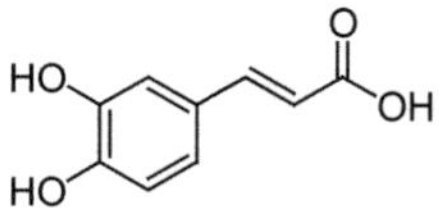
38
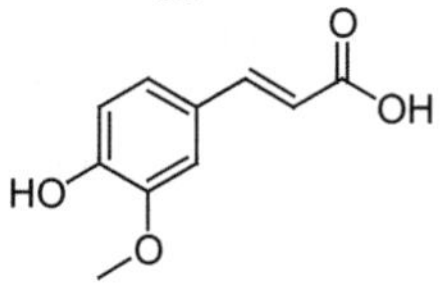
39
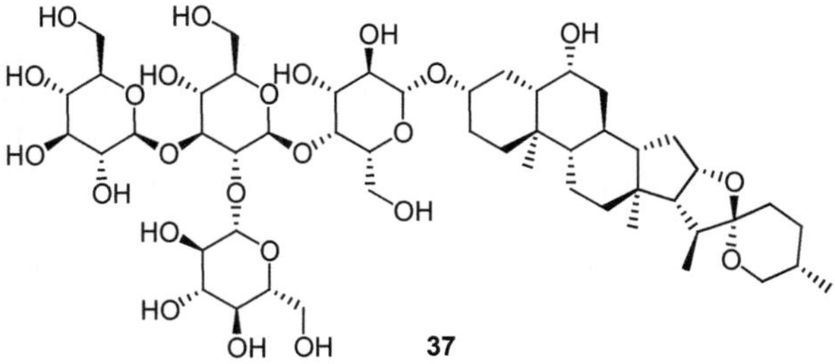
37
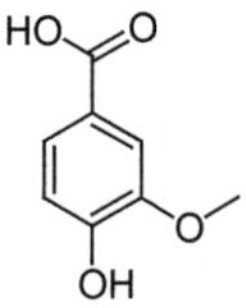
40

Fig. 3 Structures of (*E*)-ajoene (**27**), (*Z*)-ajoene (**28**), allicin (**29**), DAS (**30**), DADS (**31**), DATS (**32**), 2-vinyl-(4*H*)-1,3-dithiin (**33**), proto-eruboside B (**34**), voghieroside C1 (**35**), eruboside B (**36**), *iso*-eruboside B (**37**), caffeic acid (**38**), ferulic acid (**39**), and vanillic acid (**40**), isolated from *Allium sativum* bulb spice believed to originate from Kazakhstan, Uzbekistan, and Western China [32, 33]. Garlic is one of the oldest authenticated and most important herbs that has been used since ancient times as a traditional medicine [34].

L-Cysteine sulfoxides such as alliin (garlic), methiin (garlic, onions, leeks and shallots), propiin (shallots), isoalliinin (onions and shallots) are considered as major sulfur-containing compounds in intact plant of *Allium* species [35]. Once the plant tissue is damaged, these precursors are converted into thiosulfinates (such as allicin) through enzyme reactions, and further decomposed into other sulfur-containing compounds such as allyl sulfides, ajoenes and vinyldithiins, depending on the manufacturing process [36]. Other additional major sulfur-containing compounds detected in intact *Allium* species are γ-glutamyl peptides, which can be converted to other water-soluble organosulfur compounds, such as allyl cysteines during the extraction of garlic in aqueous solution [37].

Bulbs of *A. sativum* have been reported to contain ajoenes ((*E*)-ajoene (**27**), (*Z*)-ajoene (**28**)), thiosulfinates (allicin (**29**)), sulfides (diallyl sulfide (DAS) (**30**), diallyl disulfide (DADS) (**31**), diallyl trisulfide (DATS) (**32**)), vinyldithiins (2-vinyl-(4*H*)-1,3-dithiin (**33**)), saponins (proto-eruboside B (**34**), voghieroside C1 (**35**), eruboside B (**36**), iso-eruboside B (**37**)), and phenolics (caffeic acid (**38**), ferulic acid (**39**), and vanillic acid (**40**)) (Fig. 3) [32–34, 38, 39]. Sulfur compounds are purported to be responsible for the medicinal properties of garlic [38] and the results of pertinent biological studies are presented in Sect. 5.2.

4.3 Aloe vera

The genus *Aloe* (family Asphodelaceae) encompasses over 400 species distributed in temperate and tropical regions worldwide. The most highly commercially significant species is *Aloe vera* (L.) Burm.f. (syn: *A. barbadensis* Miller) (Plate 3) while two other species, *Aloe ferox* Miller and *Aloe arborescens* Miller, have wide use in Asia and Africa [40]. The leaves of these perennial plants are fleshy and arranged as rosettes. They can be divided into three major parts with differentiated chemical compositions, the outer green rind, the outer pulp region, and the inner colorless parenchyma ("pulp") containing the aloe mucilage.

There are different polysaccharides in *Aloe vera*, such as glucomannan and acetylated mannan or acemannan (**41**). Many health-promoting effects of *Aloe vera* have been attributed to such polysaccharides [41, 42]. Acemannan (**41**), an important component of *A. vera*, is acetylated at position β-(1→4) of the polymannose unit (Fig. 4). The inner pulp of the leaves of *A. vera* contains approximately 99% water and is also made up of polysaccharides, with one of the most abundant being acemannan (**41**), which is a partially acetylated mannan considered to be a quality control marker that may be used for characterizing *A. vera* extracts [43]. *A. vera* also contains anthraquinones and their glycosides in the outer green epidermis, phenolic compounds in the outer pulp region (anthraquinones, anthrones, chromones, coumarins, and flavonoids), pyrones and steroids. The pulp is rich in polysaccharides

Plate 3 Leaves of *Aloe vera* (Family Asphodelaceae). Online source "*Aloe vera.*" Photograph courtesy Biswarup Ganguly (https://upload.wikimedia.org/wikipedia/commons/a/a1/Aloe_vera_-_Agri-Horticultural_Society_of_India_-_Alipore_-_Kolkata_2013-01-05_2326.JPG). Accessed Oct. 27, 2022

(acemannan (**41**)) and phenolic compounds. Compound **41** is partially acetylated at the C-2 and C-3 positions, exhibiting an acetyl:mannose monomer ratio of approximately 1:1, and containing a side chain such as galactose attached to the C-6 units. Acemannan (**41**) has been found with a Man:Glc:Gal:GalA:Fuc:Ara:Xyl ratio of 120:9:6:3:2:2:1 with traces of Rha and GlcA [44–46]. Pectins and arabinogalactan have been isolated from the inner pulp of both *A. vera* and *A. ferox*, while two glucomannans, arborans A and B, have been isolated from an active polysaccharide fraction of *A. arborescens* [47].

The outer pulp region below the rind consists of vascular bundles and is from where the bitter-tasting latex of *Aloe* species is derived. This may contain an array of compounds including anthrones, anthraquinones, chromones and their glycosides [48]. Aloin is the main anthrone *C*-glycoside found in a variety of aloe species, including *A. vera*, *A ferox* and *A. arborescens*, and this occurs naturally as a mixture of two diastereomers known as aloin A (barbaloin) (10*S*) (**42**) and aloin B (isobarbaloin) (10*R*) (**43**) [49]. Aloe-emodin is considered a degradation product of aloin although some authors have reported that it does not play a key role for the characterization of commercial *Aloe* [50, 51]. Aloesin (aloeresin B) is one of the most important compounds present in the three *Aloe* species mentioned above and could be regarded as the parent compound of the *Aloe* chromones [48]. The 2′-*O*-*p*-coumaric acid ester derivative aloeresin A (**44**), is present in *A. vera* and *A. ferox* [50]. Several isomeric and substituted isomeric forms as well as chromones containing cinnamic acid moieties have also been reported [52]. Pyrones, flavonoids, and alkaloids have also been isolated from *Aloe* species. Aloenin, the *O*-glucoside of a phenyl pyrone was first isolated from *A. arborescens* but later detected in *A. ferox* and another group of *Aloe* species referred to as the aloenin chemical group [50, 53].

The *Aloe* gel from the inner leaf pulp also contains proteins (lectins), enzymes (alkaline phosphatase, amylase, carboxypeptidase, and catalase), vitamins (A, C, E, B1, B2, B6, B12, and β-carotene) and minerals [45, 54–58]. The variations in concentrations of these chemical constituents depend on the plant part used, the extraction process, the solvent employed, and the stage of growth, and plant source [54, 55]. Overall, the nutraceutical properties of *Aloe vera* have been attributed to acemannan (**41**) and several studies on its biological properties are presented in Sect. 5.3.

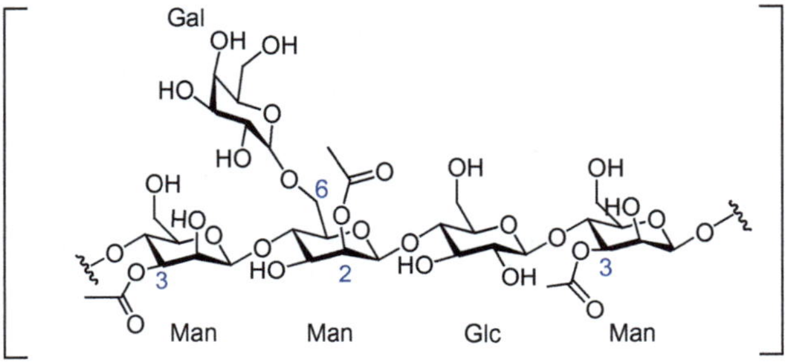

41

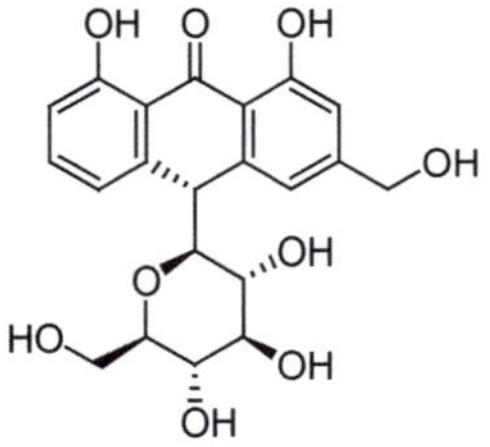

42

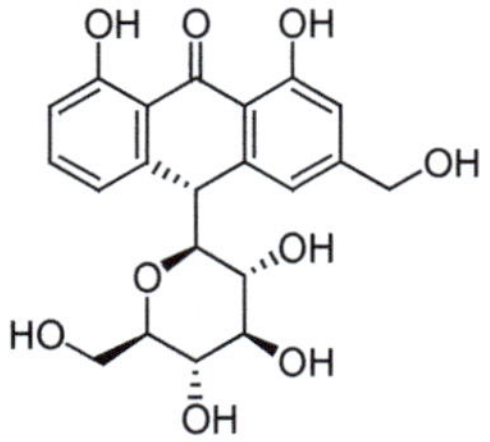

43

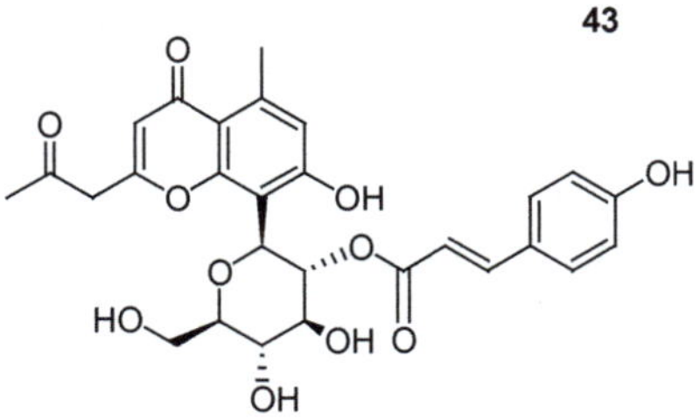

44

Fig. 4 Chemical structure of constituents of *Aloe vera*, acemannan (**41**), aloin A (**42**), aloin B (**43**), and aloeresin A (**44**)

4.4 *Cinnamomum spp. (Cinnamomum verum)*

The genus *Cinnamomum* (family Lauraceae) comprises about 250 species of evergreen trees and shrubs, collectively known as Cinnamon, which are distributed in tropical and subtropical Asia, Australia, Pacific Islands, and other regions [59]. Apart from its use as a spice, the bark of cinnamon species, as well as their essential oils and extracts obtained from these are used as components of dietary supplements and in traditional herbal medicinal products and formulations [60]. Cinnamon is mainly obtained from the dried inner bark of cultivated varieties of *Cinnamomum verum* J.Presl (Sri Lankan or true cinnamon) (syn. *Cinnamomum zeylanicum*) (Plate 4). Other arboreal *Cinnamomum* species are also referred to as cinnamon such as *Cinnamomum cassia* (L.) J.Presl (Chinese cinnamon), *Cinnamomum burmannii* (Nees & T.Nees) Blume (Indonesian cinnamon), and

Plate 4 *Cinnamomum verum* tree (Family Lauraceae). Online source "Ceylon Cinnamon (*Cinnamomum verum*)" by Rich Kostecke, CC BY-NC 4.0 (https://www.inaturalist.org/photos/143709825). Accessed Nov. 15, 2022

Cinnamomum loureiroi Nees (Vietnamese cinnamon). Cinnamon is used worldwide for different types of applications, including as a spice and a flavoring agent. Various chemical constituents have been isolated and identified from different *Cinnamomum* species including coumarins, flavonoids, lignans, phenylpropanoids, terpenes and other constituents [59]. The chemical composition of cinnamon varies depending on several factors that include the part of the plant used, the growing season, the age of the trees, location, and the methods of separation, extraction, and purification [61, 62].

Cinnamaldehyde (**46**) (Fig. 5) is considered as the dominant volatile compound from the bark of *Cinnamomum* plants and its concentration differs between species [60]. In addition to **49**, other compounds such as cinnamic acid (**47**) have also been detected not only in *C. verum*, but also in the other species mentioned above. Eugenol (**48**) occurs in the stem bark of *C. verum* [63], but is present in higher concentrations in the leaves. A rare benzyl benzoate chemotype has also been reported [64]. Regarding other species, eugenol (**48**) is either present in very low concentration levels in the bark of *C. burmannii* and *C. loureiroi* or absent as in the bark of *C. cassia* [60, 65]. Coumarin is a natural flavoring and fragrance compound that is widespread in plants and is considered hepatotoxic. It occurs at significant concentrations in *C. cassia*, *C. burmannii* and *C. loureiroi*, representing a potential risk to health. *C. verum* contains only traces of this compound [60].

Recently, a modern approach has been used to compare the metabolomic profile of the major cinnamon species. Coumarin, cinnamaldehyde, methoxycinnamaldehyde, cinnamoyl methoxyphenyl acetate, proanthocyanidins and other components were found to vary among *C. verum*, *C. cassia*, *C. burmannii* and *C. loureiroi*. Such variations have been used to develop a strategy for differentiating these four cinnamon species [66].

In traditional medicinal preparations of *Cinnamomum zeylanicum*, benzoic acid (**45**), (*E*)-cinnamaldehyde (**46**), cinnamic acid (**47**), eugenol (**48**), and *o*-methoxy-cinnamaldehyde (**49**) (Fig. 5) have been identified. All *Cinnamomum* species contain

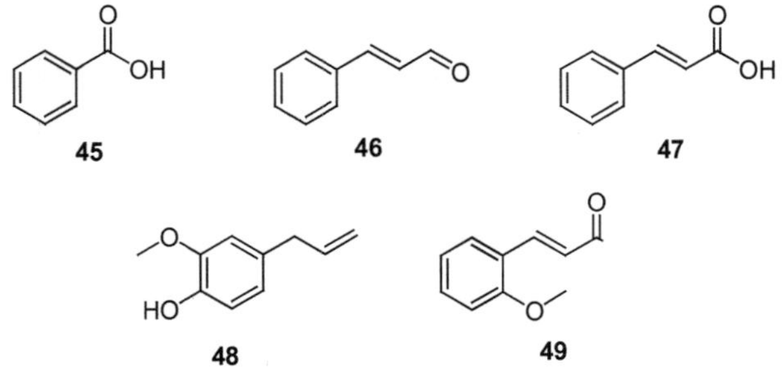

Fig. 5 Chemical structures of benzoic acid (**45**), (*E*)-cinnamaldehyde (**46**), cinnamic acid (**47**), eugenol (**48**) and *o*-methoxy-cinnamaldehyde (**49**) isolated from *Cinnamomum* spp.

the active compound cinnamaldehyde (**46**), which is found at different concentrations in all plant parts. Discussion of the purported mechanism of action of the phytochemicals in *Cinnamomum* spp. is provided in Sect. 5.4.

4.5 *Citrus aurantium*

The genus *Citrus* (family Rutaceae) comprises a large number of species, varieties, cultivars, and hybrids, including some of the most important cultivated fruit trees worldwide, such as *Citrus sinensis* (L.) Osbeck (Sweet Orange), *Citrus limon* (L.) Osbeck (lemon), *Citrus paradisi* Macfad. (grapefruit), *Citrus maxima* (Burm.) Merr. (pomelo), *Citrus reticulata* Blanco (mandarin), and *Citrus aurantium* L. (sour/bitter orange) (Plate 5) [67].

Plate 5 Fruits, leaves and flowers of *Citrus aurantium* (Family Rutaceae). Online source "*Citrus aurantium.*" Photograph courtesy Zeynel Cebeci (https://upload.wikimedia.org/wikipedia/commons/c/c9/Turun%C3%A7_-_Citrus_aurantium_02.JPG). Accessed Oct. 27, 2022

Flavonoids are the major class of phytochemicals that have been detected in the peel, pulp, and seeds of *Citrus* fruits. Hesperidin and naringenin are the two most widely observed free flavanone aglycones, in addition to isosakuranetin and eriodictyol. These flavanones are usually glycosylated by a disaccharide unit at the C-7 position, either by a neohesperidose (as in naringin (**50**) (Fig. 6)) or rutinose (as in hesperidin) moiety. Apigenin, luteolin, diosmetin and chrysoeriol flavones and their *O*-glycosides or *C*-glycosides (mostly of glucose, rutinose and neohesperidose) are a second major group of flavonoids present. Polymethoxylated flavones (PMFs) such as nobiletin and tangeretin, even though present in low concentration, are of importance due to their biological properties [68].

Coumarins and furanocoumarins (a subtype of coumarins containing an additional furan ring) are another class of compounds found in *Citrus* species (in higher concentration in the peel) that have been associated with defense against pathogens. These compounds are present as dimethyallylated and/or geranylated derivatives such as aurapten (7-geranyloxycoumarin), bergamottin (5-geranyloxypsoralen), and imperatorin (8-dimethylallyloxypsoralen) [69]. Limonoids and their glycosides are highly oxygenated tetracyclic triterpenoids characterized by the presence of a furan ring linked to the D ring at C-17. The diversification within this family of compounds mostly occurs in the A and B rings along with different oxidation levels, while the D ring is either opened or lactonized. Limonoids are often reported in the seeds and

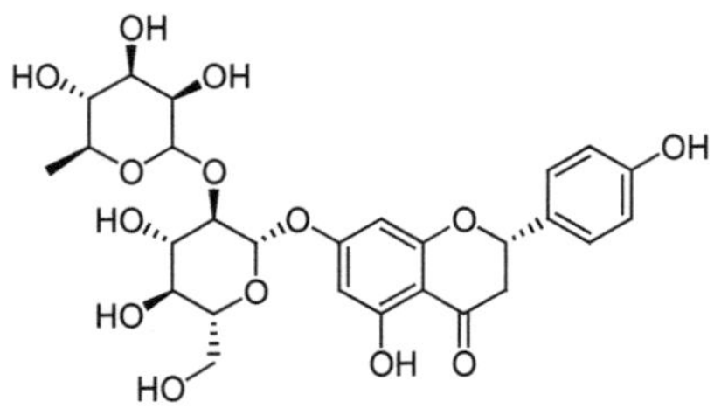

50

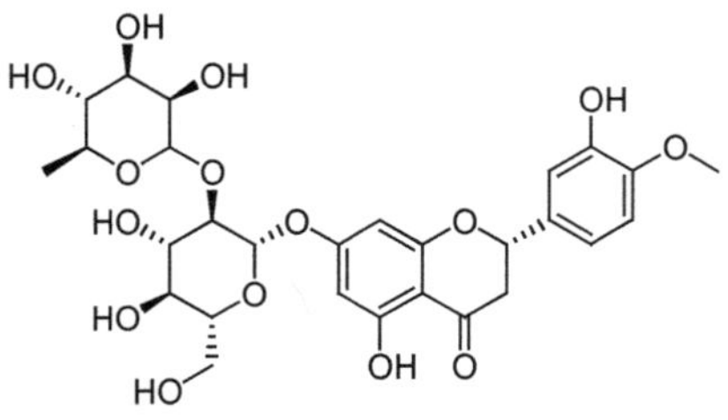

51

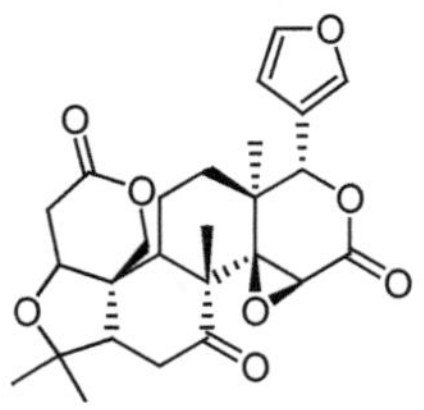

52

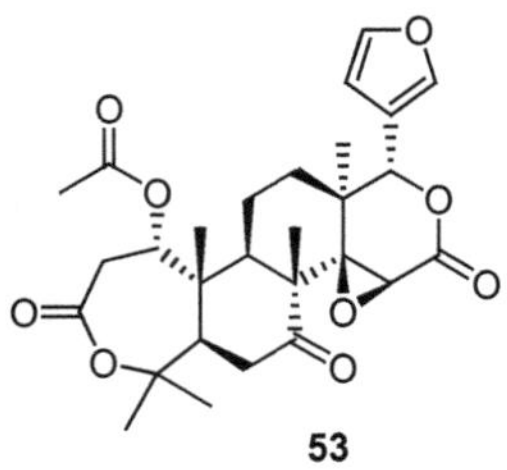

53

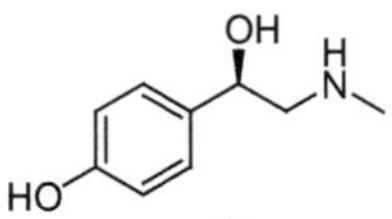

54

Fig. 6 Structures of naringin (**50**), neohesperidin (**51**), limonin (**52**), nomilin (**53**) and *p*-synephrine (**54**) isolated from *Citrus aurantium*

fruits of *Citrus* species and for most of them, limonin (**52**) and nomilin (**53**) are the most abundant aglycones [70].

Synephrine (**54**), and other phenolic phenethylamines, including octopamine and *N*-methyltyramine are present in different *Citrus* species and are particularly abundant in extracts from the unripe fruit or peel of *C. aurantium* [71, 72]. Volatile compounds account for about 0.5–5% of *Citrus* peel fresh weight. Limonene and γ-terpinene are two of the major components in the essential oils of the peels of *Citrus* spp. [71].

The phytochemical composition of *C. aurantium* responsible for health-promoting effects includes phenolic compounds, terpenoids, and alkaloids. Among the flavonoids in *C. aurantium* are naringin (**50**) and neohesperidin (**51**) (Fig. 6) [73–75]. Other classes of secondary metabolites found in this plant are limonoids, which are oxygenated triterpenoids biosynthesized from acetate-mevalonate. The main limonoids in *C. aurantium* include limonin (**52**) and nomilin (**53**) (Fig. 6) [76, 77].

Citrus aurantium also contains phenylethylamine alkaloids as *p*-synephrine (**54**) (a protoalkaloid). The structure of *p*-synephrine (**54**) has a hydroxy group in a *para* position on the benzene ring and has some structural similarity to ephedrine [73, 77]. Studies on the biological properties of this and other *Citrus* spp. constituents are presented in Sect. 5.5.

4.6 *Crataegus spp.*

The genus *Crataegus* (Rosaceae) consists of about 250 species of branched shrubs or trees distributed in temperate regions of Africa, Asia, Europe and North America [78]. Some *Crataegus* species are popularly known as "hawthorn" and according to the *European Pharmacopoeia* and the non-official *American Herbal Pharmacopoeia* (AHP), the leaves, flowers, and fruits of *Crataegus monogyna* Jacq. (Plate 6B) and *Crataegus laevigata* (Poir.) DC. (Plate 6A) or their hybrids are used in herbal preparations [79].

According to the AHP, the leaves and flower material of *C. monogyna* and *C. laevigata* must contain a minimum of 1.5% flavonoids expressed as hyperoside, while the hawthorn berries must not contain less than 1.0% procyanidins expressed as cyanidin chloride [78]. More common components such as sugar alcohols, organic acids, hydroxycinnamic acids and terpenoids are also described in the literature for this genus [78].

Among the bioactive chemical constituents of *C. monogyna* and *C. laevigata* are included the flavonoid derivatives vitexin (**55**), hyperoside (**56**), rutin (**57**), vitexin-2″-*O*-α-L-rhamnoside (**58**), (−)-epicatechin (**59**), (+)-catechin (**60**), and the oligomeric procyanidins procyanidin B2 (**61**), B4 (**62**), B5 (**63**) and C1 (**64**) as shown in Fig. 7 [80–83]. Biological studies performed on the phytochemicals identified from these species will be discussed in Sect. 5.6.

Plate 6A Leaves and flowers of *Crataegus laevigata* (Family Rosaceae). Online source "*Crataegus laevigata*." Photograph courtesy Mennerke bloem (https://commons.wikimedia.org/wiki/File:Crataegus_laevigata_002.JPG). Accessed Oct. 27, 2022

Plate 6B Leaves and flowers of *Crataegus monogyna* (Family Rosaceae). Online source "*Crataegus monogyna*." Photograph courtesy Luc. T (https://upload.wikimedia.org/wikipedia/commons/4/44/Crataegus_monogyna_%2817760108236%29.jpg). Accessed Oct. 27, 2022

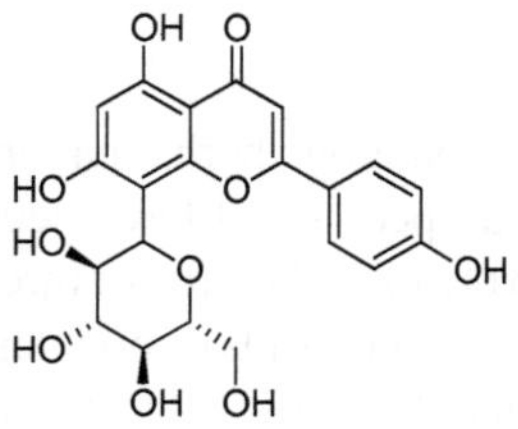
55

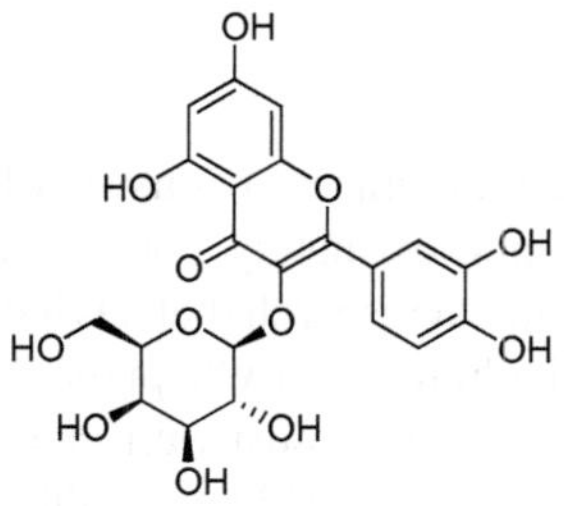
56

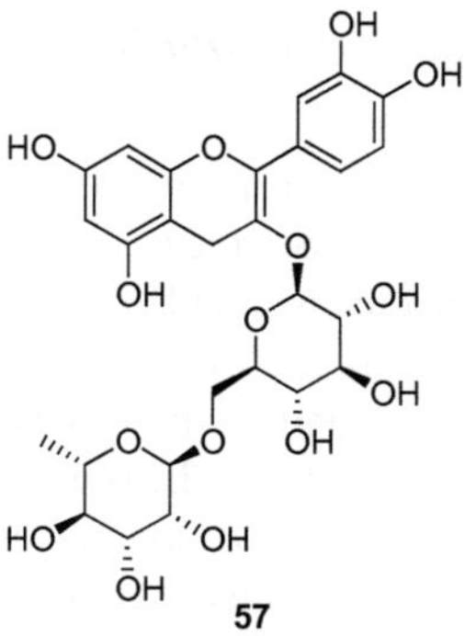
57

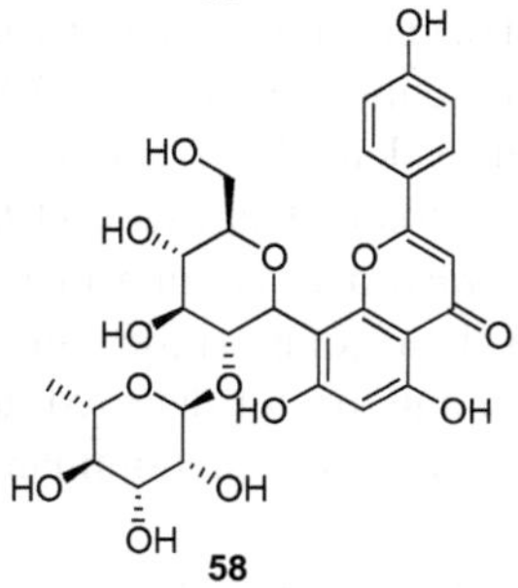
58

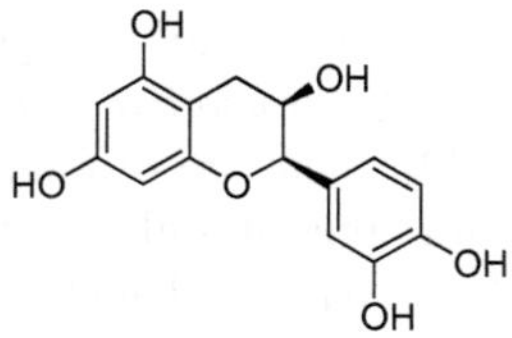
59

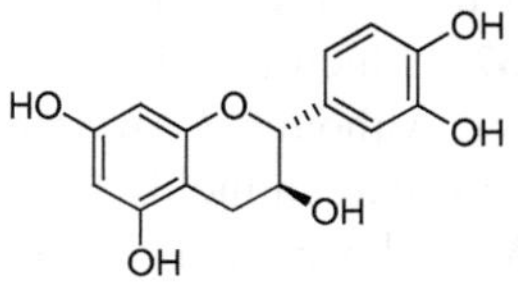
60

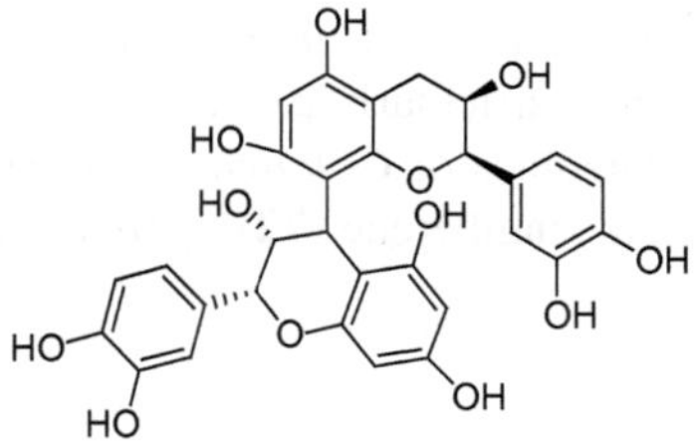
61

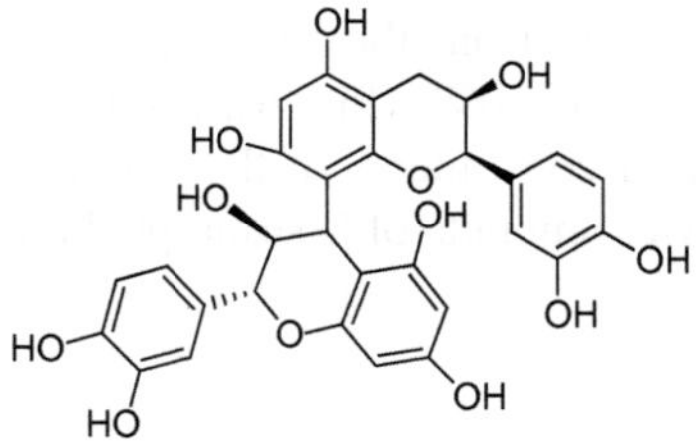
62

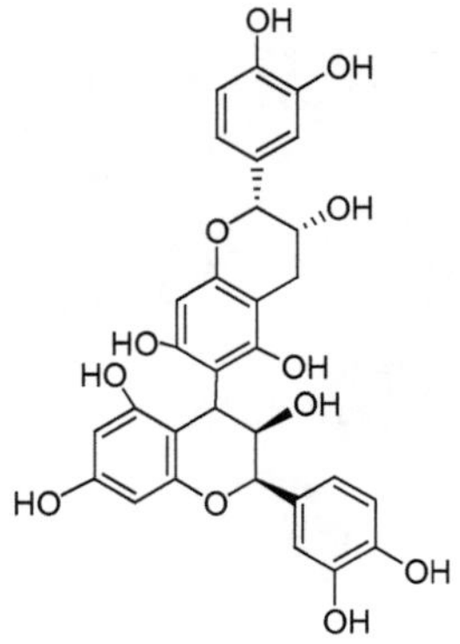
63

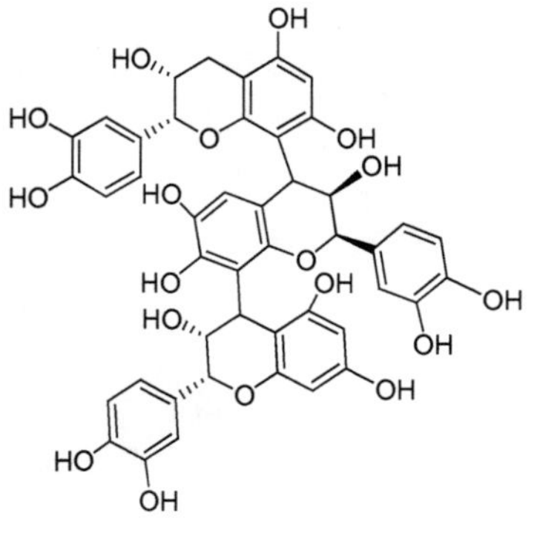
64

Fig. 7 Structures of the bioactive constituents from *Crataegus* spp. including the flavonoid derivatives vitexin (**55**), hyperoside (**56**), rutin (**57**), vitexin-2″-*O*-α-L-rhamnoside (**58**), (−)-epicatechin (**59**), (+)-catechin (**60**), and procyanidins B2 (**61**), B4 (**62**), B5 (**63**), and C1 (**64**)

4.7 *Curcuma longa*

The genus *Curcuma* (family Zingiberaceae) consists of rhizomatous herbs that are native to Southeast Asia. Of approximately 100 accepted species, 20 have been subjected to detailed phytochemical research that has led to the identification of three main groups of compounds including diarylheptanoids (curcuminoids), monoterpenoids and sesquiterpenoids [84]. The most well-investigated species is the edible *Curcuma longa* L. (known as turmeric) (Plate 7), although this can be adulterated with other traditionally used species like *Curcuma xanthorrhiza* Roxb., *Curcuma aromatica* Salisb. and *Curcuma zedoaria* (Christm.) Roscoe [85, 86].

The dried rhizomes of *C. longa* contain a variable amount of curcuminoids (Fig. 8), up to about 3% of the dry weight, with the main ones being curcumin (**65**) (up to 70% of the total curcuminoids), demethoxycurcumin (**66**), and bisdemethoxycurcumin (**67**), which together are responsible for the orange-yellow color of turmeric [19, 87]. Although *C. xanthorrhiza*, *C. aromatica* and *C. zedoaria* also contain these three curcuminoids [85, 88, 89], there are differences in their composition ratio. While bisdemethoxycurcumin is present only in trace amounts or not detected at all in *C. xanthorrhiza* and *C. aromatica*, demethoxycurcumin is the predominant curcuminoid in *C. zedoaria* [90, 91].

A variety of sesquiterpenes and monoterpenes are found in the volatile oils obtained from the rhizomes of the different species of *Curcuma*. In general, α-turmerone (**68**) (Fig. 9), β-turmerone (**69**), and particularly *ar*-turmerone (**70**) are major constituents of the volatile oil of the rhizomes of *C. longa* accompanied by lesser amounts of β-sesquiphellandrene (**71**), α-zingiberene (**72**), germacrone (**73**),

Plate 7 Roots of *Curcuma longa* (Family Zingiberaceae). Online source "Turmeric and turmeric powder on white background" by Professional Photographer, CC BY-NC-ND 2.0 (https://www.flickr.com/photos/30478819@N08/49576975953/in/photostream/). Accessed Oct. 31, 2022

terpinolene (**74**), *ar*-curcumene (**75**) and α-phellandrene (**76**) (Fig. 8) [92]. Interestingly, samples from Brazil contain primarily (*Z*)-γ-atlantone, (*E*)-γ-atlantone, and *ar*-turmerone [93].

The essential oil of *C. xanthorrhiza* rhizomes is also mainly composed of sesquiterpenes, inclusive of xanthorrhizol (**77**), which is used as marker compound, β-curcumene (**78**), and *ar*-curcumene (**75**) [94–96]. The major components reported in the literature for *C. zedoaria* are curzerenone (**79**), epicurzerenone (**80**), camphor (**81**), germacrone (**73**), and curzerene (**82**) (Fig. 9). There is greater variability regarding the composition of *C. aromatica* of different origins. Curdione (**83**), 1,8-cineole (**84**), *ar*-curcumene (**75**), xanthorrhizol (**77**) and germacrone (**73**) are frequently reported in the volatile oil of the rhizomes of this species [92].

Hence, the major chemical constituents of turmeric (*C. longa*) are the curcuminoids inclusive of curcumin (**65**), desmethoxycurcumin (**66**), and bisdemethoxycurcumin (**67**), and the sesquiterpenes α-turmerone (**68**), β-turmerone (**69**), and *ar*-turmerone (**70**) [97, 98]. Discussion of the biological aspects of the major phytochemicals in this species are provided in Sect. 5.7.

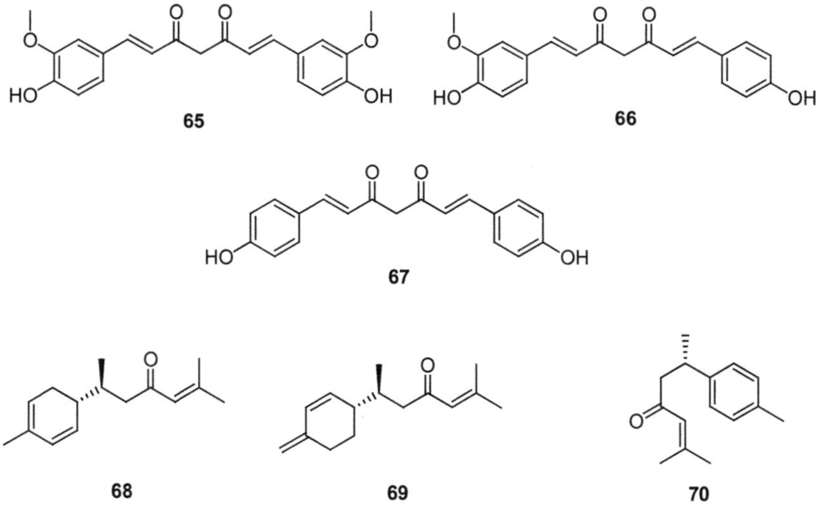

Fig. 8 Structures of characteristic secondary metabolites isolated from turmeric (*Curcuma longa* rhizomes) including curcumin (**65**), desmethoxycurcumin (**66**), bisdemethoxycurcumin (**67**), α-turmerone (**68**), β-turmerone (**69**), and *ar*-turmerone (**70**)

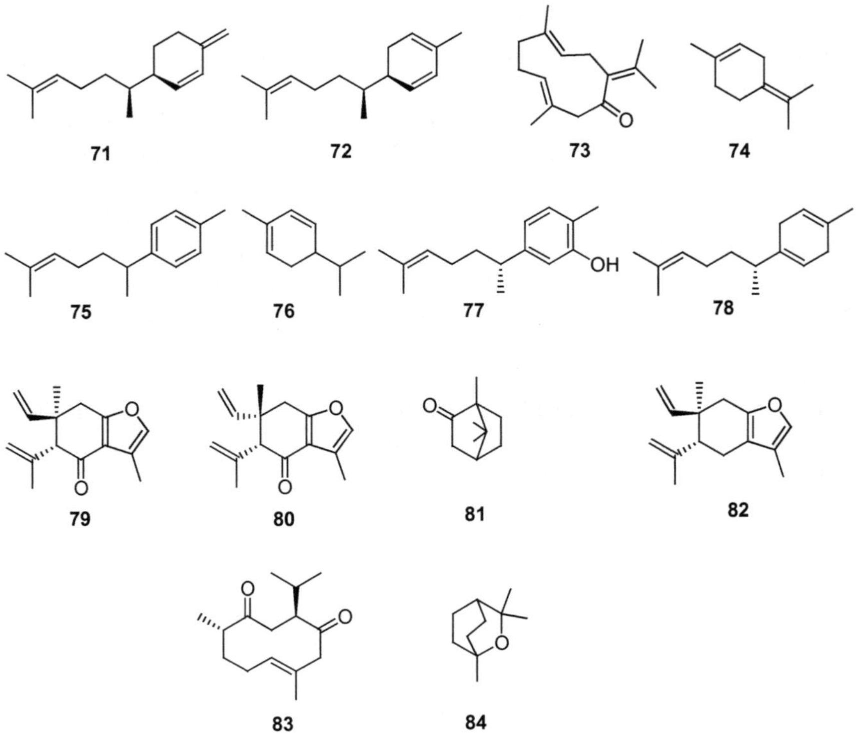

Fig. 9 Structures of additional characteristic secondary metabolites identified in *Curcuma* spp. including β-sesquiphellandrene (**71**), α-zingiberene (**72**), germacrone (**73**), terpinolene (**74**), *ar*-curcumene (**75**), α-phellandrene (**76**), xanthorrhizol (**77**), β-curcumene (**78**), curzerenone (**79**), epicurzerenone (**80**), camphor (**81**), curzerene (**82**), curdione (**83**), and 1,8-cineole (**84**)

4.8 Echinacea spp.

The genus *Echinacea* (family Asteraceae) includes nine recognized species of perennial plants that were used by Native Americans [99]. The roots and aerial parts of three of these species, *Echinacea purpurea* (L.) Moench (Plate 8), *Echinacea angustifolia* DC., and *Echinacea pallida* (Nutt.) Nutt., are commonly used in herbal supplement formulations [100].

Alkamides, caffeic acid derivatives, polysaccharides and glycoproteins, polyacetylenes and polyenes, flavonoids, and terpenoids have been reported in *Echinacea* spp. extracts. Alkamides are structures that contain a carbonyl group connected to an alkyl group (which can vary in the degree of unsaturation in the aliphatic chain), and a nitrogen atom (Fig. 10). They have been isolated primarily from the roots of *E. purpurea* and *E. angustifolia* and are major constituents present in the ethanol-water extracts [100]. Three alkamides, dodeca-(2*E*,4*E*,8*Z*,10*E*)-tetraenoic acid isobutylamide (**85**), pentadeca-(2*E*,9*Z*)-diene-12,14-diynoic acid isobutylamide

Plate 8 Leaves and flowers of *Echinacea purpurea* (Family Asteraceae). Online source "*Echinacea purpurea.*" No caption (https://www.greentecnursery.com/purple-coneflower/). Accessed Oct. 27, 2022

(**86**), and dodeca-(2*E*,4*E*)-dienoic acid isobutylamide (**87**) were identified as the main constituents of *E. angustifolia* roots. Likewise, three main alkamides have been identified in *E. purpurea* roots including undeca-(2*Z*,4*E*)-diene-8,10-diynoic acid isobutylamide (**88**), dodeca-(2*Z*,4*E*)-diene-8,10-diynoic acid isobutylamide (**89**), and dodeca-(2*E*,4*E*,8*Z*,10*E*)-tetraenoic acid isobutylamide (**85**) [100]. In a recent study, Chang et al. reported a new compound, echinalkamide (**90**), isolated from root cultures of *Echinacea purpurea* [101].

Different caffeic acid derivatives have been isolated and used as chemical markers in *Echinacea* species as they have been found in a relatively large number of plant families. Chlorogenic acid (**91**) and cynarin (**92**) are caffeoyl quinic acids. While the former is found in minimal amounts in all *Echinacea* species, the latter is predominant in *E. angustifolia* roots. Two caffeoyl tartaric acids, cichoric acid (**93**) followed by caftaric acid (**94**), are the main derivatives of *E. purpurea.* Echinacoside (**95**)

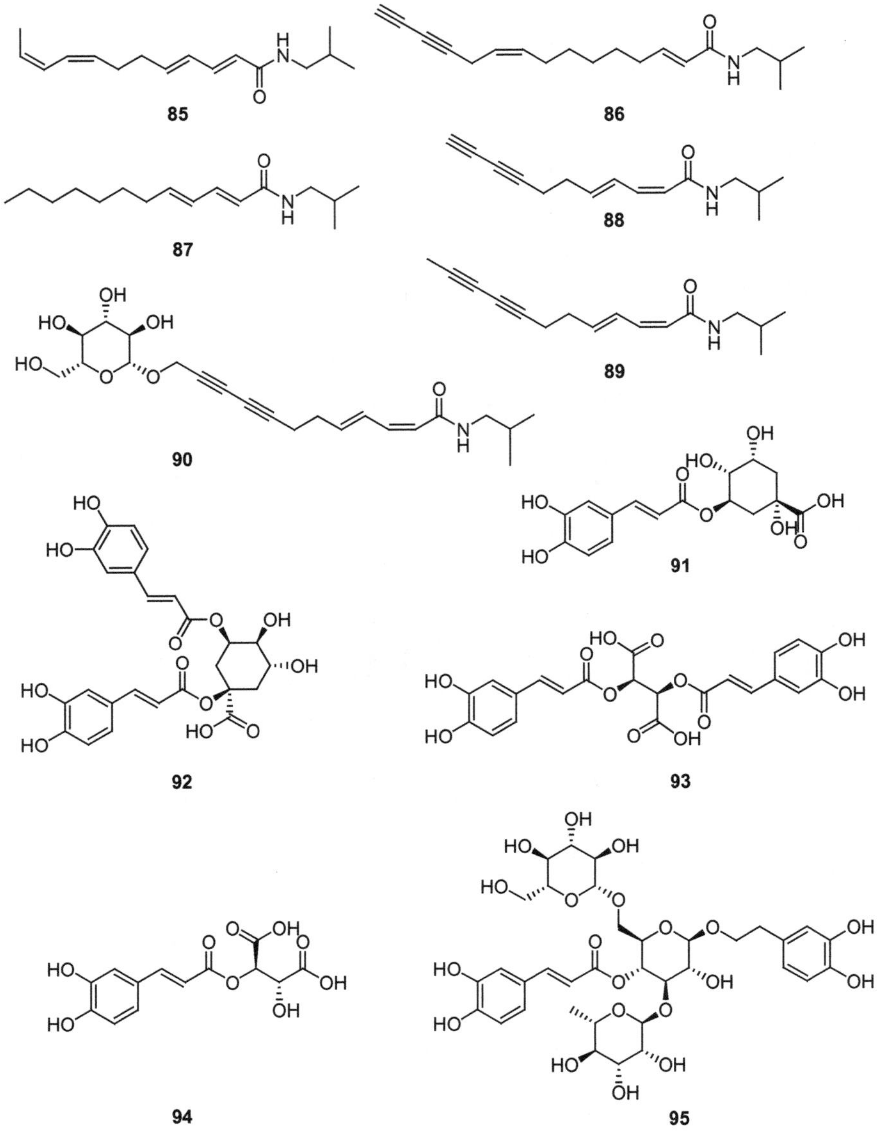

Fig. 10 Dodeca-(2*E*,4*E*,8*Z*,10*Z*)-tetraenoic acid isobutylamide (**85**), pentadeca-(2*E*,9*Z*)-diene-12,14-diynoic acid isobutylamide (**86**), dodeca-(2*E*,4*E*)-dienoic acid isobutylamide (**87**), undeca-(2*Z*,4*E*)-diene-8,10-diynoic acid isobutylamide (**88**), dodeca-(*Z*,4*E*)-diene-8,10-diynoic acid isobutylamide (**89**), echinalkamide (**90**), chlorogenic acid (**91**), cynarin (**92**), cichoric acid (**93**), caftaric acid (**94**), and echinacoside (**95**), isolated from *Echinacea* spp.

(caffeoyl glycoside) is a characteristic compound of *E. pallida* and *E. angustifolia*, but it is present only in traces in *E. purpurea* [102, 103]. *Echinacea* species also contain polysaccharides (inulin-type fructans, acidic arabinogalactans, xyloglucans and rhamnoarabinogalactans) and glycoproteins [102]. The biological effects of the phytochemicals in *E. purpurea* are discussed in Sect. 5.8.

4.9 Epimedium spp.

Epimedium is a genus of the family Berberidaceae consisting of more than 50 herbaceous species, distributed primarily in mainland China. Five of these species are included in the *Chinese Pharmacopoeia* including *Epimedium koreanum* Nakai (Plate 9), *Epimedium sagittatum* Maxim., *Epimedium brevicornum* Maxim., *Epimedium pubescens* Maxim. and *Epimedium wushanense* T.S Ying. However, more than one *Epimedium* species is used as the source of *Epidemii Folium* [104].

The major flavonoid compounds from *Epimedium* spp. (Horny Goat Weed or Barrenwort) are C-8 prenylated flavonol glycosides including icariin (**96**), and epimedins A (**97**), B (**98**), and C (**99**), which are used as quality control markers and constitute approximately 50% the total flavonoids [105]. Other flavonoids from *Epimedium* are diphylloside B (**100**), desmethylicaritin (**101**), icariside II (**102**), epimedokoreanin B (**103**), icaritin (**104**), and breviflavone B (**105**), as shown in Fig. 11. Biological studies on these constituents are discussed below in Sect. 5.9.

Plate 9 Leaves and flowers of *Epimedium koreanum* (Family Berberidaceae). Online source "Korean Barrenwort (*Epimedium koreanum*)" by 空猫 T. N, CC BY-SA 4.0 (https://www.inaturalist.org/photos/129359778). Accessed Oct. 31, 2022

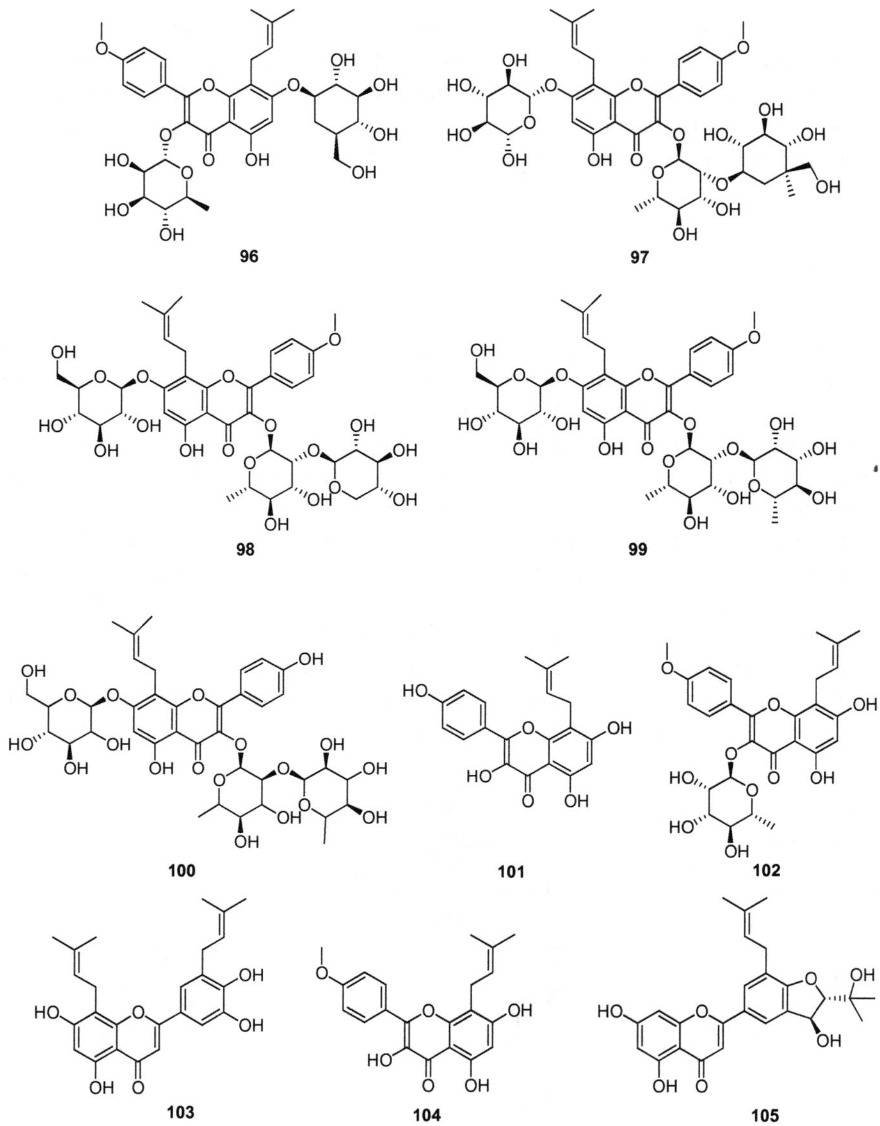

Fig. 11 Structures of the major flavonoid components found in *Epimedium* spp., namely, icariin (**96**), epimedin A (**97**), B (**98**), and C (**99**), diphylloside B (**100**), desmethylicaritin (**101**), icariside II (**102**), epimedokoreanin B (**103**), icaritin (**104**), and breviflavone B (**105**)

4.10 *Euterpe oleracea*

The genus *Euterpe* (family Arecaceae) comprises about 28 species of tropical palm trees distributed in South America [106], with three species being the most predominant. Thus, *Euterpe oleracea* Mart. (Plate 10) and *Euterpe precatoria* Mart. grow in the Amazon rainforest areas of Brazil and their fruits are known popularly as açaí, while *Euterpe edulis* Mart., a primary source of palm heart, is native to the Brazilian Atlantic Forest and its fruits are known as juçara [107].

More than 80 compounds have been reported from the berries of these tropical palms, including anthocyanins, non-anthocyanin flavonoids, phenolic acids, stilbenes, lignans and procyanidins, of which the levels vary depending on various factors such as the genotype investigated, ripening, extraction, processing, and storage conditions [107]. *Euterpe* fruits are considered abundant sources of anthocyanins and the most predominant group are the cyanidin glycosides. Cyanidin-3-glucoside (**106**) (Fig. 12) and cyanidin-3-rutinoside (**107**) are the major compounds present in the fruits of *E. oleracea*, *E. precatoria* and *E. edulis*, in addition to pelargonidin, peonidin and malvidin glycosides, which have been found at much lower concentration levels [108, 109].

As in other tropical fruits, flavonoids, and phenolic acids in their free or glycosylated forms are frequently found in *Euterpe* fruits. Epicatechin (**59**), quercetin (**108**), ferulic (**39**), gallic (**109**), protocatechuic (**110**) and vanillic (**40**) acids as well

Plate 10 Fruits of *Euterpe oleracea* (Family Arecaceae). Online source "Açaí Palm (*Euterpe oleracea*)" by Frederico Acaz Sonntag, CC BY-NC 4.0 (https://www.inaturalist.org/photos/86960923). Accessed Nov. 15, 2022

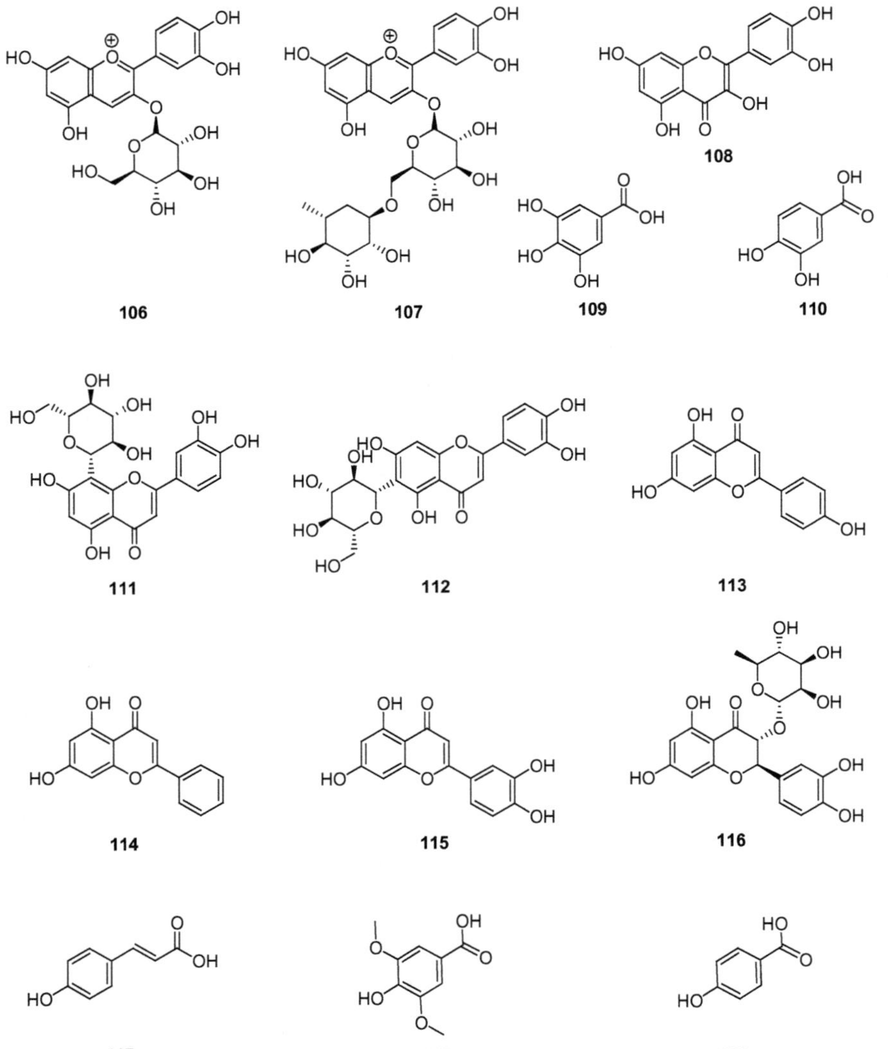

Fig. 12 Structures of cyanidin-3-glucoside (**106**), cyanidin-3-rutinoside (**107**), quercetin (**108**), gallic acid (**109**), protocatechuic acid (**110**), orientin (**111**), isoorientin (**112**), apigenin (**113**), chrysin (**114**), luteolin (**115**), taxifolin-3-rhamnoside (**116**), *p*-coumaric acid (**117**), syringic acid (**118**), and *p*-hydroxybenzoic acid (**119**) isolated from *Euterpe oleracea*

as glycosylated derivatives of apigenin, chrysoeriol and dihydrokaempferol are the main compounds present in *E. edulis.* In turn, orientin (**111**) and isoorientin (**112**) occur in *E. oleracea* together with apigenin (**113**), chrysin (**114**), epicatechin (**59**), catechin (**60**), luteolin (**115**), vitexin (**55**), taxifolin-3-rhamnoside (**116**), as well as gallic (**109**), caffeic (**38**), *p*-coumaric (**117**) and chlorogenic acids (**91**). Catechin (**60**), orientin (**111**), quercetin (**108**), and vanillic (**40**), syringic (**118**), and chlorogenic acids have been reported from *E. precatoria* [107].

A variety of phenolic acids have been identified in *E. oleracea* including vanillic (**40**), syringic (**118**), *p*-hydroxybenzoic (**119**), protocatechuic (**110**), and ferulic acids (**39**), as well as flavonoids (e.g., quercetin (**108**)) and procyanidin oligomers [cyanidin-3-glucoside (**106**) and cyanidin-3-rutinoside (**107**)] (Figs. 3 and 12) [110]. Studies on the biological testing of this botanical are discussed in Sect. 5.10.

4.11 Garcinia gummi-gutta

The genus *Garcinia* (family Clusiaceae) consists of approximately 400 species of trees or shrubs distributed in tropical Asia, Africa, New Caledonia, Polynesia, and Brazil [111, 112]. Among the edible species, *Garcinia mangostana* L., *Garcinia indica* (Thouars) Choisy, and *Garcinia gummi-gutta* (L.) N. Robson (syn. *Garcinia cambogia* Desr.) (Plate 11) are notable species from an economic point of view

Plate 11 Branches and fruits of *Garcinia gummi-gutta* (Family Clusiaceae). Online source Malabar tamarind/Kudam puli (*Garcinia gummi-gutta*)—Grafted (Plate 9-Garcinium). Accessed Oct. 19, 2022

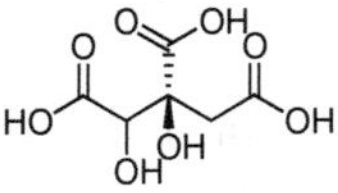

120

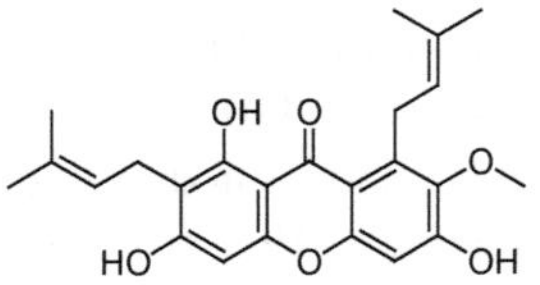

121

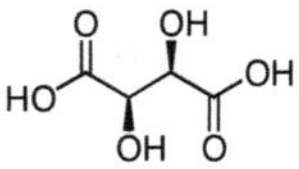

122

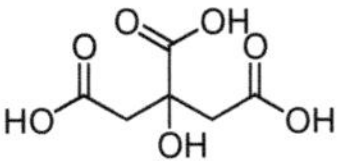

123

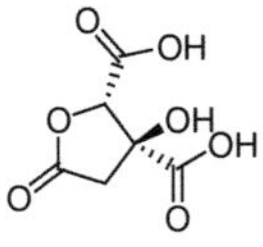

124

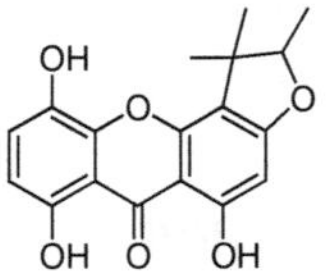

125

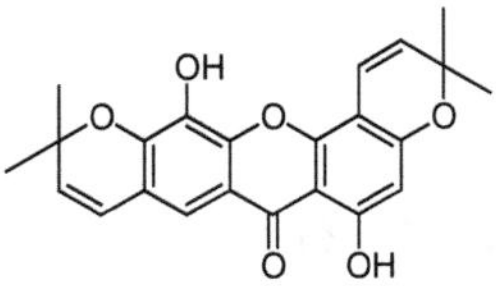

126

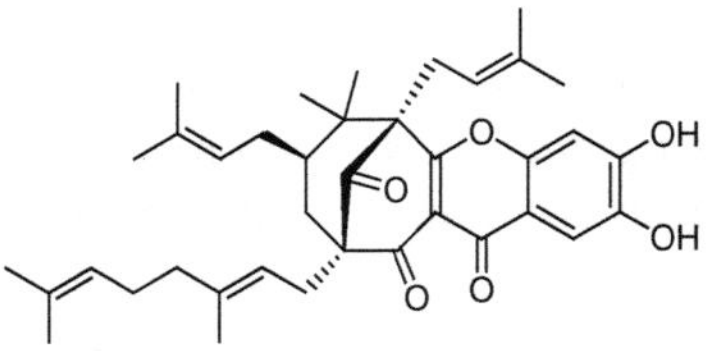

127

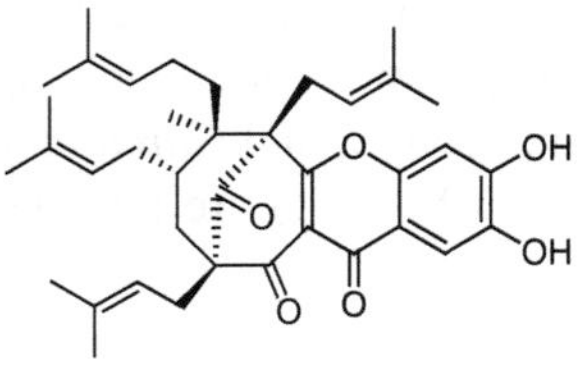

128

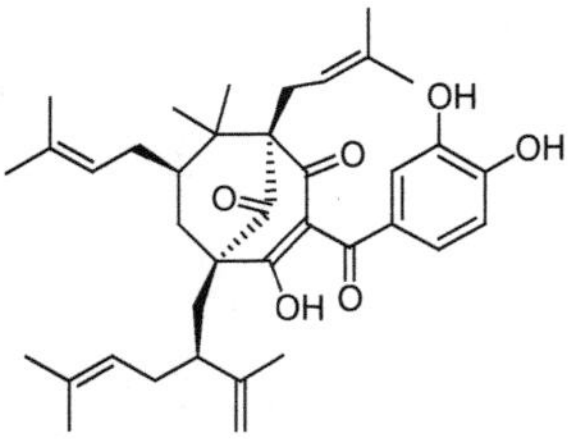

129

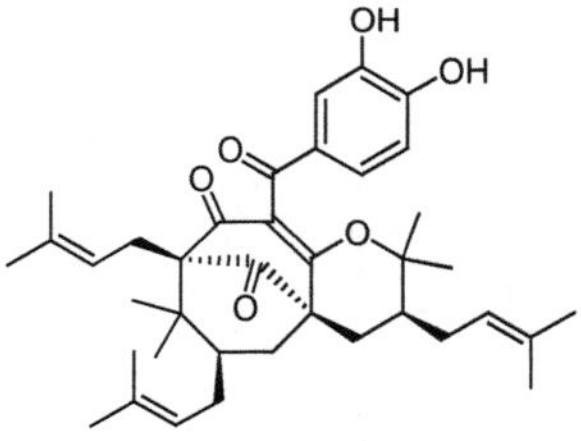

130

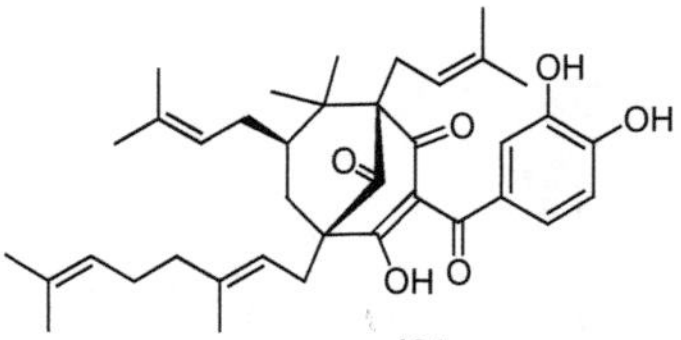

131

Fig. 13 Hydroxycitric acid (HCA) (**120**), α-mangostin (**121**), tartaric acid (**122**), citric acid (**123**), garcinia lactone (**124**), garbogiol (**125**), rheediaxanthone (**126**), oxy-guttiferone M (**127**), oxy-guttiferone K (**128**), garcinol (**129**), isogarcinol (**130**), and guttiferone M (**131**), isolated from *Garcinia gummi-gutta*

[113, 114], but several other species are also used traditionally for various culinary and medicinal purposes [113].

Garcinia species contain in their fruits various compounds, such as organic acids, polyisoprenylated benzophenones, prenylated xanthones and flavonoids [111]. Recently, a simultaneous determination of some of these bioactive compounds present in leaves of different *Garcinia* species was performed using LC-MS/MS. Hydroxycitric acid (HCA, garcinia acid, **120**) was present in higher amounts in *G. gummi-gutta* and *G. indica* when compared to other species in the genus, while α-mangostin (**121**) is a major constituent of *G. mangostana* (Fig. 13) [115].

Published phytochemical studies have revealed the presence in *Garcinia gummi-gutta* of alkaloids, flavonoids, organic acids (HCA (**120**), tartaric acid (**122**), and citric acid (**123**)), garcinia lactone (**124**), xanthones (garbogiol (**125**), rheediaxanthone (**126**), oxy-guttiferone M (**127**), oxy-guttiferone K (**128**)) and benzophenones (garcinol (**129**), isogarcinol (**130**), guttiferone M (**131**)), among other compounds isolated from various parts of the plant (Fig. 13) [116, 117].

Supplements from *Garcinia gummi-gutta* contain from 20 to 60% HCA (**120**). Compound **120** is an α,β-dihydroxy-tricarboxylic acid. The fruit contains approximately 10–30% of **120**, which can be isolated in the free form, as a mineral salt or as the lactone **124**. A discussion on the biological effects of this botanical is given in Sect. 5.11.

4.12 Ginkgo biloba

Ginkgo biloba L. (family Ginkgoaceae) (Plate 12), is a deciduous tree considered as the only living species of the Ginkgophyta division in the family Ginkgoaceae, having an ancient history of medical use dating back more than 5,000 years [118].

The major chemical constituents of *Ginkgo biloba* are flavonoids (flavonols, biflavones, and biginkgosides), terpenoid lactones (ginkgolides and bilobalide), and ginkgolic acids [119–121]. The major flavonol components are glycosides of quercetin (**108**), kaempferol (**132**), and isorhamnetin (**133**) that contain a sugar moiety at the C-3 hydroxy group, containing glucose and/or rhamnose in a variety of mono-, di-, or triglycoside substitution patterns. The biflavones ginkgetin (**134**), isoginkgetin (**135**), and bilobetin (**136**) that have been isolated from *G. biloba* are shown in Fig. 14. The flavonol glycosides quercetin 3-*O*-β-D-glucopyranoside (**137**) and quercetin 3-*O*-α-L-[6‴-*p*-coumaroyl-(β-D)-glucopyranosyl-(1→2)-rhamnopyranoside]-7-*O*-β-D-glucopyranoside (**138**) have also been found in *G. biloba* [122] (Fig. 14).

Biginkgosides are a group of minor components isolated by Ma et al. in 2016 from an extract of *Ginkgo biloba* leaves [120]. These compounds are dimers possessing a cyclobutene ring formed by a [2+2]-cycloaddition between two flavonol coumaroyl

Plate 12 Leaves of *Ginkgo biloba* (Family Ginkgoaceae). Online source "*Ginkgo biloba.*" No caption (https://www.gardenia.net/plant/ginkgo-biloba-majestic-tubifolia). Accessed Oct. 27, 2022

glucorhamnosides [120]. Shown in Fig. 15 are biginkgosides E (**139**), H (**140**) and F (**141**) [120].

Terpenoid trilactones are a unique bioactive class of compounds isolated only from *Ginkgo biloba* so far [121]. Ginkgolides are diterpenes consisting of six five-membered rings, including one spiro [4, 4] nonane carbocyclic ring, three lactone rings, and a tetrahydrofuran ring, and bilobalide (**142**), shown in Fig. 15, is a sesquiterpene lactone [121]. Ginkgolides A (**143**), B (**144**), C (**145**), J (**146**) also shown in Fig. 15, and these and bilobalide occur typically at concentration levels of 5.4–6.6% in standardized *Ginkgo biloba* extracts [121].

In the U.S., most of *G. biloba* products on the market are made from whole, cut, and powdered leaves, and dry leaf extracts and liquid leaf extracts. These ginkgo extracts are available in the form of powders, tinctures, and tablets and may be

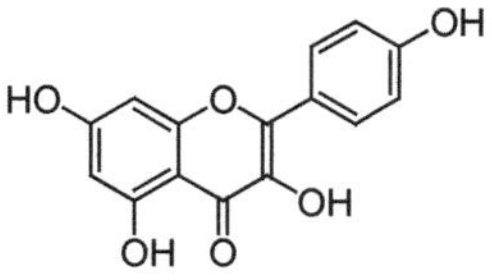

132

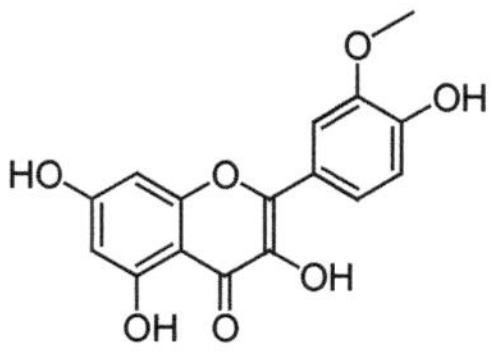

133

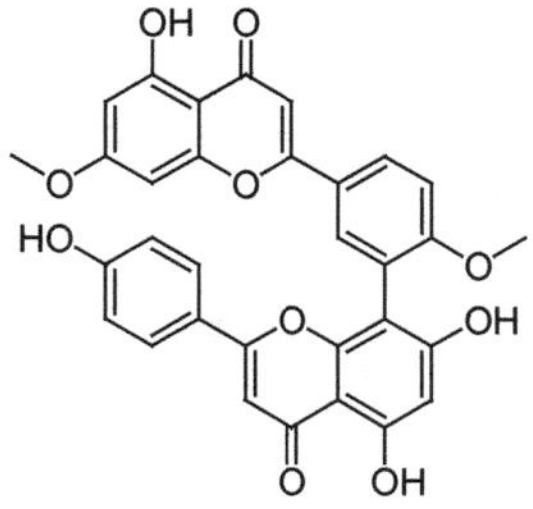

134

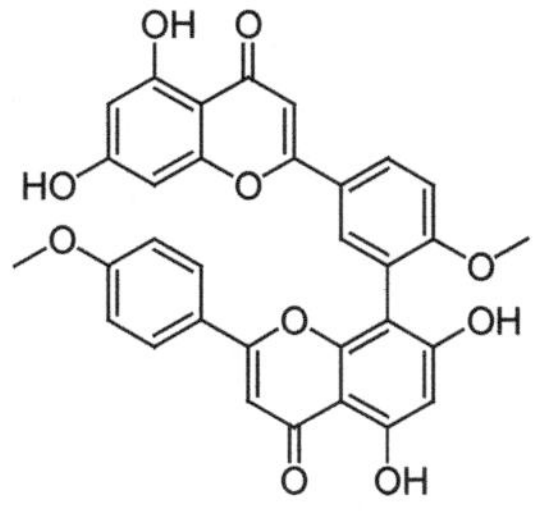

135

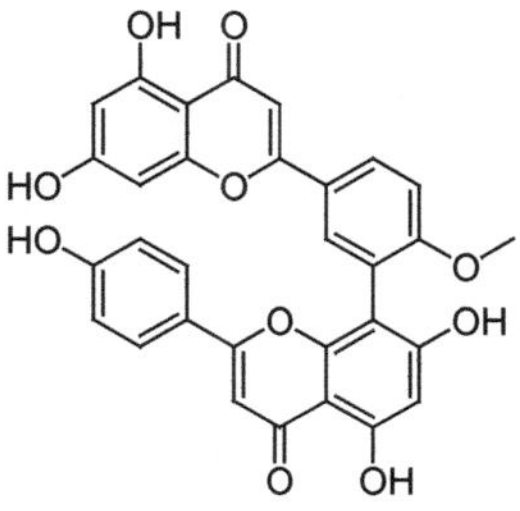

136

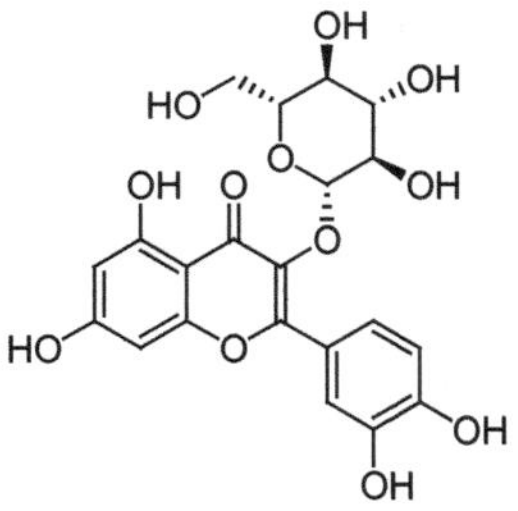

137

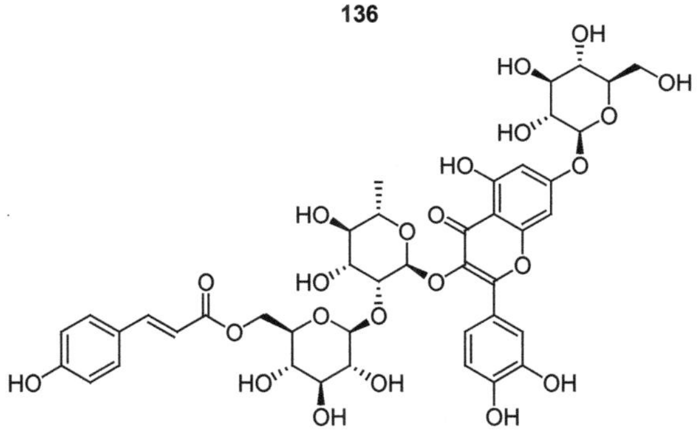

138

Fig. 14 Structures of flavonol aglycones kaempferol (**132**), and isorhamnetin (**133**), and biflavones ginkgetin (**134**), isoginkgetin (**135**), and bilobetin (**136**) isolated from *Ginkgo biloba*. Structures of the flavonoid glycosides quercetin 3-*O*-β-D-glucopyranoside (**137**) and quercetin 3-*O*-α-L-[6‴-*p*-coumaroyl-(β-D)-glucopyranosyl-(1→2)-rhamnopyranoside]-7-*O*-β-D-glucopyranoside (**138**) also isolated from *Ginkgo biloba*

standardized to contain 24% flavonol glycosides, 6% terpene trilactones and less than 5 ppm of ginkgolic acids (known to be contact allergens) [123, 124]. Selected biological studies on these isolates are presented in Sect. 5.12.

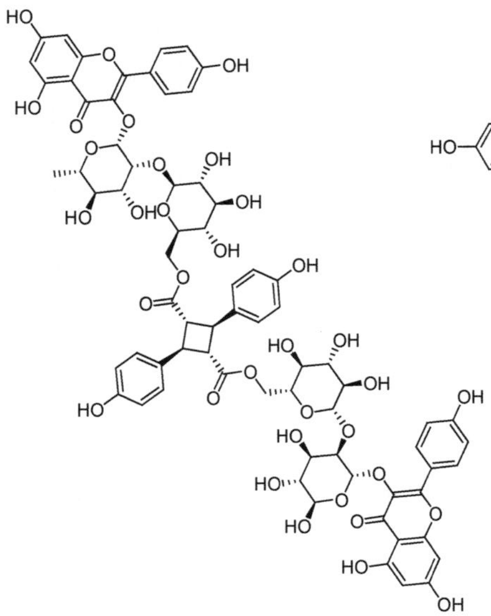

141

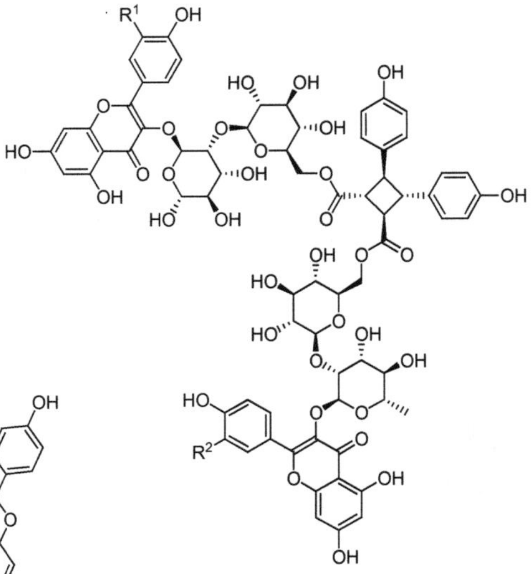

139 R¹ = OH, R² = OH
140 R¹ = H, R² = OH

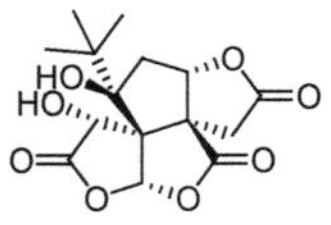

142

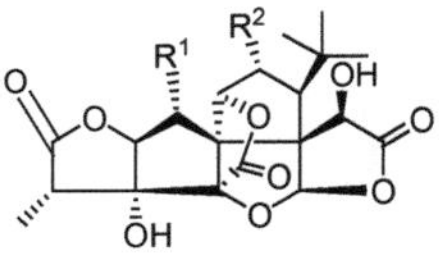

143 R¹ = H, R² = H
144 R¹ = OH, R² = H
145 R¹ = OH, R² = OH
146 R¹ = H, R² = OH

Fig. 15 Structures of biginkgosides E (**139**), H (**140**), and F (**141**) and terpenoid trilactones bilobalide (**142**), and ginkgolides A (**143**), B (**144**), C (**145**), and J (**146**) from *Ginkgo biloba*

4.13 Hypericum perforatum

St. John's wort (*Hypericum perforatum* L.) (Plate 13) is a yellow-flowered and herbaceous perennial plant native to Europe and Asia [125]. It is used in the United States as a botanical dietary supplement. *Hypericum* species are of the family Hypericaceae and consist of over 500 perennial herbs or shrubs subdivided into 36 sections in the genus [126]. Several of these species such as *Hypericum androsaemum* L., *Hypericum barbatum* Jacq., *Hypericum crux-andreae* (L.) Crantz, *Hypericum hirsutum* L., *Hypericum humifusum* L., *Hypericum maculatum* Crantz, *H. montanum* L., *Hypericum patulum* Thunb., and *Hypericum tetrapterum* Fr. have been reported as adulterants of St. John's wort (*H. perforatum*) [127]. The characteristic constituents that have been isolated from *Hypericum* species (flowering aerial

Plate 13 Leaves and flowers of *Hypericum perforatum* (Family Hypericaceae). Online source "*Hypericum perforatum.*" No caption (https://www.mygardenlife.com/plant-library/4044/hypericum/perforatum). Accessed Oct. 27, 2022

parts) are presented in Figs. 15 and 16 and include flavonoids and other phenolics, polyprenylated acylphloroglucinols (PAPs), xanthones, naphthodianthrones, and monoterpenes [125, 126].

Quercetin (**108**) and flavonoid glycosides including avicularin (**147**), guaiaverin (**148**), hyperoside (**56**), isoquercitrin (**145**), and rutin (**57**) are common constituents among several *Hypericum* species. Among the various phenolic acids found in *Hypericum* species, chlorogenic acid is the most abundant and widely identified [127, 128]. Hyperforin (**149**), a prenylated phloroglucinol, and hypericin (**150**), a naphthodianthrone, shown in Fig. 16, are regarded as the major active constituents from St. John's wort (*H. perforatum*) [125]. The content of these constituents varies considerably between and within *Hypericum* species. For example, *H. perforatum* and the main adulterant *H. maculatum* have a similar content level of hypericin, but

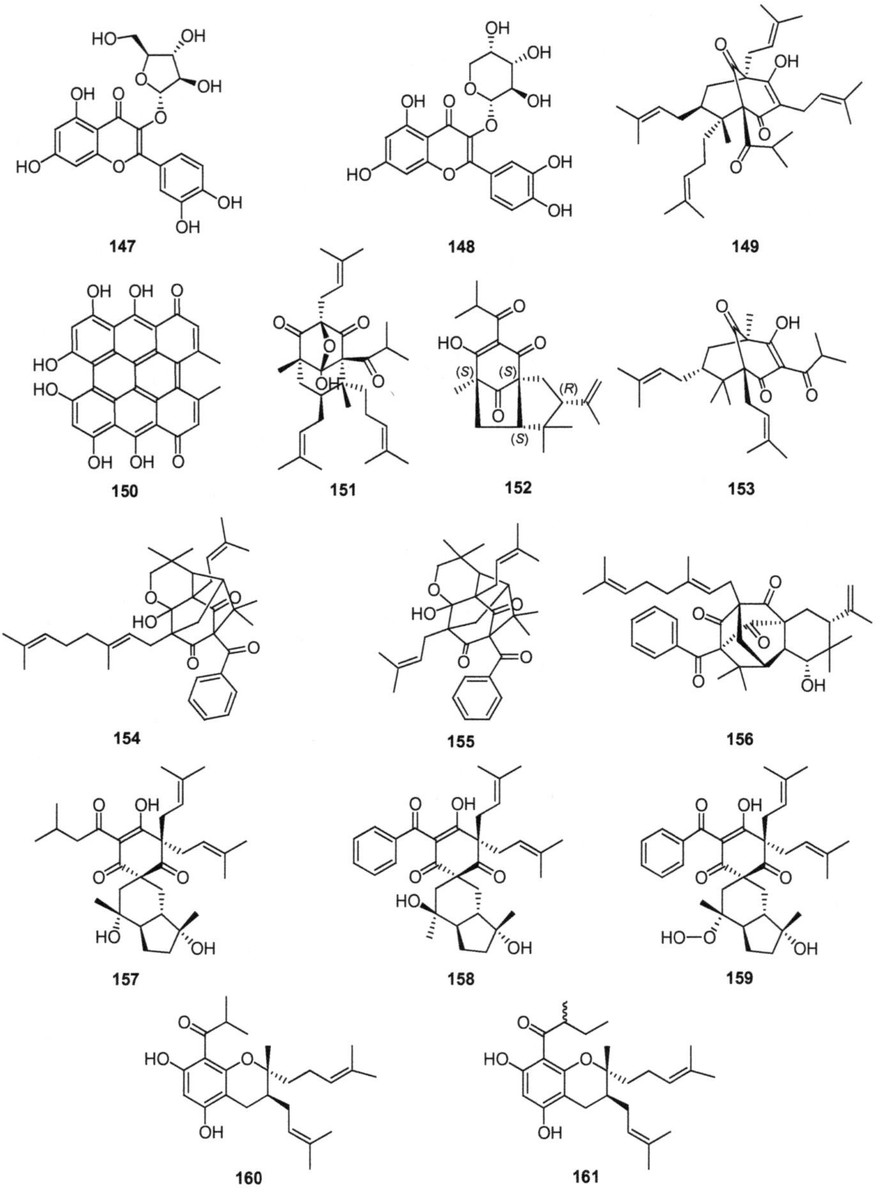

Fig. 16 Structures of constituents from *Hypericum perforatum*, including avicularin (**147**), guaiaverin (**148**), the prenylated phloroglucinol hyperforin (**149**), the napthodianthrone hypericin (**150**), and the phloroglucinol derivatives hyperibone J (**151**), ialibinone B (**152**), hyperpapuanone (**153**), sampsoniones A (**154**), B (**155**), and I (**156**), tomoeone F (**157**), hyperbeanols B (**158**) and D (**159**), and hypercalyxones A (**160**) and B (**161**)

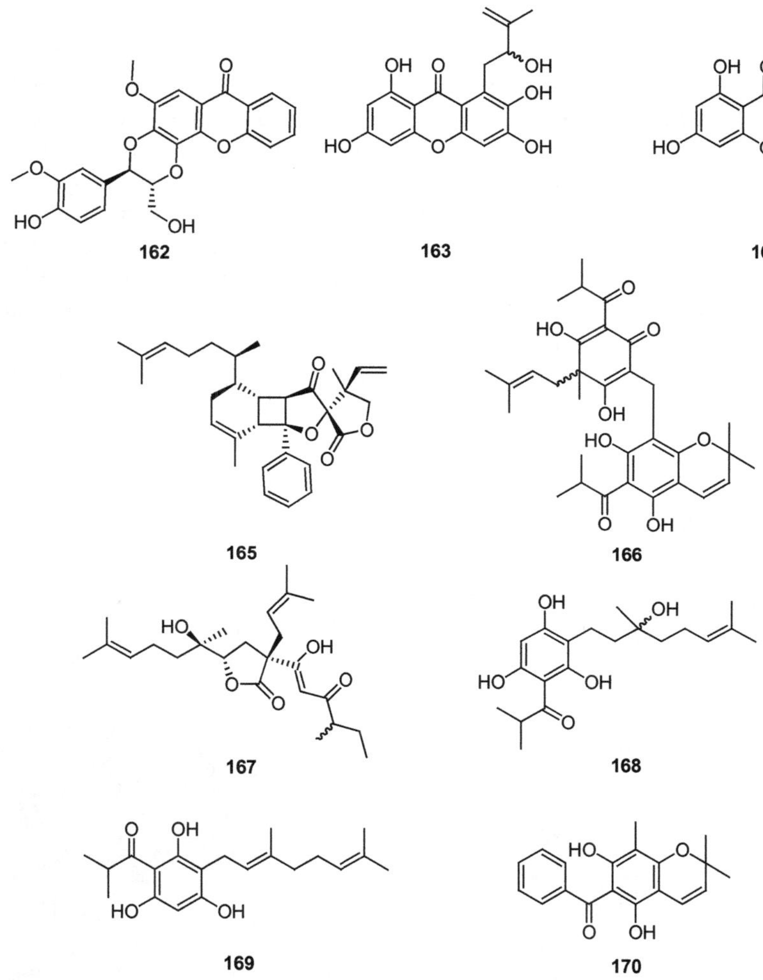

Fig. 17 Structures of additional constituents from *Hypericum perforatum* including the xanthone derivatives kielcorin (**162**) and hyperxanthones C (**163**) and E (**164**), and the further compounds biyouyanagin A (**165**), isohyperbrasilol B (**166**), yojironin A (**167**), hyperjovinols A (**168**), 3-geranyl-1-(2-methylpropanoyl)-phloroglucinol (**169**), and cariphenone A (**170**)

H. maculatum has only low concentrations of hyperforin (0.004–0.018%). Several other species including *H. olympicum* and *H. polyphyllum* have similar hyperforin (**149**) and hypericin (**150**) contents compared to *H. perforatum*, highlighting the low predictive value of these compounds in determining adulteration in *H. perforatum* preparations [127, 129].

Other phytochemicals (shown in Fig. 16) isolated from *H. perforatum* include phloroglucinol derivatives such as hyperibone J (**151**), ialibinone B (**152**), hyperpapuanone (**153**), sampsoniones A (**154**), B (**155**) and I (**156**), tomoeone F (**157**), hyperbeanols B (**158**) and D (**159**), and hypercalyxones A (**160**) and B (**161**). Sampsoniones A (**154**) and B (**155**) are classified as polyprenylated benzoylphloroglucinols possessing a novel tetracyclo-polyketone functionality with a homoadamantyl-like unit formed by complex prenyl group cyclization [130]. Additionally, the xanthone derivatives kielcorin (**162**) and hyperxanthones C (**163**) and E (**164**), and the further compounds biyouyanagin A (**165**) [130], isohyperbrasilol B (**166**), yojironin A (**167**), hyperjovinol A (**168**), 3-geranyl-1-(2-methylpropanoyl)-phloroglucinol (**169**), and cariphenone A (**170**), have been isolated from this species (Fig. 17) [130, 131]. Selected biological studies on these *H. perforatum* constituents will be presented in Sect. 5.13.

4.14 *Lepidium meyenii*

The genus *Lepidium* (Brassicaceae) consists of approximately 250 herbaceous species [132], including the Andean species *Lepidium meyenii* Walp. (Plate 14), which is known commonly as maca and is used as a botanical dietary supplement.

Plate 14 Tubers of *Lepidium meyenii* (Family Brassicaceae). Online source "Maca is a root native to the high Andes of Peru" by Holly Holmes, CC BY-NC-ND 2.0 (https://www.flickr.com/photos/129099219@N03/23657042145). Accessed Oct. 31, 2022

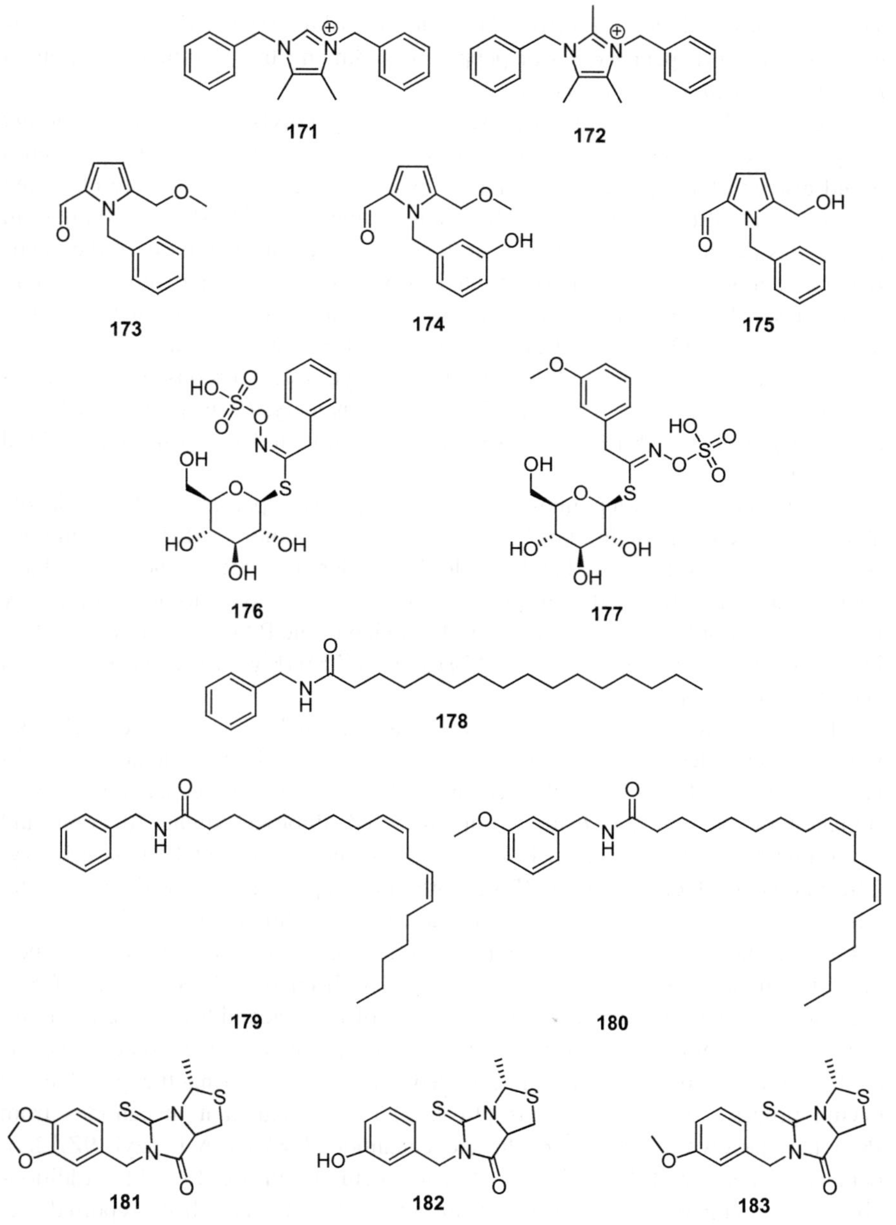

Fig. 18 Structures of constituents of *Lepidium meyenii* including the imidazole alkaloids lepidiline A (**171**) and lepidiline B (**172**), the pyrrole alkaloids macapyrrolins A (**173**), B (**174**) and C (**175**), the glucosinolates glucotropaeolin (**176**) and glucolimnanthin (**177**), the macamides *N*-benzylhexadecanamide (**178**), *N*-benzyl-(9*Z*,12*Z*)-octadecadienamide (**179**), and *N*-(3-methoxybenzyl)-(9*Z*,12*Z*)-octadecadienamide (**180**), and the thiohydantoins (+)-meyeniin A (**181**), meyeniin B (**182**), and meyeniin C (**183**)

This genus also includes a group of more widely distributed species known as peppergrass, pepperwort, pepperweed or peppercress. Among them is the cosmopolitan *Lepidium sativum* L. (garden cress).

The main class of compounds present in the hypocotyls of the genus *Lepidium* are glucosinolates (GSLs), which are sulfur- and nitrogen-containing glycosides found also in other species of the family Brassicaceae [133]. They are classified into aliphatic GSLs, aryaliphatic GSLs, and indole GSLs [134]. Even though all three types of glucosinolates have been found in different plant parts of the genus *Lepidium*, mono-, di- and tri-substituted arylaliphatic GSLs are most abundant in the seeds. Interestingly, highly substituted arylaliphatics GSLs, such as 3,4,5-trimethoxybenzyl GSL and 3,4-dimethoxybenzyl GSL, are restricted to only a few plant genera [135]. Indeed, 3,4,5-trimethoxybenzyl GSL is proposed as a chemotaxonomic marker for several *Lepidium* species, namely, *L. hyssopifolium* Desv., *L. coronopus* (L.) Al-Shebbaz, *L. sordidum* A. Gray, and *L. densiflorum* Schrad [134, 136].

Another group of compounds reported in *Lepidium* species are the imidazole alkaloids. Seven dimeric imidazole alkaloids lepidine, lepidine AK [137], lepidines B–F, and two monomeric imidazole alkaloids, semilepidinosides A and B [138] have been isolated and identified from the seeds of *L. sativum.* Also, atypical quaternary imidazole alkaloids (e.g. lepidiline A (**171**) and lepidiline B (**172**)) and pyrrole alkaloids [e.g. macapyrrolins A (**173**), B (**174**) and C (**175**)) (Fig. 18) have been isolated and identified from maca [139–141].

Thus, the main secondary metabolites isolated from *L. meyenii* include glucosinolates, macamides, macaenes, thiohydantoins and alkaloids, as presented in Fig. 18 [142, 143]. The GSLs characterized are *N*-hydroxy sulfates consisting of a *β*-D-glucopyranose unit, an *O*-sulfated anomeric (*Z*)-thiohydroximate function, and a variable aglycone side chain [135, 142]. The most abundant GSL is a benzyl GSL known as glucotropaeolin (**176**), representing approximately 80% of the total GSLs, followed by *m*-methoxyglucotropaeolin (glucolimnanthin, **177**) (Fig. 18) [142]. The biological activities of these compounds are also attributed to their degradation products [135]. Additionally, it has been postulated that the GSLs present in *L. meyenii* may serve as precursors of the macamides. These are long-chain alkyl *N*-benzylamides present only in this species and they could be generated, together with macaenes (unsaturated fatty acids), during the post-harvest drying process of the hypocotyls [144]. The most abundant macamides from Peruvian maca which are *N*-benzylhexadecanamide (**178**) and *N*-benzyl-(9*Z*,12*Z*)-octadecadienamide (**179**), which are shown structurally in Fig. 18 [142]. Additionally, the 3-methoxy derivative of the macamide **179** has also been reported, i.e. *N*-(3-methoxybenzyl)-(9*Z*,12*Z*)-octadecadienamide (**180**) [145].

Thiohydantoins, sulfur analogs of hydantoin (imidazolidine-2,4-diones), are also present in maca and several of these structures have been reported including meyeniins A (**181**), B (**182**), and C (**183**) [142]. The content of the secondary metabolites present in *L. meyenii* varies depending on the drying method [144], the area of cultivation, and the color of the hypocotyls (with black, purple, red and yellow varieties) [146]. The hypocotyl color may be a less important factor for the variation of the resultant chemical composition [147]. A summary of biological studies on *L. meyenii* is presented in Sect. 5.14.

4.15 Linum usitatissimum

The genus *Linum* (family Linaceae) includes about 200 species of annual or perennial species that are distributed in temperate and subtropical areas. This genus includes the agronomically important species *Linum usitatissimum* L. (common flax) (Plate 15), of which the fiber is used to produce linen and the seeds to produce linseed oil [148].

Plate 15 Leaves and flowers of *Linum usitatissimum* (Family Linaceae). Online source "*Linum usitatissimum.*" Photograph courtesy Roser1954 (https://upload.wikimedia.org/wikipedia/commons/b/b2/Common_Flax_or_Linseed_%28Linum_usitatissimum%29_flower._Chapeltoun_North_Ayrshire.jpg). Accessed Oct. 27, 2022

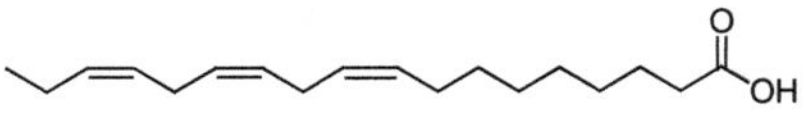

184

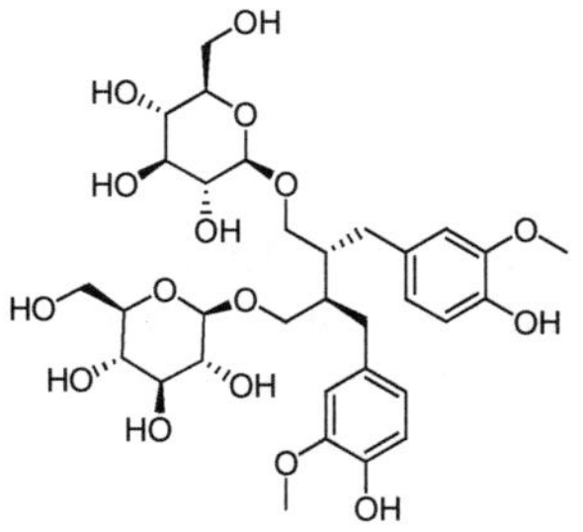

185

Fig. 19 α-Linoleic acid (**184**) and secoisolariciresinol diglucoside (**185**) isolated from *Linum usitatissimum*

The fatty acid profiles for various *Linum* species have been reported. The predominant chain length is 18 carbons with either linolenic or linoleic acid being the principal fatty acid [149]. *L. usitatissimum* contains a high content of the ω-3 fatty acid, α-linolenic acid (**184**), in particular (Fig. 19).

Lignans, phenolic acids, phytosterols, cyanogenic glycosides, and flavonoids as well as various miscellaneous compounds have been also reported in the genus *Linum*. The principal types of lignans belong to the dibenzyl butane group, including secoisolariciresinol diglucoside (**185**) (Fig. 19), and the aryltetralin group (e.g. 5-methoxypodophyllotoxin) and the arylnaphthalene group (e.g. justicidin B) [149]. The phytoestrogenic lignan, secoisolariciresinol diglucoside (**185**) is the principal dietary lignan present in flaxseed. The C-6 hydroxy of the glucose of compound **185** may be esterified to the hydroxymethylglutaric unit [150–152]. Currently available biological data supporting the use of flax seed is presented in Sect. 5.15.

4.16 Lycium spp.

The genus *Lycium* (family Solanaceae) contains approximately 80 species of shrubs or small trees distributed mainly in Asia, South America, and southern Africa. Two of the primary edible and medicinal species that are sold as dietary supplement on the U.S. are *Lycium barbarum* L. (Plate 16) and *Lycium chinense* Mill. [153]. The common names of these species are goji or wolfberry, respectively. These species are used interchangeably in traditional Chinese medicine (TCM), traditional Japanese Kampo medicine, and traditional Korean medicine [153]. Phytochemical studies of these species have focused on the leaves, fruits and root bark and have described the presence of polysaccharides, carotenoids, flavonoids, alkaloids, amides and terpenoids [154].

Polysaccharides isolated from the *L. barbarum* fruits comprise 5–8% of the dried fruit and consist of a complex mixture of highly branched and partly characterized polysaccharides and proteoglycans [155, 156]. The six main monosaccharides present in the *L. barbarum* polysaccharides are fucose, arabinose, xylose, glucose, mannose, and galactose [156, 157]. Saccharides from *L. barbarum* that have been shown with biological properties as described in Sect. 5.16.

Plate 16 Fruits of *Lycium barbarum* (Family Solanaceae). Online source "*Lycium barbarum* (Goji berries)" by Lotus Johnson, CC BY-NC 2.0 (https://www.flickr.com/photos/ngawangchodron/15121728286). Accessed Oct. 31, 2022

4.17 *Marrubium vulgare*

The genus *Marrubium* (family Lamiaceae) consists of about 40 species of herbaceous plants native to temperate regions of Europe and Asia, and currently distributed throughout the world [158]. Characteristic constituents of the genus *Marrubium* are diterpenes, phenylethanoid glycosides, and flavonoids [159]. The characteristic diterpenes can be classified in pre-furanoid, e.g., premarrubiin (**186**) and furanoid labdanoids including marrubiin (**187**), a marker compound for the genus [160].

Marrubium vulgare L. (horehound) (Plate 17) is a widespread and most known representative species of the genus indigenous to Central and Western Asia, North Africa, Europe, and South America [161, 162]. Some of the constituents present in *M. vulgare* (obtained mainly from the aerial parts), include monoterpenoids, diterpenoids as well as phenolic compounds [161]. Specific constituents isolated from the *M. vulgare* herb include the substituted monoterpenoid, marrubic acid (**188**), the diterpenoids marrubiin (**187**), 12(*S*)-hydroxymarrubiin (**189**), 11-oxomarrubiin (**190**), marrubenol (**191**), pre-marrubiin (**186**), and marruliba-acetal (**192**), and the phenylethanoid glycoside, marruboside (**193**), which are shown in Fig. 20 [161]. Studies on the bioactive constituents of this species are discussed in Sect. 5.17.

Plate 17 Leaves of *Marrubium vulgare* (Family Lamiaceae). Online source "*Marrubium vulgare*" by Eugene Zelenko, CC BY-SA 2.0 (https://www.flickr.com/photos/eugenezelenko/401586185). Accessed Oct. 31, 2022

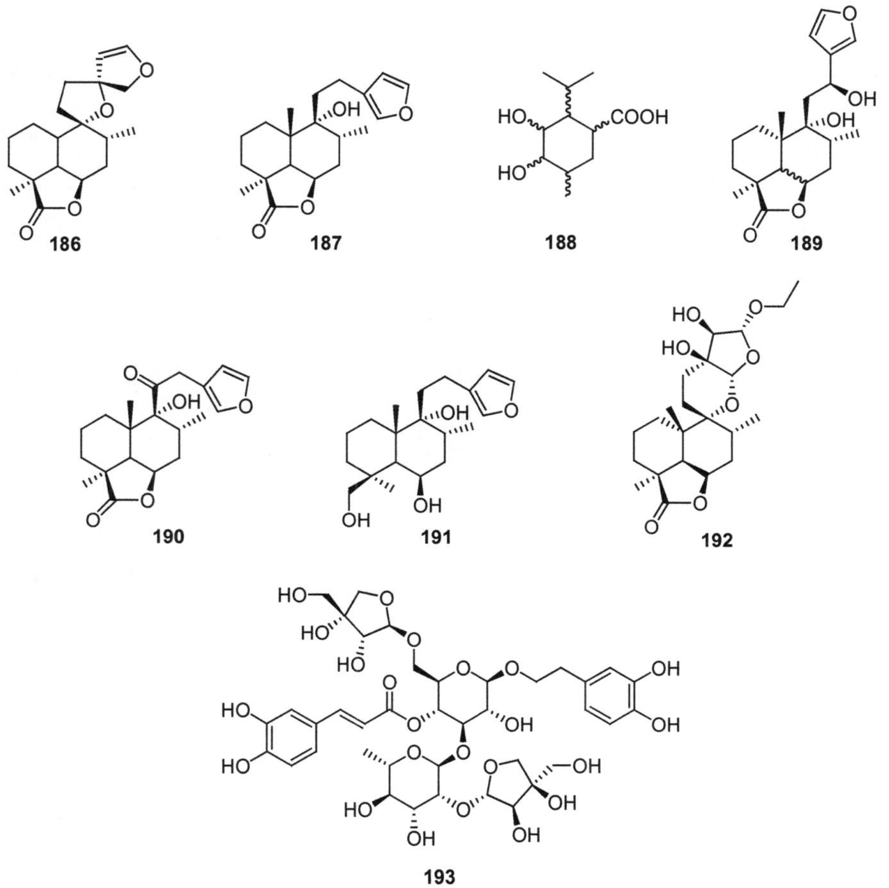

Fig. 20 Structures of constituents from *Marrubium vulgare* including the substituted monoterpenoid marrubic acid (**188**), the diterpenoids pre-marrubiin (**186**), marrubiin (**187**), (12*S*)-hydroxymarrubiin (**189**), 11-oxomarrubiin (**190**), marrubenol (**191**), and marruliba-acetal (**192**), and the phenylethanoid glycoside, marruboside (**193**)

4.18 Matricaria chamomilla

The genus *Matricaria* (family Asteraceae) comprises a small number of flowering plant species native to North America, Eurasia, and North Africa that have also been introduced in the southern hemisphere [163]. *Matricaria chamomilla* L. (syn: *Chamomilla recutita* (L.) Rauschert) (Plate 18) is the most widely distributed representative of the genus and is known as chamomile or German chamomile [164, 165]. The composition of *M. chamomilla* essential oil and extracts has been investigated widely and more than 120 constituents have been identified [166]. The essential oil (0.3–1.5% total yield) is comprised mainly of sesquiterpenoids such as α-bisabolol

Plate 18 Flowers of *Matricaria chamomilla* (Family Asteraceae). Online source "German Chamomile (*Matricaria chamomilla*)" by Adam Makhauri, CC BY-NC 4.0 (https://www.inaturalist.org/photos/197755254). Accessed Nov. 15, 2022

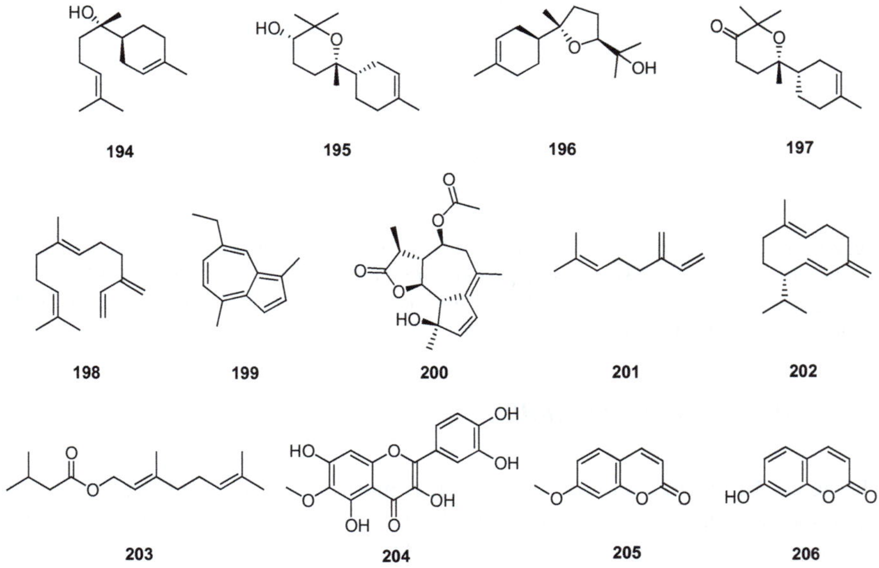

Fig. 21 α-Bisabolol (**194**), bisabolol oxide A (**195**), bisabolol oxide B (**196**), bisabolone oxide A (**197**), β-farnesene (**198**), chamazulene (**199**), matricin (**200**), myrcene (**201**), germacrene D (**202**), geranyl isovalerate (**203**), patuletin (**204**), herniarin (**205**) and umbelliferone (**206**) obtained from *Matricaria chamomilla* and related species

(**194**), bisabolol oxides A (**195**), and B (**196**), bisabolone oxide A (**197**) and β-farnesene (**198**) (Fig. 21). Additionally, the blue color typical of the majority of chamomile oil samples originates from chamazulene (**199**), an artefact of the distillation process derived from the nonvolatile sesquiterpene lactone matricin (**200**). The essential oil also contains spiroethers [167, 168].

According to the content of the main components of the essential oil, from two to six chemotypes have been distinguished. The *European Pharmacopoeia* differentiates between only two chemotypes, with one rich in α-bisabolol (**194**) (10–65%) and the other dominated by bisabolol oxides (29–81%), although their concentrations are different in the volatile oils from various regions of the world [166, 167]. The composition of the volatile oil of other *Matricaria* species has been also investigated. Bisabolol oxide A (**195**) is the predominant compound from the flower heads of *Matricaria aurea* (Loefl.) Sch.Bip, while the major constituents of the essential oil from aerial parts of *Matricaria discoidea* DC. are myrcene (**201**), β-farnesene (**198**), germacrene D (**202**), geranyl isovalerate (**203**), and spiroethers [169].

Chamomile flower extracts contain the flavonoids apigenin (**113**), luteolin (**115**), quercetin (**108**), and patuletin (**204**), lactones like matricin (**200**), coumarins like herniarin (**205**) and umbelliferone (**206**), and phenolic acids such as caffeic acid (**38**) and chlorogenic acid (**91**). The constituents of their essential oils are mainly chamazulene (**199**), (*E*)-β-farnesene (**198**), and α-bisabolol (**194**) (Figs. 3, 10, 12 and 21) [164, 170, 171]. Selected biological studies performed on these phytochemicals are presented in Sect. 5.18.

4.19 Momordica charantia

Momordica (family Cucurbitaceae) is a genus of about 60 species of annual or perennial climbing herbs, including edible tree crops such as *Momordica balsamina* L. (African pumpkin), *Momordica charantia* L. (bitter melon) (Plate 19), and *Momordica foetida* Schumach. (bitter cucumber), which are distributed predominantly in warm tropical regions of Africa and Southeast Asia [172]. These species are characterized by a wide diversity of compounds including phenolic acids, flavonoids, carotenoids, cucurbitane triterpenoids, and phytosterols isolated from different parts of the plant [173].

Cucurbitacins are bitter tasting, and are tetracyclic triterpenoids isolated from different *Momordica* species and other plants of the Cucurbitaceae family. They possess a 19(10→9β)-abeo-10α-lanost-5-ene skeleton and exhibit with a variety of oxygenated functional groups and positions of unsaturation. Momordicosides A–U, charantosides, goyaglycoside, kuguasaponins, and karavilosides are some cucurbitane-type triterpenes from *M. charantia* [174]. Certain cucurbitane-type triterpenoids from *M. balsamina* have unusual oxidation patterns. They have been divided in several groups, namely, those of the balsaminol, balsaminoside, balsaminagenin, karavilagenin, and cucurbalsaminol groups [175].

Plate 19 Fruits of *Momordica charantia* (Family Cucurbitaceae). Online source "*Momordica charantia*" by Eric Hunt, CC BY-NC-ND 2.0 (https://www.flickr.com/photos/ericinsf/1086530814). Accessed Oct. 31, 2022

Phytochemical studies performed on *M. charantia* suggest that *M. charantia* polysaccharides found in various parts of the plant have diverse biological activities [176, 177]. The heteropolysaccharide contents vary in their composition depending on their environmental conditions, but the polysaccharide content ranges from 5.91 to 10.62% of the dry powder and consists of arabinose, glucose, mannose, galactose, and rhamnose [176]. Correspondingly, the biological activities of these polysaccharides with different molecular weights and monosaccharide compositions can vary [176]. Cucurbitane-type compounds from *M. charantia* are considered to be responsible for the major bioactivities of this botanical and include cucurbitacins B (**207**), E (**208**), and I (**209**), charantoside A (**210**), momordicoside I (**211**), karavilagenin C (**212**), and kuguaglycoside A (**213**), as shown in Fig. 22 [175, 178]. Studies on the bioactivity of these constituents are discussed in Sect. 5.19.

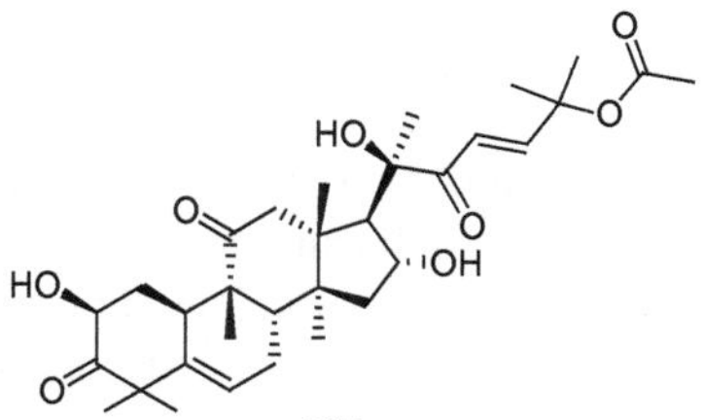

207

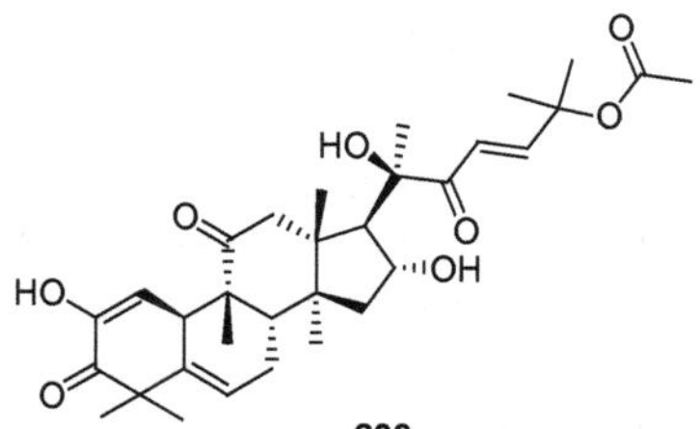

208

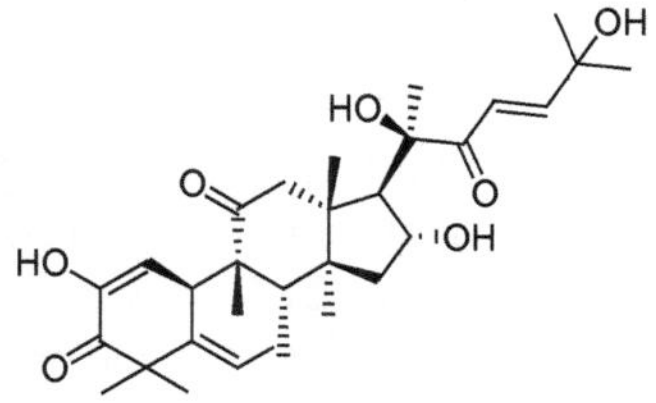

209

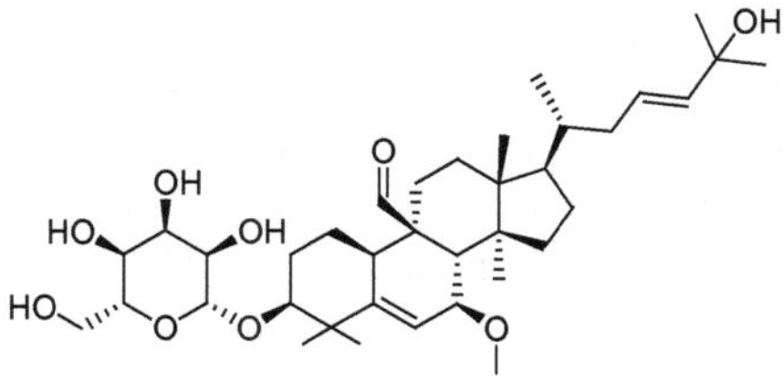

210

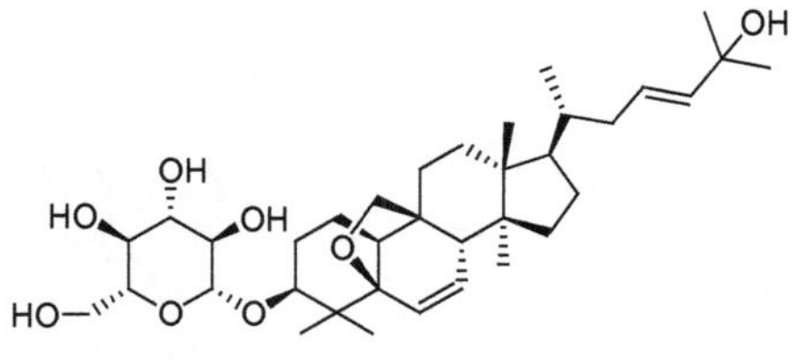

211

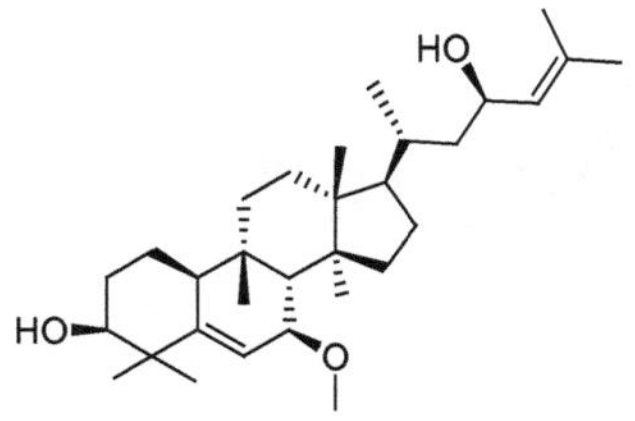

212

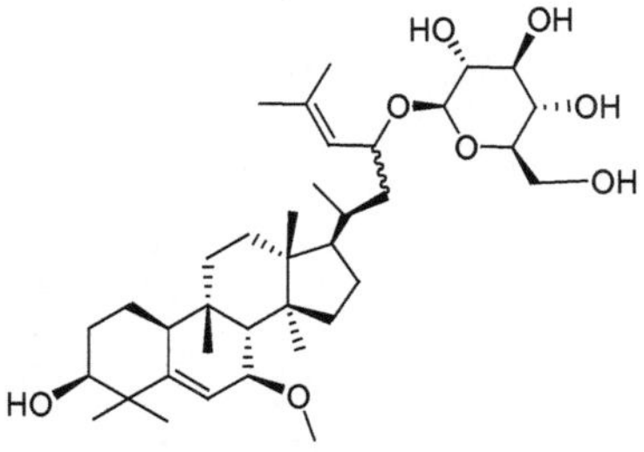

213

Fig. 22 Cucurbitacin B (**207**), cucurbitacin E (**208**), cucurbitacin I (**209**), charantoside A (**210**), momordicoside I (**211**), karavilagenin C (**212**), and kuguaglycoside A (**213**) isolated from *Momordica charantia*

4.20 *Nigella sativa*

Nigella is a small genus (family Ranunculaceae) of annual herbaceous plants native to the Mediterranean region from West Asia to northern India and includes about 26 accepted species [179]. Two *Nigella* species primarily have gained attention in the scientific literature, the ornamental species *Nigella damascena* L., and the popular spice *Nigella sativa* L. (black cumin) (Plate 20).

Plants from this genus are mainly consumed for their seeds and seed oil, which besides fatty acids and other nutrients, contain alkaloids, dolabellane diterpenoids, triterpenes and phenolic compounds [180]. *N. sativa* is characterized by a variable volatile composition that present different chemotypes according to the major chemical constituents (Fig. 23), including thymoquinone (**214**) (absent in *N. damascena*), the phenylpropanoid anethole (**215**), α-pinene (**216**), *p*-cymene (**217**), carvacrol (**218**) and thymol (**219**) [180]. *Nigella damascena* seed oil is characterized by the presence

Plate 20 Flowers and leaves of *Nigella sativa* (Family Ranunculaceae). Online source "Nigella Sativa (black cumin)-cultivated" by Arthur Chapman, CC BY-NC-SA 2.0 (https://www.flickr.com/photos/arthur_chapman/4848335080). Accessed Oct. 31, 2022

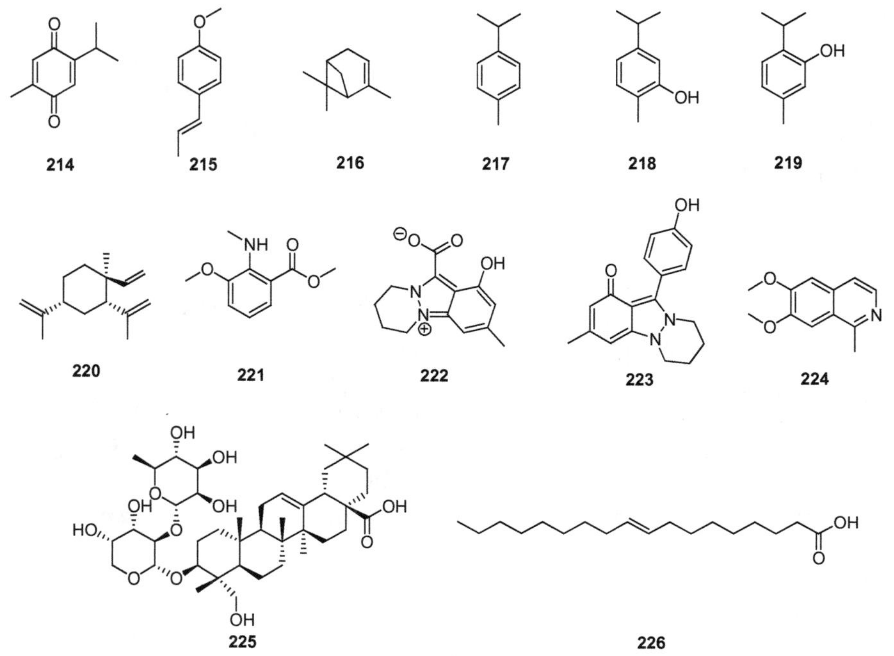

Fig. 23 Structures of the volatile oil components thymoquinone (**214**), anethole (**215**), α-pinene (**216**), *p*-cymene (**217**), carvacrol (**218**), thymol (**219**), and β-elemene (**220**), and of damascenine (**221**), nigellicine (**222**), nigellidine (**223**), nigellimine (**224**), the triterpenoid saponin, α-hederin (**225**), and the fatty acid oleic acid (**226**) from *Nigella sativa* and *Nigella damascena*

of almost 100% sesquiterpenes, of which *β*-elemene (**220**) was the most representative, although this compound also may be present in high concentrations in the essential oils of in *N. orientalis* and *N. arvensis* [180]. Additionally, a major constituent of the essential oil of *N. damascena* seeds is damascenine (**221**), a blue-fluorescing nitrogen-containing compound that is absent in *N. sativa* [179].

Also found in *N. sativa* are the pyrazole alkaloids nigellicine (**222**) and nigellidine (**223**), the isoquinoline alkaloid nigellimine (**224**), the triterpenoid saponin *α*-hederin (**225**), and the fatty acids linoleic acid (**184**) and oleic acid (**226**) [181]. Selected biological studies performed on these phytochemicals are presented in Sect. 5.20.

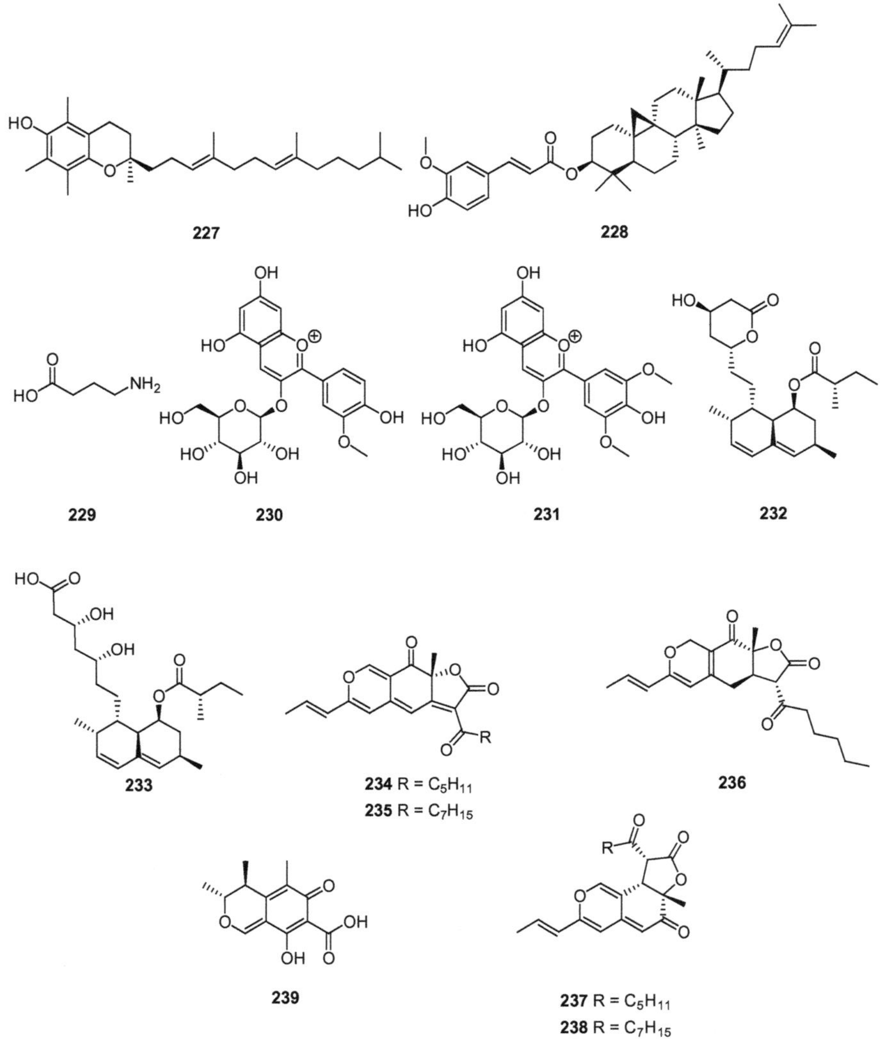

Fig. 24 Structures of constituents of *Oryza sativa* inclusive of red yeast rice, γ-tocotrienol (**227**), *γ*-oryzanol (**228**), γ-aminobutyric acid (GABA, **229**), peonidin-3-glucoside (**230**), malvidin-3-glucoside (**231**), monacolin K (in the lactone (**232**) and acid form (**233**)), rubropunctatin (**234**), monascorubrin (**235**), monascin (**236**), monasfluore A (**237**), monasfluore B (**238**), and citrinin (**239**)

4.21 *Oryza sativa*

Rice (*Oryza sativa* L., family Poaceae) is one of the most widely cultivated species for human consumption of the 22 domesticated species of the genus *Oryza* species. The *O. sativa* can be classified into three widely cultivated ecological varieties (subspecies), namely, *indica*, *japonica*, and *javanica*. White grain varieties (having a brown color before milling) are the most common; however, some have a different phenotype (black and red rice) [182]. In addition, these pigmented rice varieties have attracted attention as a source of vitamins, sterol derivatives and phenolic compounds, which are primarily located in the bran layer of the rice grain [183]. Phenolic acids, anthocyanins, and proanthocyanidins have been reported from rice. In general, white rice contains mainly phenolic acids, red rice contains procyanidins, whereas black rice is characterized by the presence of anthocyanins [184].

Major components of pigmented rice include essential amino acids, dietary fiber, folate, lignin, minerals and other phytochemicals [185]. The latter include γ-tocotrienol (a type of vitamin E, **227**), γ-oryzanol (**228**), γ-aminobutyric acid (GABA, **229**), the phenolic acids ferulic acid (**39**) and *p*-coumaric acid (**117**) and

Plate 21 Red yeast rice. Online Source "Red Yeast Rice (Hongqu) a dried culture of *Monascus purpureus*". Photograph courtesy ssbandit CC BY-SA 3.0 (https://upload.wikimedia.org/wikipedia/commons/1/15/Hongqu_3.jpg). Accessed Oct. 19, 2022

the anthocyanidins cyanidin-3-glucoside (**106**), peonidin-3-glucoside (**230**), and malvidin-3-glucoside (**231**), which are presented in Fig. 24 [185–187].

Red yeast rice is an Asian traditional medicine and food additive made by fermenting cooked rice by artificial inoculation of *Monascus purpureus* (Plate 21) [188]. Most dietary supplement products contain red yeast rice, instead of plain rice. Phytochemical analysis has led to the identification of monacolin K (occurring in both a lactone (**232**) and acid form (**233**)) and its analogs rubropunctatin (**234**), monascorubrin (**235**), monascin (**236**), monasfluore A (**237**), and monasfluore B (**238**), as well as organic acids, amino acids, sterols, decalin derivatives, flavonoids, lignans, coumarins, terpenoids, and polysaccharides [189].

More than twenty monacolins have been isolated from red yeast rice including monacolin K (**230**), the first, but not the only reported lipid-lowering compound isolated from *M. purpureus* [188]. This can be used as a marker of quality of red yeast rice-containing dietary supplements. However, the U.S. FDA has stated that products containing monacolin K are similar to a drug, so they may be subject to regulation [189].

Finally, citrinin (**239**) is a polyketide-derived mycotoxin that is produced by several fungi, including *M. purpureus.* It is a potentially toxic fermentation by-product in red yeast rice that may cause nephrotoxicity in humans [190]. Biological activities of these compounds are presented in Sect. 5.21.

4.22 Panax spp.

The genus *Panax* (family Araliaceae) consists of approximately 18 species of perennial herbaceous plants distributed from the subtropical to temperate regions in Asia and eastern North America. Sixteen of these species are found in Asia and two in North America [191]. Among them, the most highly investigated species are *Panax ginseng* C.A. Mey. (Plate 22) (Asian ginseng or Korean ginseng), *Panax quinquefolius* L. (American ginseng), *Panax notoginseng* (Burkill) F.H. Chen (Chinese ginseng), *Panax japonicus* (T. Nees) C.A. Mey (Japanese ginseng), and *Panax vietnamensis* Ha & Grushv. (Vietnamese ginseng).

Korean red (*Panax ginseng*) and American ginseng (*Panax quinquefolius*) are among the most popular *Panax* species consumed as dietary supplements [192]. *Panax ginseng* includes both white and red ginseng, which differ in the way the harvested roots are processed, namely, by air-drying or steaming at 95–100°C, respectively [193]. *Panax ginseng* has been used in traditional medicine to help prevent and alleviate a broad range of diseases [194]. *Panax quinquefolius* is traded as American ginseng [193]. Commercial herbal products are often adulterated with *Panax* species other than that specified on the label. There are also reports of the substitution of

Plate 22 Ginseng roots, *Panax ginseng* (Family Araliaceae). Online source "Ginseng" by Eugene Kim, CC BY 2.0 (https://www.flickr.com/photos/eekim/5152038565). Accessed Nov. 15, 2022

ginseng roots, the medicinally recommended part of the plant, by the leaves, stems or flowers [195].

Various phytochemicals have been identified from *P. ginseng*, including polysaccharides, alkaloids, glycosides, and phenolic acids, but the most-well characterized constituents from ginseng species are the ginsenosides [196]. This class of compounds are saponins, and more specifically triterpenoidal dammarane glycosides, and can be further classified into the protopanaxadiol- or protoanaxatriol-type based on the positioning of the sugar moiety at C-3 or -6 [197]. Ginsenosides Rb1 (**240**), Rb2 (**241**), Rc (**242**), Rd (**243**), Rg3 (**244**), Rh2 (**245**) are grouped in either the (20*S*)- or (20*R*)-protopanaxadiol types, because of the location of the sugar moiety at carbon-3 (Fig. 25). In turn, ginsenosides Re (**246**), Rg1 (**247**), Rg2 (**248**), and Rh1 (**249**) are characterized as of the (20*S*)- or (20*R*)-protopanaxatriol types, and their structures shown in Fig. 26 [198]. Ginsenoside Rf (**250**) is also a protoanaxatriol-type ginsenoside and is only detected in *P. ginseng*, whereas (24*R*)-pseudoginsenoside F11 (**251**), an ocotillol-type ginsenoside, is mainly detected in *P. quinquefolius* [196]. Studies on the biological activities of the ginseng saponins are presented in Sect. 5.22.

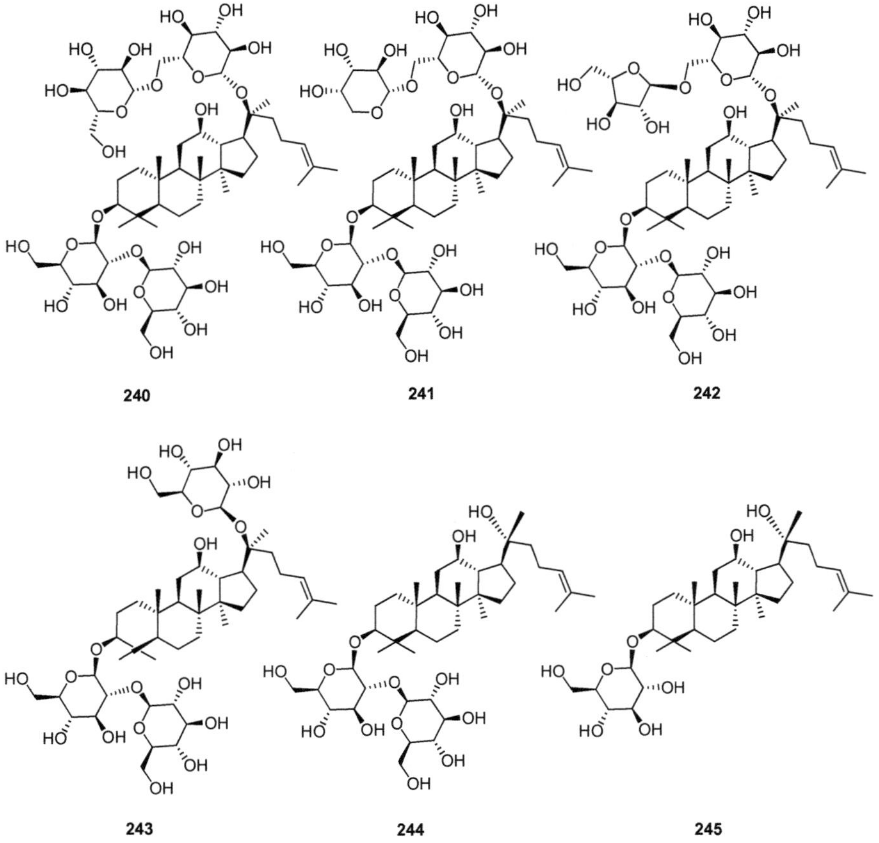

Fig. 25 Structures of the (20*S*)-protopanaxadiol-type ginsenosides Rb1 (**240**), Rb2 (**241**), Rc (**242**), and Rd (**243**), and the (20*R*)-protopanaxadiol-type ginsenosides Rg3 (**244**) and Rh2 (**245**)

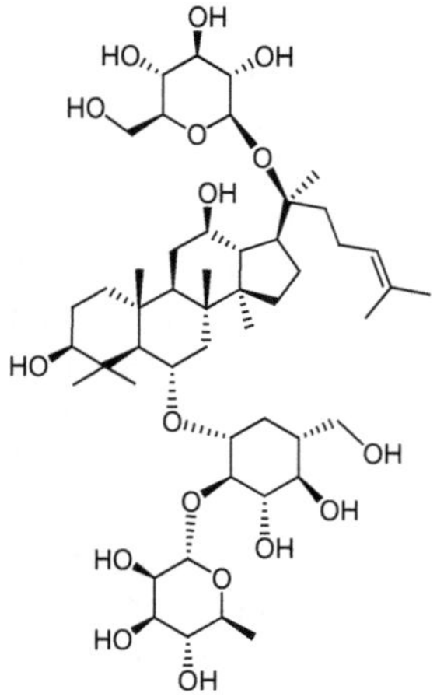

246

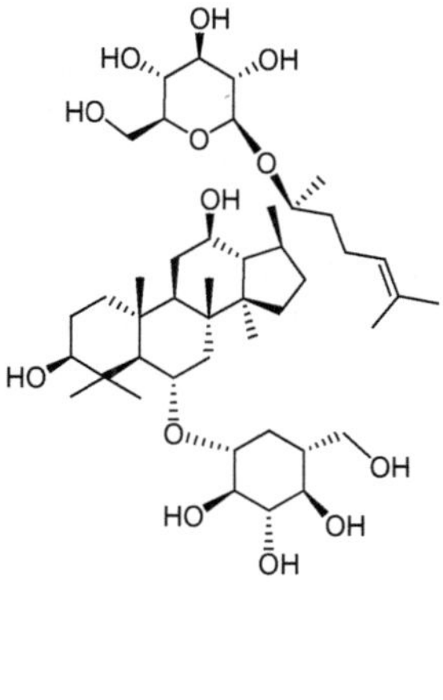

247

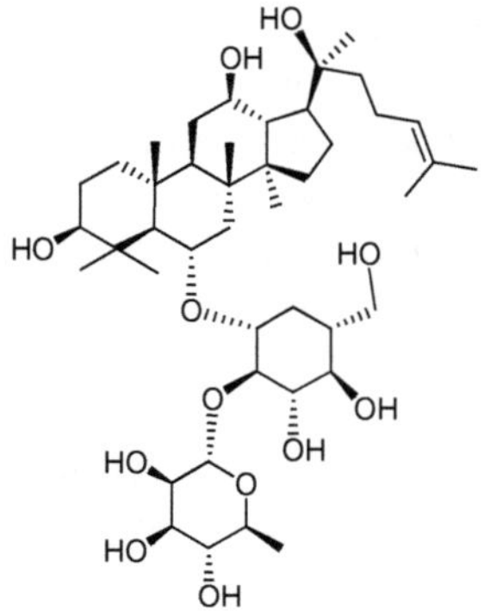

248

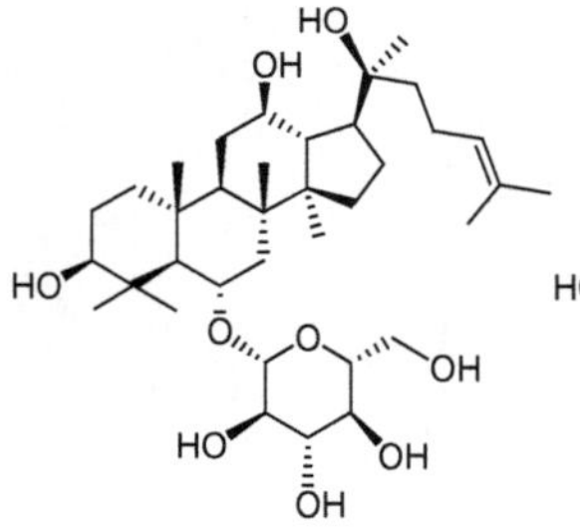

249

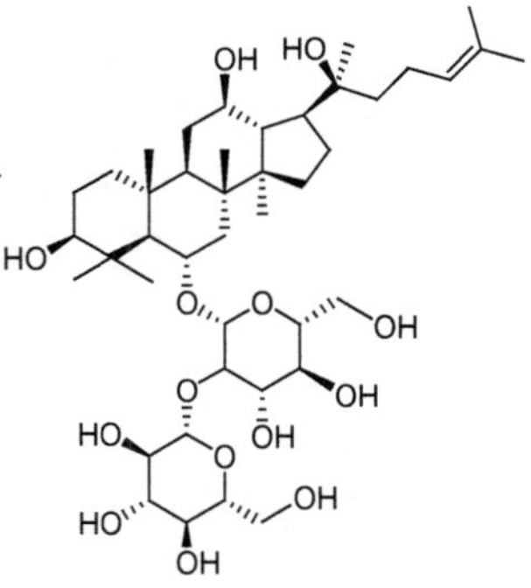

250

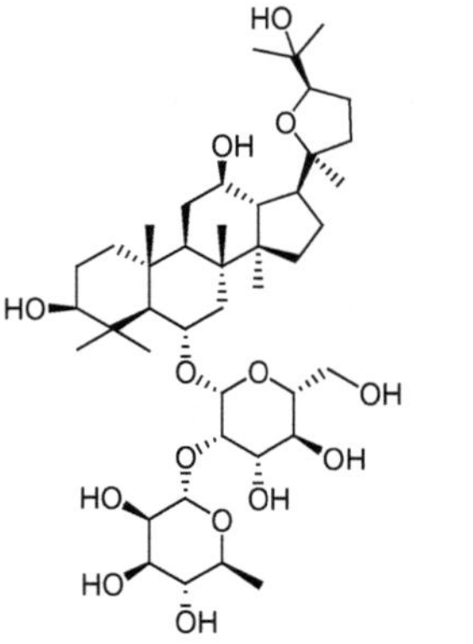

251

Fig. 26 Structures of the (20*S*)-protopanaxatriol-type ginsenosides Re (**246**), Rg1 (**247**), Rg2 (**248**), Rh1 (**249**) and ginsenoside Rf (**250**) and the ocotillol-type ginsenoside (24*R*)-pseudoginsenoside F11 (**251**)

4.23 *Pausinystalia johimbe*

Pausinystalia johimbe (K. Schum.) Pierre ex Beille (syn. *Corynanthe johimbe* K.Schum.) (Plate 23) is an evergreen tree of the flowering plant family Rubiaceae, native to West and Central Africa. It is commonly known as yohimbe and it has been utilized traditionally as an aphrodisiac [199].

Yohimbine (**252**), shown in Fig. 27, is the most abundant and active indole alkaloid found in the stem bark of yohimbe that has had use medicinally as a treatment for erectile dysfunction [200]. This compound has been categorized as a prescription drug in the United States where it has been included as a drug in the "United States Pharmacopeia". However, the bark of the yohimbe plant and its extract containing yohimbine (**252**) are commercially available as dietary supplements in the U.S. [201, 202].

Other less studied indole alkaloids in yohimbe include α-yohimbine (**253**), β-yohimbine (**254**), Ψ-yohimbine (**255**), corynantheine (**256**), corynanthine (**257**), allo-yohimbine (**258**), and yohimbic acid (**259**), and are shown structurally in Fig. 27 [203]. Studies on their biological properties are presented in Sect. 5.23.

Plate 23 Fruits and leaves of *Corynanthe johimbe* (Family Rubiaceae). Online source "*Corynanthe johimbe* K. Schum." by Nicolas Texier, CC BY-NC-ND 3.0 (http://legacy.tropicos.org/Image/100592592). Accessed Nov. 15, 2022

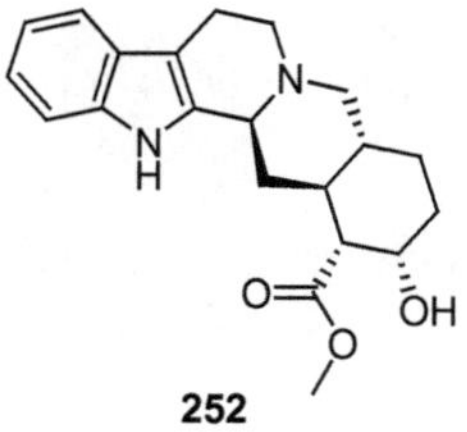

252

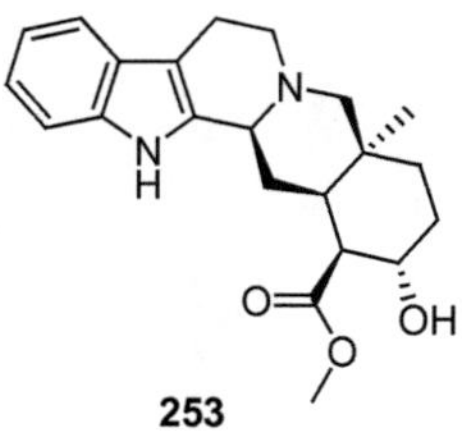

253

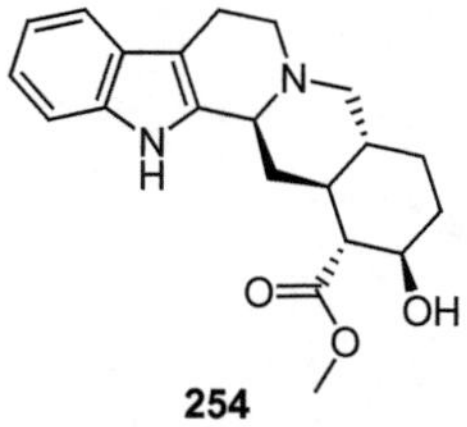

254

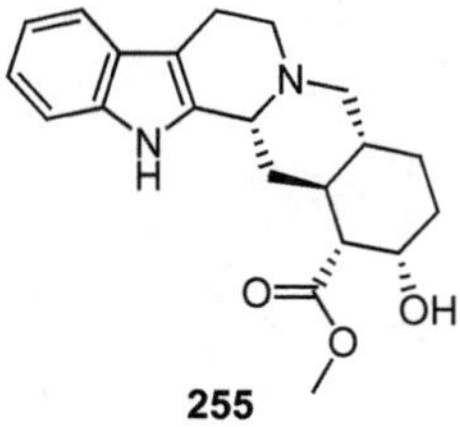

255

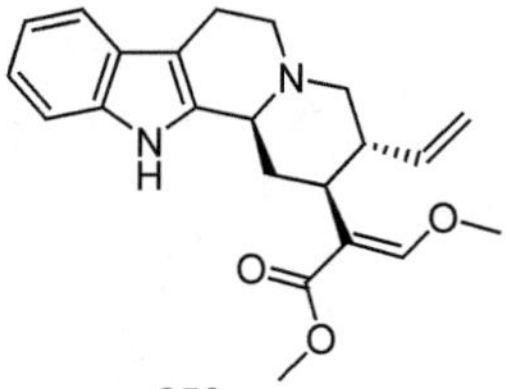

256

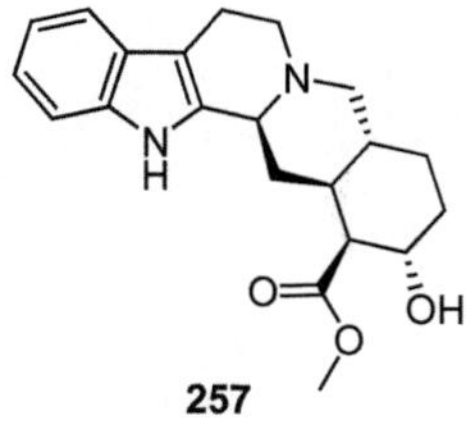

257

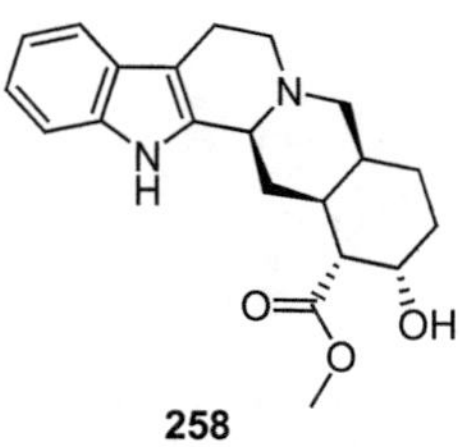

258

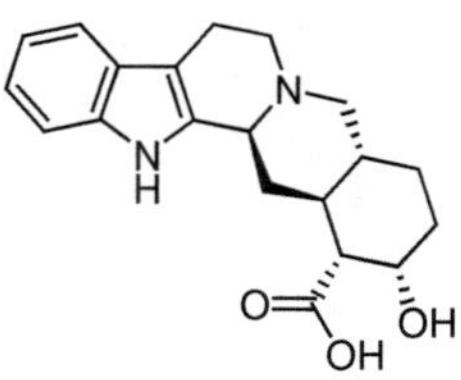

259

Fig. 27 Structures of the indole alkaloids isolated from the bark of *Pausinystalia johimbe* including yohimbine (**252**), α-yohimbine (**253**), β-yohimbine (**254**), Ψ-yohimbine (**255**), corynantheine (**256**), corynanthine (**257**), *allo*-yohimbine (**258**), and yohimbic acid (**259**)

4.24 *Sambucus nigra* and *S. canadensis*

The genus *Sambucus* (family Adoxaceae) consists of shrubs or smalls trees of complex taxonomy, with up to 30 species being recognized [204]. Two of these species are the European elderberry (*Sambucus nigra* L.; syn: *Sambucus nigra* subsp. *nigra*) (Plate 24) and the American elderberry (*Sambucus canadensis* L.; syn: *Sambucus nigra* subsp. *canadensis* (L.) Bolli). While both increasingly are produced and consumed in North America, most elderberry products sold in the United States are produced in Europe [205].

Plate 24 Flowers and leaves of *Sambucus nigra* (elderberry). Online source "Blooming Black elder (*Sambucus nigra*)". Photograph courtesy Jan Helebrant (https://www.flickr.com/photos/96541566@N06/52103106091/in/photolist-2nob3cn-wTzWZb-wTzeBw-osSMBL-xV2quK-oeWZHk-oeUKhw-xNzmDW-2m7wZEg-xgSC9G-xAuvQm-wMuECD-xCCSV2-xTux5C-xZjM9y-xvkSUs-oeXEse-r7JrTJ-U29rJ2-rpJFbQ-oupJYS-wLKDq4-y2x8Pz-xmThne-qtQcAj-oey3Wb-raGWN2-oxREzi-xv5vFV-raLY39-xkqf79-r8RnSp-xZzeTC-ouojbC-rpUvio-y1psEr-xQX9bf-x4yNMR-x3PfkZ-y56F6B-owuwqH-xHMzNN-2m7G5BA-oeYADy-xNLJaw-xb4VcW-vdVMV7-xSqGL2-xP8oNB-xZL4kG). Accessed Oct. 20, 2022

The fruits and flowers of *S. nigra* and *S. canadensis* are used medicinally and have been studied phytochemically. Some of the constituents are given in Fig. 28. Both species contain phenolic substances for which their concentration levels may vary depending on intrinsic and extrinsic factors [206]. Important factors include the cultivar grown, environmental conditions (light, temperature, amount, and frequency of rainfall, fertilization, cultivation methods), the processing method, and storage conditions [207]. Anthocyanins, mainly cyanidin glycosides, are the predominant compounds present in unprocessed *S. nigra* fruits and products [208, 209]. Of these, cyanidin 3-*O*-sambubioside (**260**) and cyanidin 3-*O*-glucoside (**106**) are the major representatives, whereas cyanidin 3-*O*-sambubioside-5-*O*-glucoside (**261**), cyanidin 3,5-*O*-diglucoside (**262**) and some non-cyanidin-based anthocyanins are present at lower concentrations [210]. These four cyanidin glycosides have been identified in *S. canadensis* fruits along with more stable acylated anthocyanins [210]. Cyanidin 3-glucoside (**106**) and cyanidin 3-*O*-sambubioside (**260**) were also detected in the fruits of another important European species, *Sambucus ebulus* L. (Dwarf elder) [208, 211].

Other phenolic compounds that are produced are three cinnamic acids 3-caffeoylquinic acid (**91**), 4-caffeoylquinic acid (**263**), and 5-caffeoylquinic acid

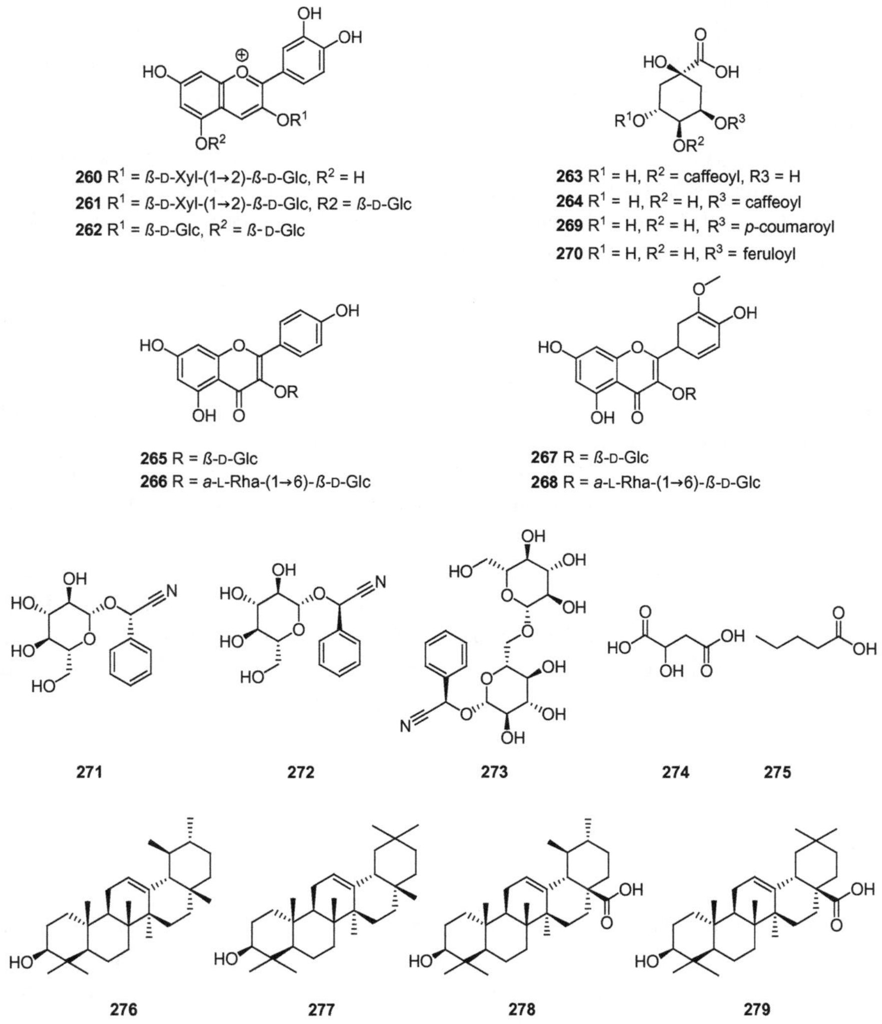

Fig. 28 Cyanidin 3-*O*-sambubioside (**260**), cyanidin-3-*O*-sambubioside-5-glucoside (**261**), cyanidin-3,5-*O*-diglucoside (**262**), 4-caffeoylquinic acid (**263**), 5-caffeoylquinic acid (**264**), kaempferol 3-*O*-glucoside (**265**), kaempferol 3-*O*-rutinoside (**266**), isorhamnetin 3-*O*-glucoside (**267**), isorhamnetin 3-*O*-rutinoside (**268**), 5-*p*-coumaroylquinic acid (**269**), 5-feruloylquinic acid (**270**), sambunigrin (**271**), prunasin (**272**), amygdalin (**273**), malic acid (**274**), valeric acid (**275**), tartaric, α-amyrin (**276**), β-amyrin (**277**), ursolic acid (**278**), and oleanolic acid (**279**) identified in *Sambucus nigra* and related species

(**264**) and the six flavonol glycosides quercetin 3-*O*-rutinoside (**57**), quercetin 3-*O*-glucoside (**137**), kaempferol 3-*O*-glucoside (**265**), kaempferol 3-*O*-rutinoside (**266**), isorhamnetin 3-*O*-glucoside (**267**) and isorhamnetin 3-*O*-rutinoside (**268**). These have been reported in both the fruits of *S. nigra* and *S. canadensis*, but their proportions differ between these species [210]. Several of these substances are also present in *S. ebulus*, and, compared to other *Sambucus* species, this species accumulates the highest levels of cinnamic acid (**47**), catechin (**60**), and epicatechin (**59**) [210, 212].

Elderberry flowers are also a source of phenolic compounds and several studies have described their composition [213–215]. The main compounds present in extracts of the flowers obtained with different solvents include 3-caffeoylquinic acid, 4-caffeoylquinic acid, 5-caffeoylquinic acid, 5-*p*-coumaroylquinic acid (**269**), 5-feruloylquinic acid (**270**), and other dicaffeoylquinic acid derivatives, while the principal flavonols found are quercetin 3-rutinoside (**57**), quercetin 3-glucoside (**145**), kaempferol 3-rutinoside (**266**) and isorhamnetin 3-rutinoside (**268**).

In addition, the presence of harmful cyanogenic glycosides has been reported in several species of the genus *Sambucus*. However, the presence and levels of cyanogenic glycosides vary widely among different specimens of *S. canadensis* [216]. Sambunigrin (**271**), prunasin (**272**), and amygdalin (**273**) have been found in the leaves, flowers and raw berries of *S. nigra*, *S. ebulus*, and *S. racemosa* and the highest content of total cyanogenic glycosides was demonstrated in the leaves of *S. nigra*, followed by their flowers [217]. However, the levels of these harmful compounds, as well as those of certain potentially toxic lectins [218, 219], may be greatly reduced to safer levels after heat treatment [220].

Other secondary metabolites reported to be present in *S. nigra* are vanillic acid (**40**), ferulic acid (**39**), chlorogenic acid (**91**), caffeic acid (**38**), *p*-coumaric acid (**117**), gallic acid (**109**), apigenin (**113**), luteolin (**115**), organic acids [malic (**274**), valeric (**275**), tartaric (**122**)], and triterpenoids (α-amyrin (**276**), β-amyrin (**277**), ursolic acid (**278**), oleanolic acid (**279**)) [206, 207, 221–223]. Relevant biological studies will be summarized in Sect. 5.24.

4.25 Serenoa repens

Serenoa repens (W. Bartram) Small (Plate 25) is part of the Arecaceae (palm) family and is the only species classified in the genus *Serenoa* [224]. Its common name, saw palmetto, is derived from its appearance as a small palm tree with sharp-edged leaves, and the fruits are used as dietary supplements [225].

The constituents of *S. repens* include the phytosterols, β-sitosterol (**280**), stigmasterol (**281**), and campesterol (**282**), the triterpenoid, lupeol (**283**), free fatty acids, inclusive of caprylic acid (**284**), capric acid (**285**), lauric acid (**286**), linoleic acid (**287**), myristic acid (**288**), palmitoleic acid (**289**), palmitic acid (**290**), stearic acid (**291**), and two monoacylglycerides, 1-monolaurin (**292**), and 1-monomyristin (**293**)

Plate 25 Flowers of *Serenoa repens* (saw palmetto). Online source "Saw Palmetto flowers". Photograph courtesy Big Cypress National Preserve (https://www.flickr.com/photos/bigcypres snps/30909396694/in/photolist-P6mJLE-SxdhS2-Q9KRwu-P9efDH-xh5yav-xPa4zn-otekSY-8Qs7SS-yrmtyx-xwzchy-wPcjF3-xG7fMm-owci5W-wNUSV7-otBSFr-oeY4cy-ovhgDX-ove y7G-tAUWrT-wZgTzk-ow5XEK-oeYiZP-SovG7b-xtKwzz-wjyN1j-tDvTvS-ouBkXr-osviCm-sDJAD8-owuCT2-u5eRXT-xK7nvD-tASqAK-oy9MRD-oeves8-owgcS5-xG47w9-xwegLe-owr akT-tm9MWY-sDEyno-osA4fo-tAHn9g-rtaA9J-otcL66-oeSYvu-ovbcKt-sDwFmQ-ow27o6-xvQ CuR/). Accessed Oct. 20, 2022

(Fig. 29) [226, 227]. Therefore, these phytosterols and fatty acid components from *S. repens* are standardized in dietary supplement preparations available in the market, including *S. repens* and its commercial standardized extract Permixon. Studies on the biological activities of these constituents are given in Sect. 5.25.

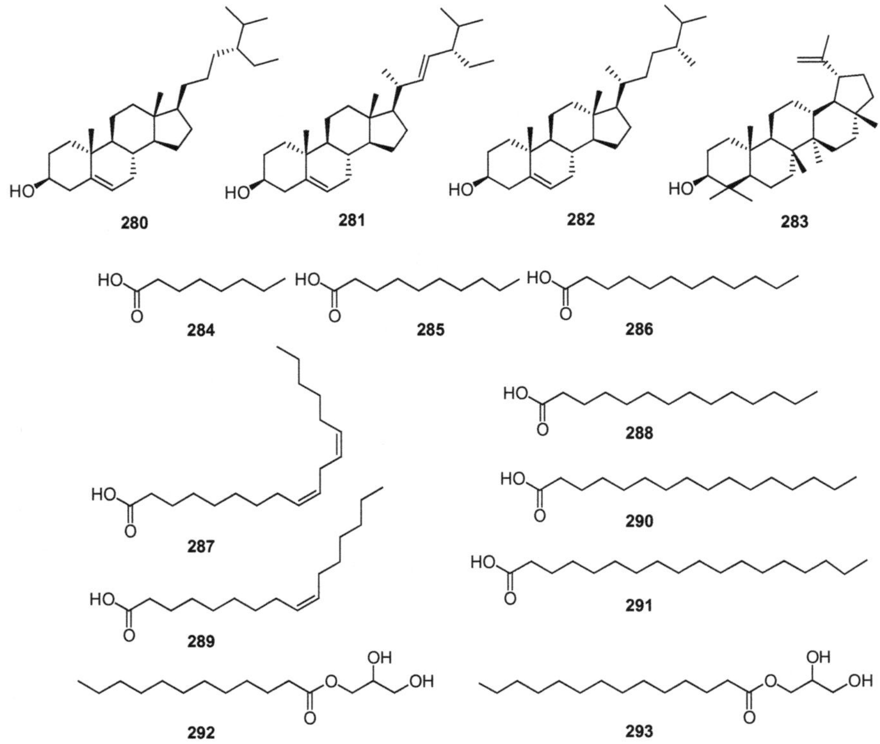

Fig. 29 Structures of the *Serenoa repens* constituents β-sitosterol (**280**), stigmasterol (**281**), campesterol (**282**), lupeol (**283**), caprylic acid (**284**), capric acid (**285**), lauric acid (**286**), linoleic acid (**287**), myristic acid (**288**), palmitoleic acid (**289**), palmitic acid (**290**), stearic acid (**291**), 1-monolaurin (**292**), and 1-monomyristin (**293**)

4.26 Silybum marianum

The genus *Silybum* (family Asteraceae) contains annual or biennial spiny herbaceous plants native to the Mediterranean basin that have now spread widely to other parts of the world [228]. The only two species that have been included in the genus are *Silybum marianum* (L.) Gaertn. (Plate 26) and *Silybum eburneum* Coss. & Durieu [229]. The fruits (often referred to as seeds) of *S. marianum* (milk thistle) have been used for medicinal purposes for over 2000 years in certain countries bordering the Mediterranean Sea [229].

Studies on *S. marianum* fruits have enabled the characterization of a diverse group of flavonolignans. Silymarin is a standardized extract of *S. marianum* fruits composed of a mixture of silybins A (**294**) and B (**295**), which make up approximately 30–65% of the total secondary metabolites present [230]. Similar flavonolignans containing a

Plate 26 Leaves and a flower of *Silybum marianum* (milk thistle). Online source "*Silybum marianum*, Parque Pozo Norte, Puertollano, Spain". Photograph courtesy Javier Martin (https://commons.wikimedia.org/wiki/File:Sylibum_marianum_Puertollano.jpg). Accessed Oct. 20, 2022

benzodioxane moiety are isosilybin A (**296**), isosilybin B (**297**) and silydianin (**298**). Other compounds that may occur in this extract include flavonolignans containing a benzofuran moiety such as silychristin A (**299**), silychristin B (**300**) and isosilychristin (**301**). Analysis of the composition of *S. marianum* from different origins has identified one chemotype with a high silybin content and another with high silydianin content. The structure of these flavonolignans as well as that of the flavonoid taxifolin (**302**), which is also present in milk thistle, is shown in Fig. 30 [229, 231]. Studies on the bioactivity of these flavonolignans are discussed in Sect. 5.26.

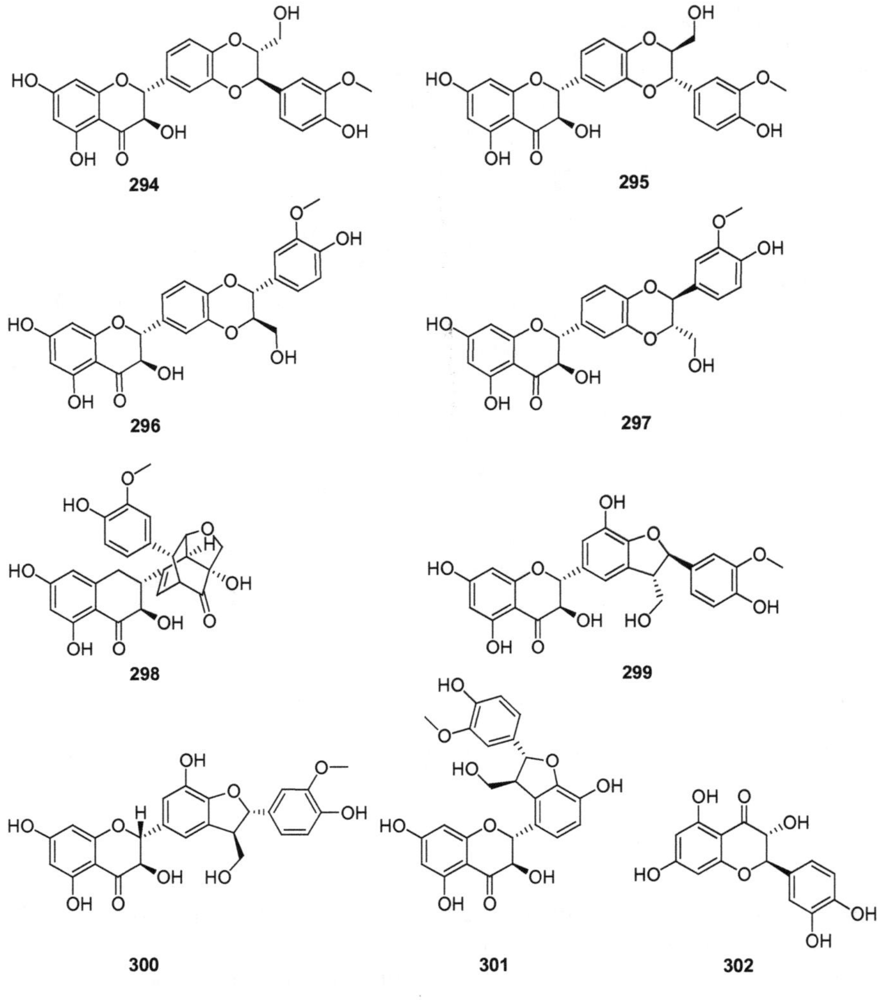

Fig. 30 Structures of constituents in the standardized *Silybum marianum* extract, silymarin, namely, the flavonolignans silybins A (**294**) and B (**295**), isosilybin A (**296**), isosilybin B (**297**), silydianin (**298**), silychristin A (**299**), silychristin B (**300**), isosilychristin (**301**), and the flavonoid taxifolin (**302**)

4.27 Trigonella foenum-graecum

The genus *Trigonella* (family Fabaceae) includes annual or perennial aromatic herbs characterized by trifoliate leaves. *Trigonella* species are distributed widely in the dry regions of the Mediterranean area, Africa, North America, and Australia [232]. According to "The World Flora Online" [233], this genus includes 114 accepted species including *Trigonella foenum-graecum* L., a traditional spice crop and medicinal plant used in Indian Ayurvedic and Sino-Tibetan traditional practices [232].

Trigonella foenum-graecum (fenugreek) seeds (Plate 27) contain a variety of phytochemicals extracted from different parts of the plant. These include flavonoids (quercetin (**108**), naringenin (**303**), rutin (**57**), vitexin (**55**)), nitrogen-containing compounds (trigonelline (**304**), choline (**305**), carpaine (**306**)), saponins (protodioscin (**307**), dioscin (**308**)), sapogenins (diosgenin (**309**)), phenolics (chlorogenic acid (**91**), *p*-coumaric acid (**117**), vanillin (**310**)), a lactone (sotolone (**311**)), amino acids (4-hydroxy isoleucine (**312**), arginine (**313**)), and polysaccharides (galactomannan (**314**)) (Figs. 7, 10, 12, 31) [234–237].

Saponins are the most abundant type of compounds in fenugreek seeds [235]. Fenugreek is one of very few known natural sources of diosgenin (**309**) [235, 238]. Diosgenin (**309**) is also an extensively studied steroidal sapogenin, and structurally is a C_{27} steroid with a double bond at C-5, a spiroacetal moiety at C-22, and a hydroxy group at C-3.

Fenugreek is a rich source of a number of nitrogen-containing compounds [235], including the plant hormone, trigonelline (**304**) that has also been found in other *Trigonella* species [239]. Fenugreek is also a good source for dietary fiber [235].

Plate 27 Seeds of *Trigonella foenum-graecum* (fenugreek). Online source "Methi dana". Photograph courtesy Miansari66 (https://commons.wikimedia.org/wiki/File:Fenugreek_seed.JPG). Accessed by Oct. 20, 2022

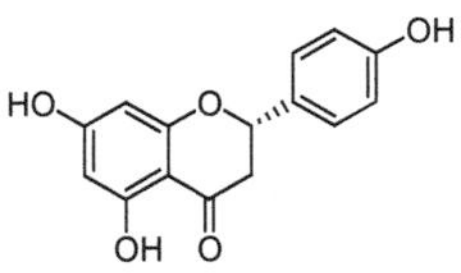
303

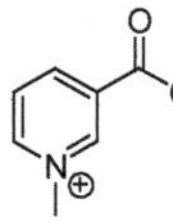
304

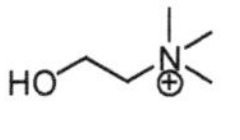
305

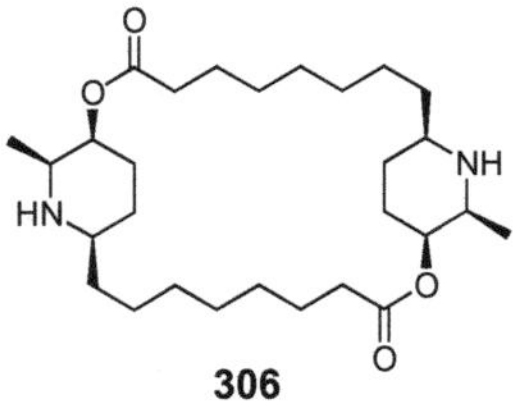
306

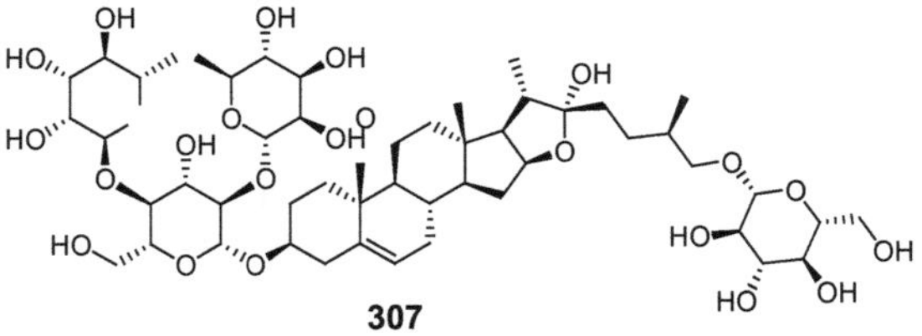
307

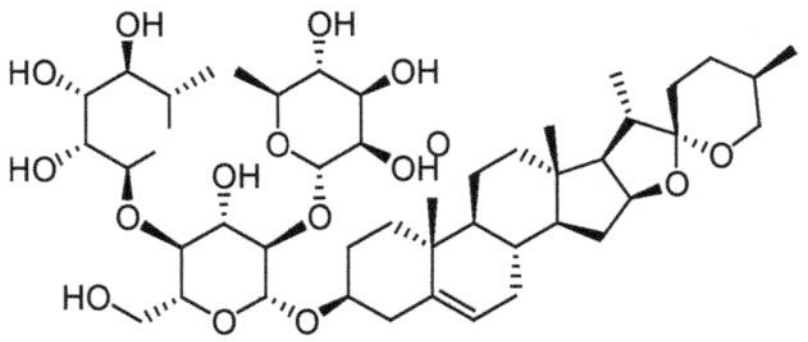
308

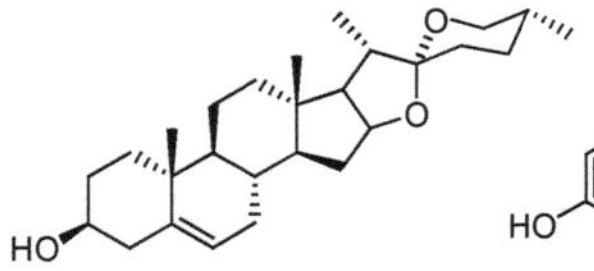
309

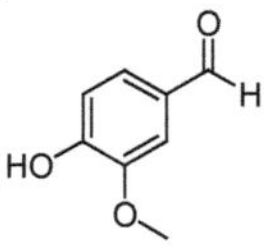
310

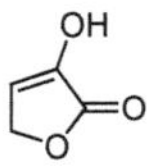
311

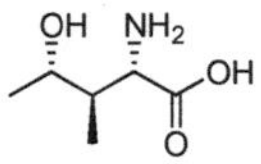
312

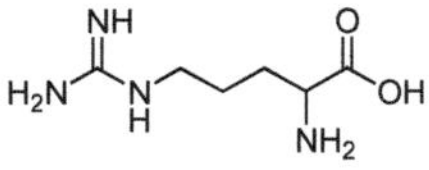
313

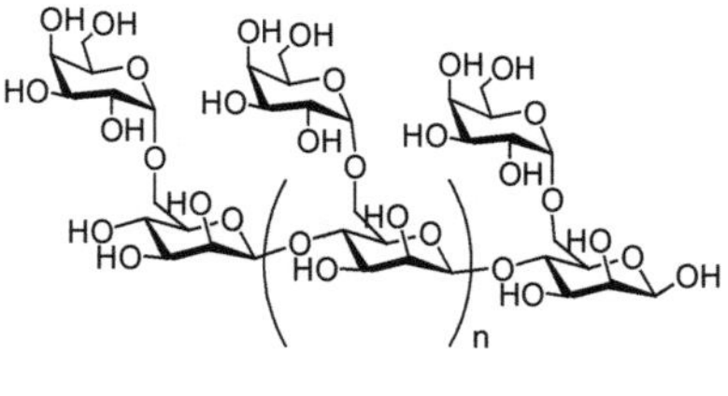
314

Fig. 31 Naringenin (**303**), trigonelline (**304**), choline (**305**), carpaine (**306**), protodioscin (**307**), dioscin (**308**), diosgenin (**309**), vanillin (**310**), sotolone (**311**), 4-hydroxy-isoleucine (**312**), arginine (**313**), and galactomannans (**314**) isolated from *Trigonella foenum-graecum*

Galactomannan (**314**) is a carbohydrate (soluble fiber) fraction representing 45–60% of the fenugreek seed. The major part of it is a polysaccharide component with 1:1 or 1:2 galactose to mannose ratio [238]. Biological studies on phytochemicals from *T. foenum-graecum* are discussed in Sect. 5.27.

4.28 *Vaccinium macrocarpon*

The genus *Vaccinium* (family Ericaceae) consists of more than 450 species of shrubs with a wide geographic distribution in the Northern Hemisphere [240]. Fruits of several species of *Vaccinium* (cranberry) were consumed by various indigenous tribes of North America and are now considered to be health-promoting and high value commercial crops [241]. There are four species of cranberry, with the most important being *Vaccinium macrocarpon* Aiton (large cranberry or American cranberry) (Plate 28) and *Vaccinium oxycoccos* L. (small cranberry or European cranberry) [240]. Often confused with the cranberry is the wild *Vaccinium vitis-idaea* L., known as lingonberry [242].

Major anthocyanins common to *V. macrocarpon*, *V. oxycoccos* and *V. vitis-idaea* are cyanidin 3-*O*-galactoside (**315**), cyanidin 3-*O*-glucoside (**106**), and cyanidin 3-*O*-arabinoside (**316**). *Vaccinium macrocarpon* and *V. oxycoccos* contain also peonidin 3-*O*-galactoside (**317**), peonidin 3-*O*-glucoside (**230**) and peonidin 3-*O*-arabinoside

Plate 28 Berries and leaves of *Vaccinium macrocarpon* (cranberry). Online source "Highbush Cranberry Fruit". Photograph courtesy Joanna Gilkeson/USFWS Midwest Region (https://www.flickr.com/photos/usfwsmidwest/14878337909/in/photolist-oEKnvp-xyy2jF-wAvEoW-xxdMij-wAKstf-xyAW82-xxJo7n-wAWcJH-xyANWD-xg7FXe-xg8kHy-xxXmgH-xvisM7-xym8pa-xxWqoT-xxAfsF-xxeZSq-xy9q96-xwFLaQ-xx4s8d-xga4Zh-xgd4gL-xfUiGs-wAww5Q-xycc5e-xvrebd-xg1Ach-xxdZeU-xgnNun-xxyqPg-xxNRZ8-xvoMGm-xyjCeZ-xxxU6n-xwVAkm-xytPip-xxA5Hv-xgfyNZ-xvcJ4G-xxTEv6-wAN7zb-xveb57-xfU62W-wBcKeB-xypti2-xxPdPe-wAQXL3-xxeLGN-xvxE2o-xgmaYj). Accessed Oct. 20, 2022

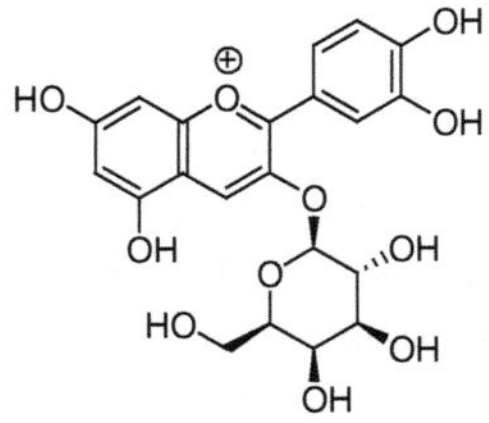

315

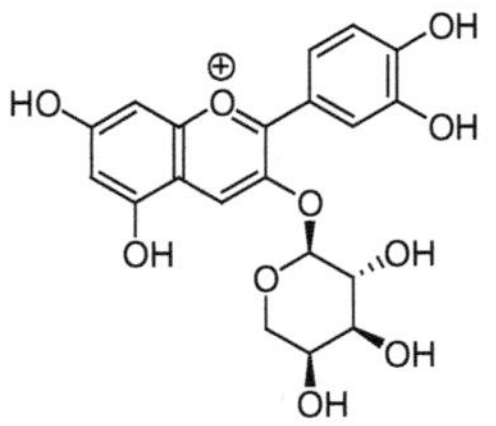

316

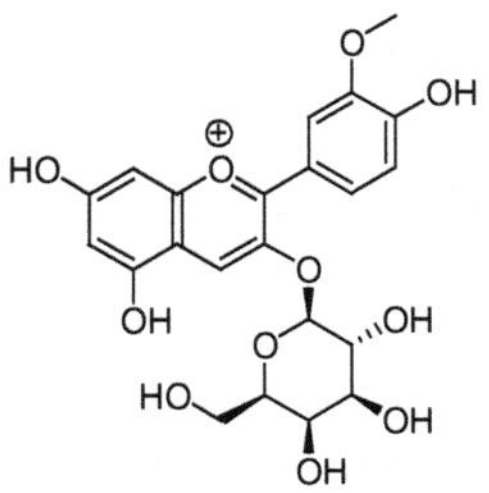

317

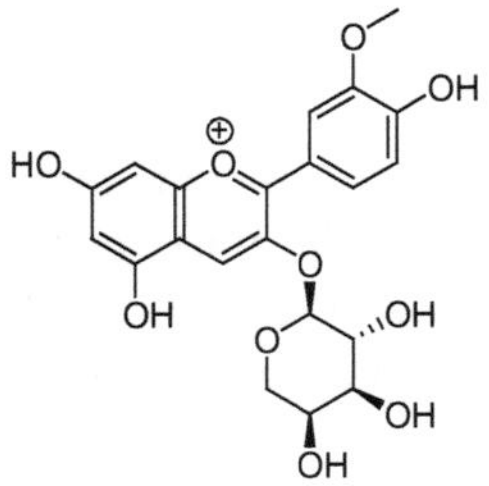

318

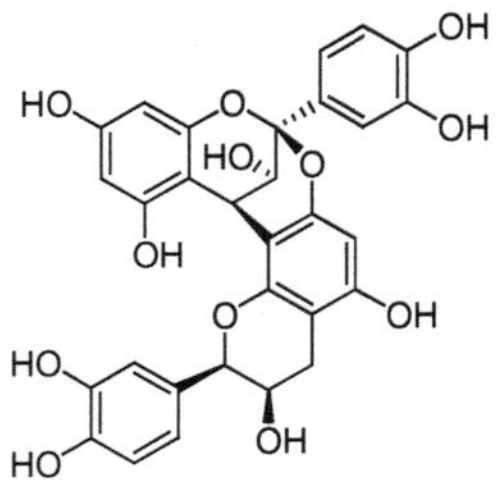

319

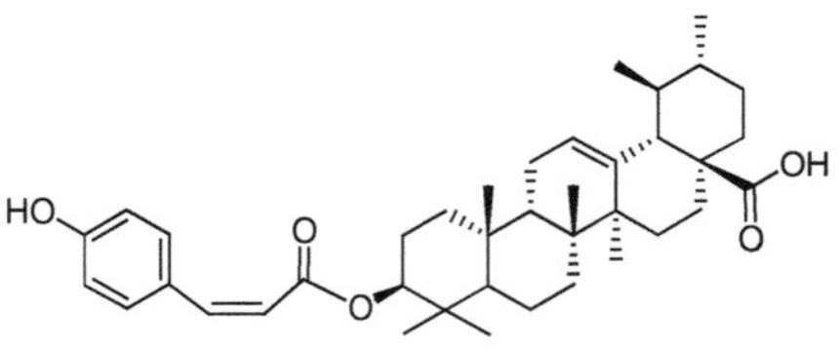

320

Fig. 32 Structures of cyanidin 3-*O*-galactoside (**315**), cyanidin 3-*O*-arabinoside (**316**), peonidin 3-*O*-galactoside (**317**), peonidin 3-*O*-arabinoside (**318**), an A-type procyanidin (**319**), and *cis*-3-*O*-*p*-hydroxycinnamoyl ursolic acid (**320**), as identified in *Vaccinium macrocarpon*

(**318**) [243], although *V. vitis-idaea* was found not to contain measurable amounts of these compounds [244, 245]. Secondary metabolites from these species are shown structurally in Fig. 32. Cranberry proanthocyanidins (PACs), also known as condensed tannins, are oligomers and polymers of flavan-3-ols units (having primarily epicatechin units). There are two common series of PAC dimers. The "B-type" series are dimers linked either at the C-4–C-6 or C-4–C-8 positions, whereas the "A-type" series are dimers linked at the C-4–C-8 positions with an additional C-2–O–C-7 ether linkage. Cranberry PAC oligomers with a degree of polymerization greater than two may incorporate both "A-type" and "B-type" interflavan linkages. "A-type" PACs are referred as those that contain one or more "A-type" interflavan bonds in their structure whereas PAC oligomers that contain only "B-type" interflavan bonds are referred to as the "B-type" PACs [246]. *Vaccinium* species contain both "A-type" and "B-type" procyanidins but compared to other fruits, the "A-type" are predominant and are associated with the preventive and therapeutic effects of cranberry against urinary tract infections [247].

Flavonols are also abundant in cranberries, and they include mainly rhamnose, arabinose, and galactose derivatives of myricetin as well as rhamnose, galactose and glucoside derivatives of quercetin. Common phenolic acids, triterpenoids and phytosterols are also reported as constituents of different cranberry species [248]. Hence, cranberries are particularly abundant in phenolic compounds, triterpenoids, and carotenoids [248, 249]. The content of phenolic compounds in cranberries is influenced by aspects such as the cultivar used, agricultural practices, geographical area where the plant is grown, weather conditions, the degree of ripeness, the harvesting time, and storage conditions.

American cranberry is characterized by the presence of flavan-3-ols ((−)-epicatechin (**59**)), A-type procyanidins ((A-type PACs), (**319**)), anthocyanins, benzoic acid (**45**), and (*Z*)-3-*O*-*p*-hydroxycinnamoyl ursolic acid (**320**). (−)-Epicatechin (**59**) is the predominant constitutive unit in cranberry PACs (**319**). The building blocks of **319** can be condensed either via a single C–C bond between C-4 of the upper unit and C-8, with an additional ether-type bond between C-2 of the upper unit and the hydroxy group at C-7 of the lower unit [246, 249]. Biological studies on the constituents of cranberry species are discussed in Sect. 5.28.

4.29 *Valeriana officinalis*

The genus *Valeriana* (Valerianaceae) encompasses more than 250 species distributed in different regions of the world [250]. The main constituents present in species of this genus include valepotriates, valerenic acid (and its derivatives), as well as monoterpenes and sesquiterpenes present in their essential oils. Secondary metabolites from these species are presented in Figs. 33 and 34. Valepotriates (epoxy iridoid esters) are unstable compounds that are found in varying amounts in *Valeriana officinalis* L. (Plate 29) and in other commercially used species such as the Mexican *Valeriana edulis* Nutt. and the Indian *Valeriana jatamansi* Jones (syn: *V. wallichii* DC.) [251]. The main diene-type valepotriates common to these three species are valtrate (**321**) and isovaltrate (**322**) [252], of which the degradation products are baldrinal (**323**), homobaldrinal (**324**), valeric acid (**275**) and isovaleric acid (**325**), with the latter compound giving the characteristic odor of dried valerian roots [253]. Sesquiterpene acids such as acetoxy-valerenic acid (**326**), hydroxy-valerenic acid (**327**) and especially valerenic acid (**328**) are considered as chemical markers of *V. officinalis* for quality assurance purposes. Earlier studies have also detected valerenic acid (**328**) in *V. edulis* and *V. jatamansi* [254–256].

Plate 29 Flowers and stems of *Valeriana officinalis* (valerian). Online source "*Valeriana officinalis* L. 1753". Photograph courtesy T. Lefort, Val Def (https://www.flickr.com/photos/194071409@N05/51732086369/in/photolist-2mJ1XHx-xCHXdq-xLH3em-xBihHY-2mPpGpF-2mPpHcN-xWdwq4-x5ijkp-y1HHbm-oeJgXo-wGvUCS-y1MY1o-xspiG2-x4BU1W-wHq6CJ-x4JAq2-xnPASN-2mPopts-2mPk4No-2mPpDph-2mPosXB-xKU6EA-x7AbLf-x5owpe-xKKBzK-xBfxF7-oeWRB2-xo4pAo-xKwSwr-y1oUqy-otZowY-x7ekdY-xkV9MJ-oeHa1s-x55SUA-x3ph21-y3fZ5R-xDZiHc-waoXkz-xhobBW-x7rhwi-x7eXV5-x3e27S-xbUawT-wHq5HN-otpNsm-wskG3e-x7ffAG-wEg59h-xBh4gp/) Accessed Oct. 20, 2022

The essential oil of valerian has a variable composition, depending on the species of origin [250]. The oil from the roots of *V. officinalis* consists mainly of valeric acid (**275**), isovaleric acid (**325**), and monoterpenes such as bornyl acetate (**329**), myrtenyl isovalerate (**330**), myrtenyl acetate (**331**), camphene (**332**), myrtenol (**333**) and borneol (**334**). Also found are sesquiterpenes of which the structures are based on the skeletons of kessane, valeranone and valerenic acid [250]. The major sesquiterpenes are valerenal (**335**), valeranone (**336**), valerenol (**337**), valerenyl acetate (**338**), valerenyl isovalerate (**339**) and valerenic acid (**328**) [257]. The main sesquiterpenes present in the essential oil of the roots of different varieties of *V. jatamansi* are patchouli alcohol (**340**), α-patchoulene (**341**) and β-patchoulene (**342**), *α*-guaiene (**343**) and β-guaiene (**344**) and β-gurjunene (**345**) [258], while the most prominent volatile components of *V. edulis* root are methyl isovalerate (**346**) and isobutyl isovalerate (**347**) [259].

Certain lignans and flavonoids have also been isolated and identified from different *Valeriana* species. Three lignans, pinoresinol-4,4′-di-*O*-*β*-D-glucoside (**348**), 8-hydroxypinoresinol-4′-*O*-*β*-D-glucoside (**349**) and pinoresinol-4-*O*-*β*-D-glucoside (**350**) were detected in five *Valeriana* species, including *V. officinalis* [260]. In the same study, linarin (**351**) (acacetin-7-*O*-rutinoside) was identified as a flavonoid

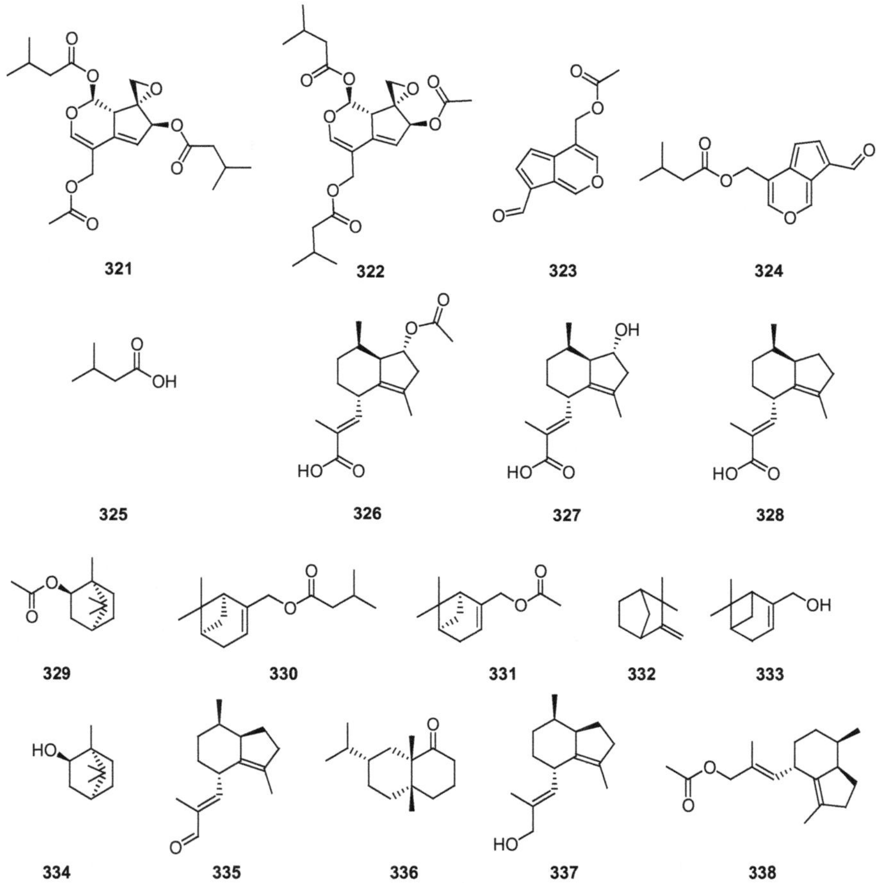

Fig. 33 Structures of constituents of *Valeriana officinalis* and related species including valtrate (**321**), isovaltrate (**322**), baldrinal (**323**), homobaldrinal (**324**), isovaleric acid (**325**), acetoxy-valerenic acid (**326**), hydroxy-valerenic acid (**327**), valerenic acid (**328**), bornyl acetate (**329**), myrtenyl isovalerate (**330**), myrtenyl acetate (**331**), camphene (**332**), myrtenol (**333**), borneol (**334**), valerenal (**335**), valeranone (**336**), valerenol (**337**), and valerenyl acetate (**338**)

common to all species investigated, but was present in higher concentration in *V. jatamansi* [260]. The pyridine alkaloids valerianine (**352**), and actinidine (**353**) have been also described for the genus *Valeriana* [261]. Studies on the bioactive constituents of *Valeriana officinalis* are discussed in Sect. 5.29.

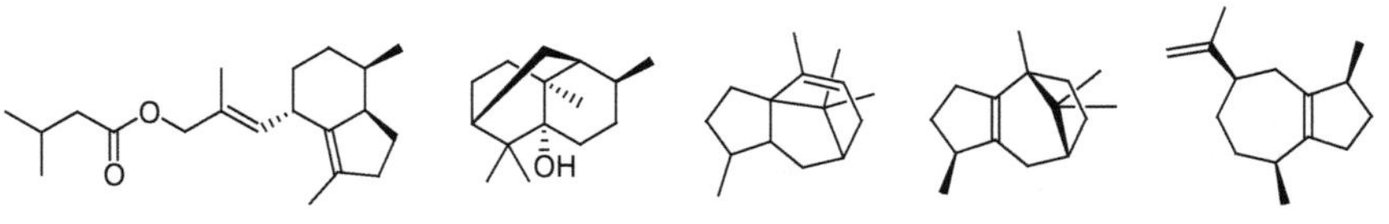

339 340 341 342 343

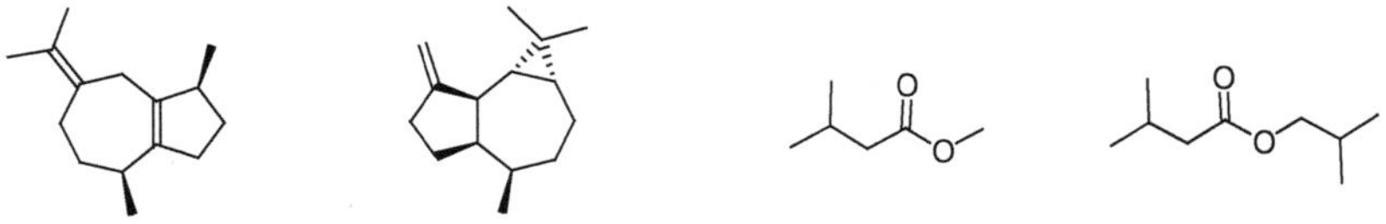

344 345 346 347

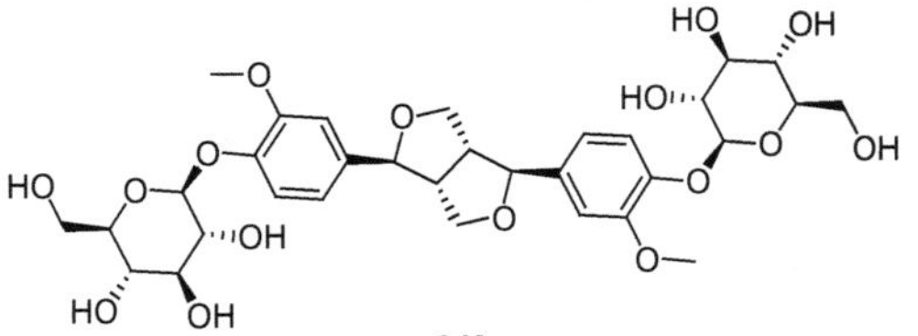

348

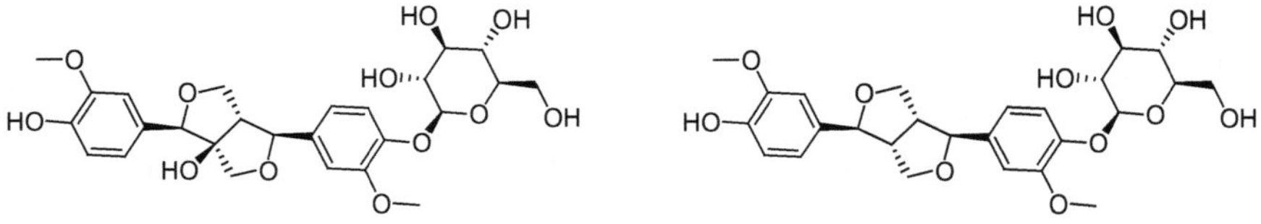

349 350

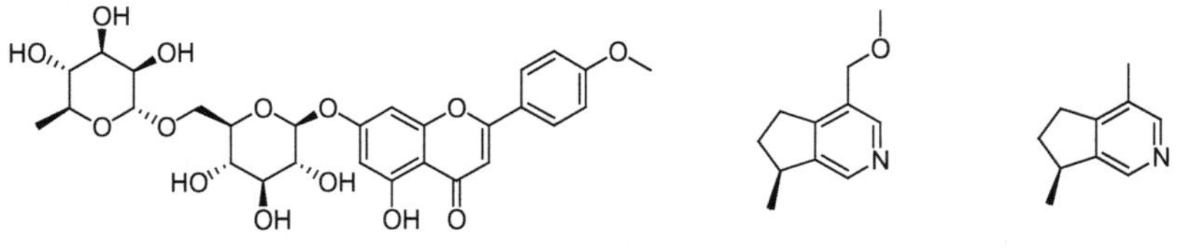

351 352 353

Fig. 34 Structures of additional constituents of *Valeriana officinalis* and related species including valerenyl isovalerate (**339**), patchouli alcohol (**340**), α-patchoulene (**341**) and β-patchoulene (**342**), α-guaiene (**343**) and β-guaiene (**344**) and β-gurjunene (**345**), methyl isovalerate (**346**), isobutyl isovalerate (**347**), pinoresinol-4,4′-di-*O*-*β*-D-glucoside (**348**), 8-hydroxypinoresinol-4′-*O*-*β*-D-glucoside (**349**), pinoresinol-4-*O*-*β*-D-glucoside (**350**) linarin (**351**), valerianine (**352**), and actinidine (**353**)

4.30 Withania somnifera

The genus *Withania* (family Solanaceae) comprises 23 species of shrubs distributed in the Mediterranean region that also occur in South Asia. Among them, two species, *Withania somnifera* (L.) Dunal ("Ashwagandha") (Plate 30) and *Withania coagulans* (Stocks) Dunal, are economically significant and have been used in Ayurvedic and other indigenous systems of medicine [262].

The major phytoconstituents isolated from *W. somnifera* and *W. coagulans* are withanolides present mainly in their roots and leaves. Withanolides are C_{28} steroidal lactones with an ergostane core and a C-8 or C-9 side chain [263]. Most of these compounds have a five- or six-membered C-22/C-26 lactone/lactol side chain (**354**) (Fig. 35) and a carbonyl functionality at C-1 [264, 265]. This intact or rearranged ergostane framework involves C-22 and C-26 being appropriately oxidized to form a five- or six-membered lactone ring that is normally attached to C-17 (**354**) (Fig. 35)

Plate 30 *Withania somnifera*. Photograph courtesy Hari Prasad Nadig under Creative Commons 2.0

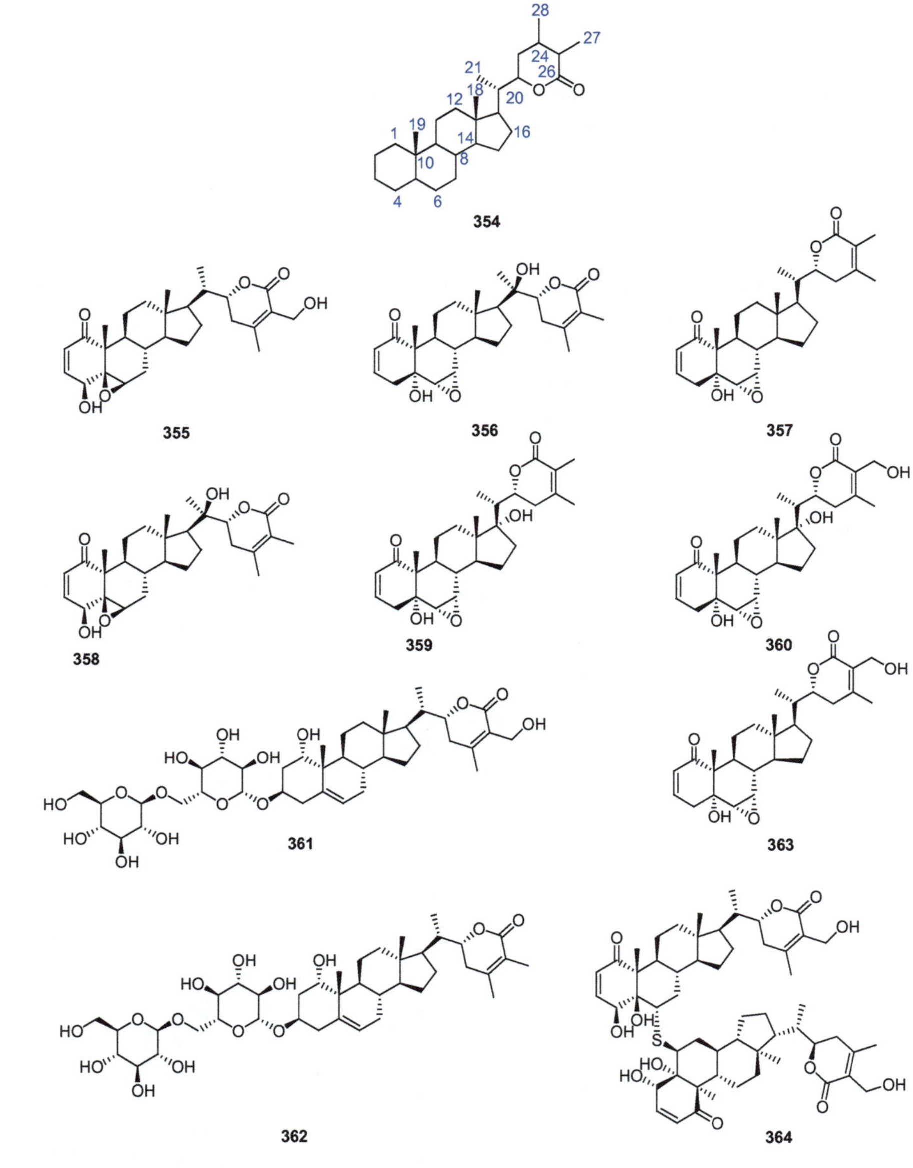

Fig. 35 Basic withanolide skeleton (**354**), and structures of withaferin A (**355**), withanolide A (**356**), withanolide B (**357**), withanolide D (**358**), withanone (**359**), 27-hydroxywithanone (**360**), withanoside-IV (**361**), withanoside-V (**362**), 12-deoxywithastramonolide (**363**), ashwagandhanolide (**364**), as isolated from *Withania somnifera*

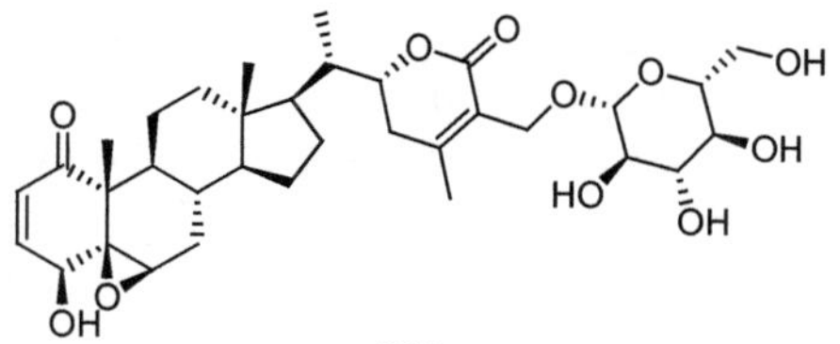

365

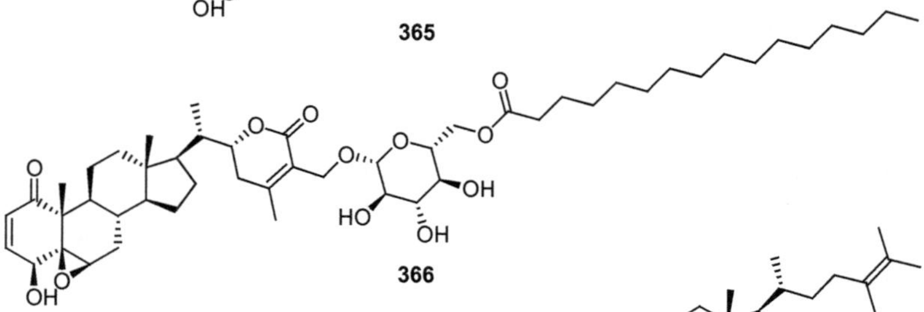

366

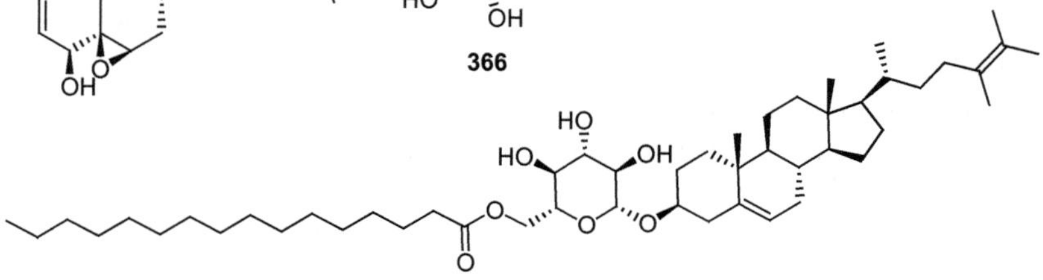

367

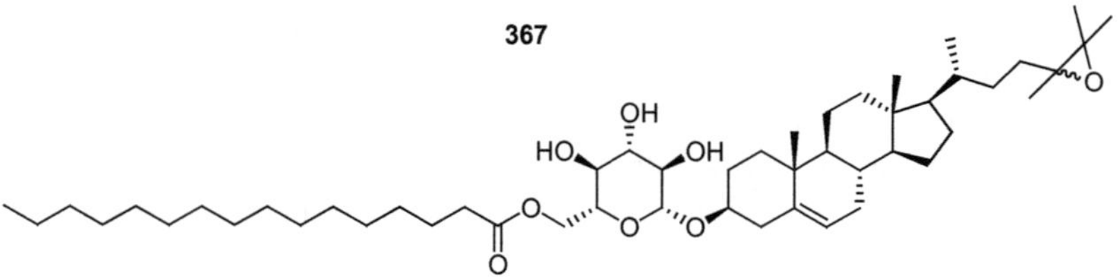

368

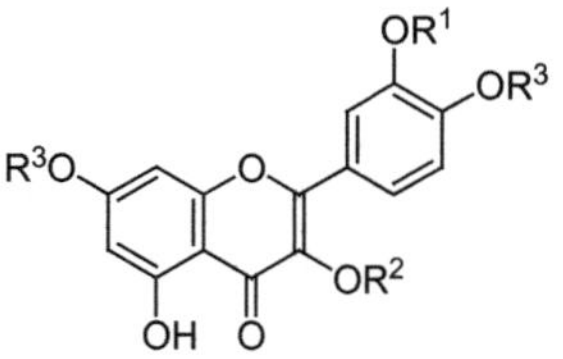

369 R^1 = OH, R^2 = α-L-Rha-(1→6)-ß-D-Gal, R^3 = ß-D-Glc

370 R^1 = OH, R^2 = α-L-Rha-(1→6)-ß-D-Glc, R^3 = ß-D-Glc

371 R^1 = H, R^2 = α-L-Rha-(1→6)-ß-D-Gal, R^3 = ß-D-Glc

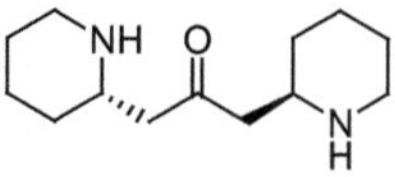

372

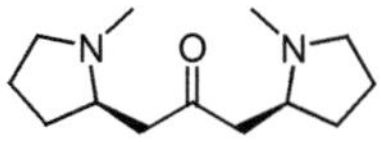

373

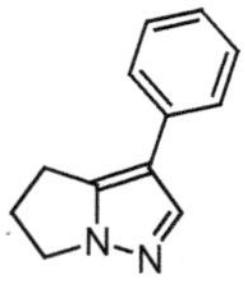

374

Fig. 36 Structures of sitoindosides IX (**365**), X (**366**), VII (**367**) and VIII (**368**), quercetin 3-*O*-robinobioside-7-*O*-glucoside (**369**), quercetin 3-*O*-rutinoside-7-*O*-glucoside (**370**), kaempferol 3-*O*-robinobioside-7-*O*-glucoside (**371**), anaferine (**372**), cuscohygrine (**373**), and withasomnine (**374**) as isolated from *Withania somnifera*

[266, 267]. The withanolide skeleton may be defined as a 22-hydroxyergostan-26-oic acid-26,22-lactone. This lactone ring may be fused with the carbocyclic part of the molecule through a carbon-carbon bond or through an oxygen bridge [266–268]. Withaferin A (**355**) and withanolide A (**356**) (Fig. 35) are two of the main representatives of this family of compounds that have been isolated from *W. somnifera* [269] and detected in *W. coagulans* [270]. *Withania somnifera* root extracts have also yielded withanolide B (**357**), withanolide D (**358**), withanone (**359**), 27-hydroxywithanone (**360**), withanosides (withanolide glycosides as e.g. withanoside-IV (**361**) and withanoside-V (**362**)), 12-deoxywithastramonolide (**363**) and ashwagandhanolide (**364**) [263, 271, 272]. Examples of other related compounds present in *W. somnifera* are sitoindosides IX (**365**) and X (**366**) (Fig. 36, withanolides containing a glucose moiety at C-27) and sitoindosides VII (**367**) and VIII (**368**) (steroidal saponins containing an additional acyl group) [269]. Coagulins and withacoagulins are important constituents of *W. coagulans* aerial parts [273].

Some flavonoids and other phenolic compounds have been detected in various plant parts of *W. somnifera* [274, 275] and *W. coagulans* [276]. Interestingly, three flavonol glycosides, quercetin 3-*O*-robinobioside-7-*O*-glucoside (**369**), quercetin 3-*O*-rutinoside-7-*O*-glucoside (**370**) and kaempferol 3-*O*-robinobioside-7-*O*-glucoside (**371**) have been isolated the aerial parts of *W. somnifera*, but were absent from the root sample [277].

Other chemical classes of constituents reported in *W. somnifera* are alkaloids and withanamides. A variety of alkaloids, such as anaferine (**372**) (piperidine alkaloid), cuscohygrine (**373**) (a pyrrolidine alkaloid) and withasomnine (**374**) (a pyrazole alkaloid), were isolated or detected mainly from the roots of *W. somnifera* [278]. Additional purine, indole and isoquinoline alkaloids were detected in *W. somnifera* by LC-MS [275]. The presence of alkaloids in *W. coagulans* has been confirmed only by qualitative phytochemical screening [279]. Withanamides are serotonin glycosides with a hydroxyfatty acyl side chain attached via an amide linkage. This class of compounds has been isolated from the fruits of *W. somnifera* [280]. Biological studies on the withanolide constituents of *Withania somnifera* are presented in Sect. 5.30.

4.31 *Zingiber officinale*

The genus *Zingiber* is one of the largest in the plant family Zingiberaceae, which comprises 150–200 species. Species in this genus are mostly aromatic rhizomatous herbs that are distributed widely in tropical to subtemperate regions of Asia. Common ginger rhizomes (*Zingiber officinale* Roscoe) (Plate 31) is the best-known and economically most important species in this genus and is used in the traditional Chinese and Ayurvedic systems of traditional medicine [281].

Phenolic compounds and terpene oils (essential oils) are the main constituents of ginger rhizome. Gingerols, the major components in the fresh rhizome, are phenolic compounds that differ from each other by the length of their alkyl side chains. They include 4-gingerol (**375**), 6-gingerol (**376**), 8-gingerol (**377**), and 10-gingerol (**378**)

Plate 31 Stems and rhizomes of *Zingiber officinale* (Family Zingiberaceae). Online source "*Zingiber officinale*." No caption (https://www.picturethisai.com/wiki/Zingiber_officinale.html). Accessed Oct. 27, 2022

(Fig. 37). These compounds are thermally labile and are transformed at high temperatures to shogaols. Shogaols, the dehydrated form of gingerols, are the principal pungent constituents in dried ginger rhizomes and include 6-shogaol (**379**), 8-shogaol (**380**) and 10-shogaol (**381**). Paradols, 6-paradol (**382**) and 8-paradol (**383**), and zingerone (**384**) [282, 283] are also phenolic components of ginger rhizomes. These compounds have been reported to be present in related species including *Zingiber zerumbet* (L.) Smith and *Aframomum melegueta* K. Schum. (Zingiberaceae) [284].

Volatile isoprenoids such as α-zingiberene (**72**), β-bisabolene (**385**), α-curcumene (**75**), α-farnesene (**386**), and β-sesquiphellandrene (**71**) are considered the main constituents of ginger essential oils (Fig. 37) [283, 285]. Other constituents include, capsaicin (**387**), 6-gingerdiol (**388**), and galanolactone (**389**) [286]. Biological studies on the constituents of *Zingiber officinale* are presented in Sect. 5.31.

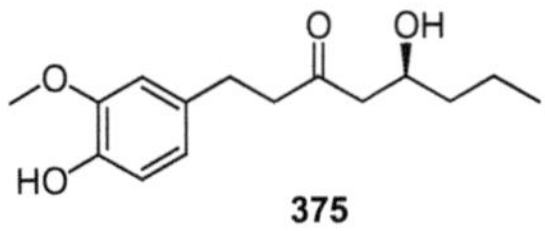

375

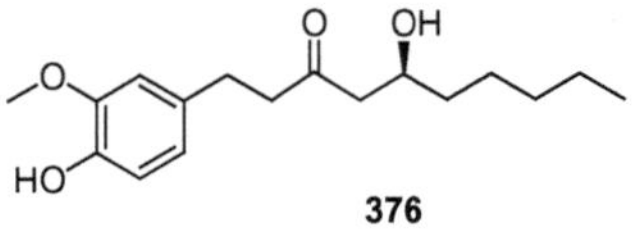

376

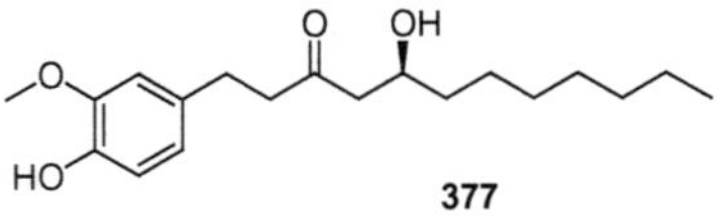

377

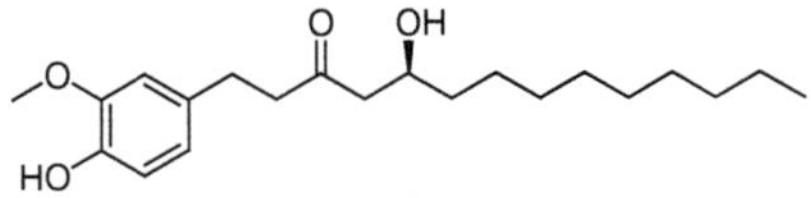

378

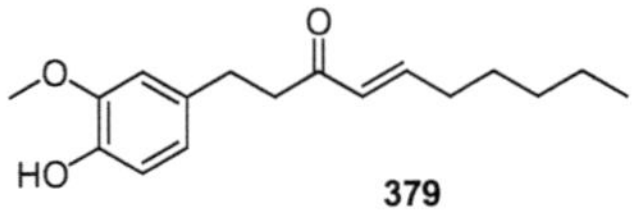

379

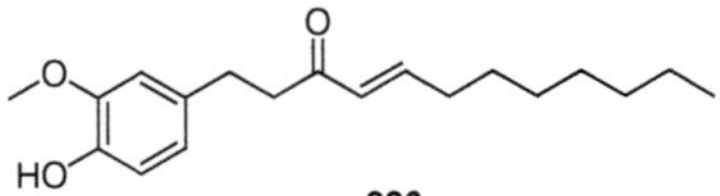

380

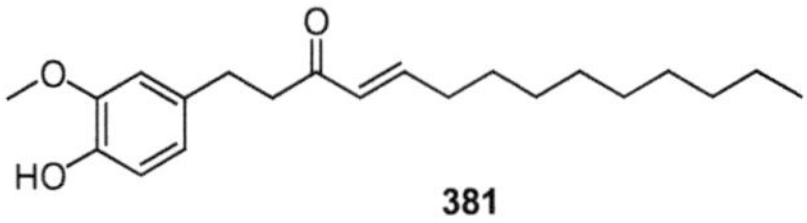

381

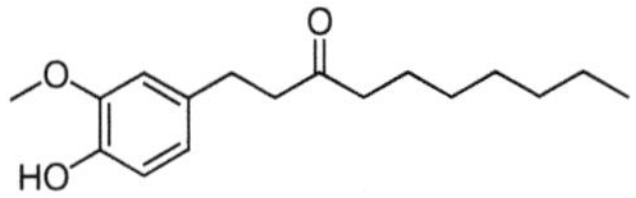

382

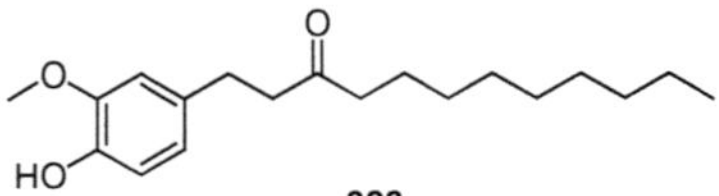

383

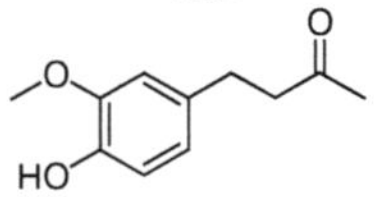

384

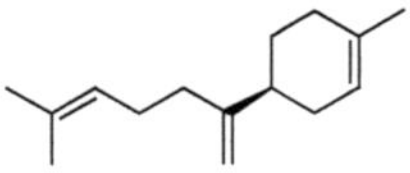

385

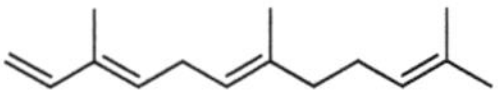

386

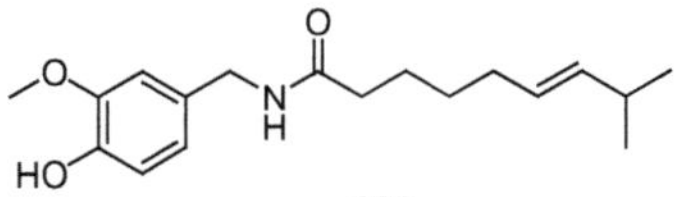

387

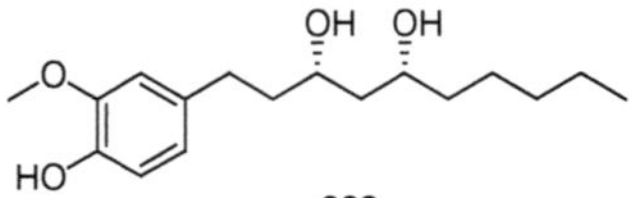

388

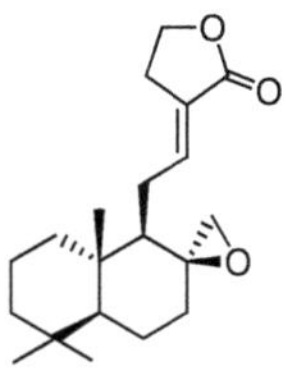

389

Fig. 37 4-Gingerol (**375**), 6-gingerol (**376**), 8-gingerol (**377**), 10-gingerol (**378**), 6-shogaol (**379**), 8-shogaol (**380**), 10-shogaol (**381**), 6-paradol (**382**), 8-paradol (**383**), zingerone (**384**), β-bisabolene (**385**) and α-farnesene (**386**) capsaicin (**387**), 6-gingerdiol (**388**), galanolactone (**389**), isolated from *Zingiber officinale*

5 Biological Evaluation in Validation of Use

Dietary supplements in the United States include botanical products containing structurally identified secondary metabolites that have been shown to be responsible for their observed biological activities. A description of in vitro, in vivo and clinical studies of extracts and secondary metabolites of a selected number of botanicals based on the source of origin is presented below. For most of the positive studies conducted today, further evidence is needed to confirm the efficacy shown.

5.1 Actaea racemosa

Recent biological studies have shown that triterpenoids from *Actaea racemosa* (black cohosh), such as actein (**4**) and 23-*epi*-26-deoxyactein (**5**), promote cell growth and differentiation of osteoblasts as well as osteocalcin production, causing elevation of alkaline phosphatase activity, collagen synthesis, and decreased production of reactive oxygen species (ROS) [21]. The results of a clinical study performed on 62 selected postmenopausal women using the extract AR BNO 1055, containing 40 mg of an *A. racemosa* preparation, conjugated estrogen (CE) (0.6 mg) or placebo for 12 weeks indicated the positive effects of the plant-containing sample on bone metabolism. A menopause rating scale was used for menopausal symptom measurements and diary levels of Cross Laps (a marker of bone degradation). It was shown that treatment with AR BNO 1055 extract for 12 weeks increased the amount of bone alkaline phosphatase. These reported biological activities correlated with beneficial effects of *A. racemosa* against postmenopausal osteoporosis [28, 287].

The cycloartane triterpenes (**4** and **5**), cimigenol-3-*O*-β-D-xyloside (**1**), and cimiracemoside A (**24**) (Fig. 1) showed growth inhibition against MCF-7 breast cancer cells with IC_{50} values of 14, 21, 22 and 41 $\mu g/cm^3$, respectively [288]. Actein (**4**) has also been shown to significantly inhibit cell proliferation, reduce the migration and motility of endothelial cells, and suppress the protein expression of VEGFR1, pJNK, and pERK, as part of the JNK/ERK pathway. The results of in vivo testing showed that oral treatment of actein (**4**) (10–15 mg/kg) for 28 days decreased tumor size and metastasis when tested in a murine xenograft model implanted with mouse syngeneic mammary tumor 4T1 cells [289]. 25-*O*-Methylcimigenol-3-*O*-α-L-arabinopyranoside (**25**) and 25-*O*-methylcimigenol-3-*O*-β-D-xylopyranoside (**26**) exhibited cytotoxic activity for three different myeloma cell lines (NCI-H929, OPM-2, and U-266) [24, 29].

Black cohosh has been associated with the alleviation of premenstrual discomfort, dysmenorrhea, and perimenopausal symptoms. Commercial products of *Actaea racemosa* are available for treatment of menopausal depression, as well as premenstrual depression. The mechanism of action for this type of activity of black cohosh remains unclear. Studies have shown that an increase in luteinizing hormone (LH) leads to estrogen level elevation and can reduce the severity of menopausal disorders. To determine the effects of a commercial preparation of *Actaea racemosa* (Remifemin®) on the secretion of LH and follicle-stimulating hormone (FSH), a placebo-controlled investigation was performed on 110 postmenopausal women (52 ± 2 years). Two months after daily administration of 8 mg of this extract, the FSH concentration was similar to placebo subjects, while the LH secretion was decreased significantly in patients treated with the extract. This suggests an estrogenic effect of this *A. racemosa* product [28, 290].

5.2 *Allium sativum*

Garlic is best known as a flavoring for food. However, over the years, this botanical has been used medicinally to prevent or treat a wide range of diseases and conditions [236]. Garlic has been attributed with therapeutic properties including the alleviation of lung disorders, whooping cough, stomach disorders, colds, earache, and may assist in preventing cardiovascular disease. Thus, an extract of *A. sativum*, prepared from aged garlic is a folk herbal remedy that has been postulated to enhance the immune system [291]. According to the National Library of Medicine, National Institutes of Health (NIH), USA, garlic is widely used for several conditions linked to the vascular system and heart, like hypercholesterolemia, hyperlipidemia, atherosclerosis, thrombosis, hypertension, and cardiac injury [236]. Sulfur compounds are purported to be responsible for the medicinal properties of garlic [38]. Allicin (**29**) (diallyl thiosulfinate), a sulfenic acid thioester, is the most biologically active sulfur-containing compound of garlic, and also is responsible for its smell and taste [34, 39].

Garlic extracts (aqueous, methanol and ethanol) have been reported to be active against a panel of bacteria including *S. dysenteriae*, *S. aureus*, *E. coli*, *Streptococcus* spp., *Salmonella* spp., and *P. mirabilis*, with minimum inhibitory concentration (*MIC*) values of 0.05–1.0 mg/cm^3. Among these extracts, the garlic aqueous extract was the most effective [32, 33, 292]. Allicin (**29**) has been established as a potent antimicrobial, antiviral, antifungal and antiprotozoal compound. Ajoenes **27** and **28** have exhibited broad-spectrum antimicrobial activity against Gram-positive and Gram-negative bacteria [32, 39]. Eruboside B (**36**) has been described as an inhibitor of the growth of *Candida albicans*, and its activity was comparable to allicin (**29**) (MIC 25 μg/cm^3) [293]. Voghieroside C1 (**35**) (1000 ppm), exhibited weak antimicrobial effects against *Botrytis cinerea* and *Trichoderma harzianum* [294].

Garlic paste and oil as well as the purified compound **29** reduced cholesterol biosynthesis in rat hepatocytes, as a result of the inhibition of 3-hydroxy-3-methylglutaryl (HMG)-coenzyme A (COA)-reductase and 14α-demethylase. Allicin (**29**) exhibited EC_{50} values $= 28 \pm 4$ $\mu M$and 17 ± 2, respectively [295]. In clinical trials, Sobenin et al. revealed that administration of Allicor, garlic powder tablets, at doses of 600 and 300 mg/day for 4 weeks and 12 months, respectively, decreased total cholesterol, total triglyceride, and low density lipoprotein, while increasing high density lipoprotein levels [296, 297].

Many in vitro and in vivo studies have suggested possible cancer chemopreventive effects of garlic preparations and their respective constituents. The anticarcinogenic effect of garlic is induced by the presence of bioactive sulfur compounds like allicin (**29**), DAS (**30**), DADS (**31**), and DATS (**32**) [298]. Shang et al. reported that the cancer-related mechanisms of action of garlic extracts may be attributed to regulating the metabolism of carcinogenic substances, inducing apoptosis, suppressing cell growth and proliferation, angiogenesis, and inhibiting cell invasion and migration [33]. Allicin (**29**) has some antitumor potential, and showed an inhibitory effect against glioma cells (U251), human hepatic adenocarcinoma cells (SK-HEP-1), human hepatocellular carcinoma cells (BEL-7402), human lung cancer cells (SK-MES-1), and human colorectal adenocarcinoma cells (DLD-1) with IC_{50} values in the range of 1.39–41.97 $\mu g/cm^3$ [298]. Also, (*Z*)-ajoene (**28**) (IC_{50} 2.5–5 μM) has inhibited the growth of glioblastoma multiforme cancer stem cells (GBM CSC), and has exhibited antiproliferative activities against other types of cancer cells [299].

5.3 Aloe vera

Aloe species increasingly are being incorporated into different cosmetic products, health drinks, foods, and beverages, and used as dietary supplement in the United States. *Aloe vera* has also been used as an ingredient of functional foods such as yogurt or for the preparation of healthy drinks, including teas. The European Commission (EC), according to Annex I of Regulation No. 1831/2003, states that *Aloe vera* can be used by the animal feed industry as sensory additive functional group "flavoring compounds", to increase the smell or palatability of feedstuff products [44].

Many health-promoting effects of *Aloe vera* have been attributed to the polysaccharides present [41, 42]. The immunomodulatory activity of one of these, acemannan (**41**), has been confirmed using in vitro and in vivo studies. Acemannan (**41**) can enhance the lymphocyte response to alloantigen in a concentration-dependent manner (2.6×10^{-7} to 2.6×10^{-9} M) and the mechanism may be related to the release of interleukin-1 (IL-1) from nuclear cells under the protection of alloantigen [44, 46, 300].

The antitumor activity of acemannan (**41**) was examined in mice implanted with sarcoma cells. Thus, a sample enriched with acemannan (**41**) was injected into the peritoneum of an Institute of Cancer Research (ICR) mouse strain at 1 mg/0.028 kg a day for six consecutive days. The results confirmed that the macrophage-activating activity of acemannan (**41**) shown in vitro could be correlated with the antitumor activity determined in vivo [46, 57, 301].

Acemannan (**41**) induces the growth of bacteria such as *Bifidobacterium* and *Lactobacillus* species. The inhibitory effect of acemannan (**41**) on the bacterial populations of human feces and on the synthesis of short-chain fatty acids produced during fermentation of human feces, has been determined using qPCR. Therefore, acemannan (**41**) may have a prebiotic potential [302].

The polysaccharides in *A. vera* are mainly responsible for their antimicrobial, antiviral, and antifungal activities [44, 303–305]. Also, *A. vera* gel has shown growth inhibitory effects against both Gram-positive and Gram-negative bacteria (*Staphylococcus aureus*, *Enterococcus faecalis*, *Escherichia coli*, and *Pseudomonas aeruginosa*). The saccharide unit mannose present might interfere with the cell wall growth of bacteria. The structure of acemannan (**41**) suggests that it might block bacterial adherence to the host epithelium [42, 44, 45, 304]. As an antifungal agent, it was demonstrated that short-term exposure of macrophages to acemannan (**41**) upregulates phagocytosis and displays activity against *Candida parapsilosis*, *Candida krusei*, and *Candida albicans* [42, 44, 45]. Acemannan (**41**) also showed activity in vitro against Newcastle disease virus, influenza virus, and human immunodeficiency virus (HIV). Acemannan (**41**) possesses antiviral activity by modifying glycosylation of both virally infected cells and the glycoprotein coats of viruses, thus inhibiting virus replication and infectivity [42, 306].

Aloe vera is a traditional remedy for diabetes mellitus worldwide. In particular, polysaccharides present in *A. vera* have been attributed to its significant antidiabetic activity determined [42, 44]. The efficacy and safety of aloe leaf gel was evaluated (300 mg capsule every 12 h for two months) in the treatment of type 2 diabetic patients that were resistant to oral synthetic antihyperglycemic drugs and required insulin. The only bioactive substance that was identified and quantified in the aloe gel used in this trial was acemannan (**41**). Acemannan (**41**) significantly lowered the blood levels of fasting glucose and glycosylated hemoglobin, without any major effects in a liver and or kidney function test [307]. Several clinical trials showed positive effects for *A. vera* gel in the treatment of patients with type 2 diabetes mellitus, but the mechanisms involved in the antihyperglycemic and antihyperlipidemic actions of **41** remain unidentified [42, 45, 57, 308]. Thus, more clinical trials are needed to confirm the therapeutic properties of acemannan (**41**) and to elucidate the exact mechanism of action.

5.4 Cinnamomum spp. (Cinnamomum verum)

Cinnamon has many different reported health properties, such as anti-inflammatory and antioxidant properties and an effect on diabetes and microbes due to the bioactive components evident. The pharmacological properties of this botanical are associated with constituents including various phenolic acids, proanthocyanidins, as well as coumarins and essential oil components. All types of cinnamon contain the active constituent cinnamaldehyde (**46**), which is found in different concentrations in all plant parts. Cinnamaldehyde (**46**), a major compound in the bark oil (65–80%) from both *C. zeylanicum* and *C. verum* showed a positive effect in reducing plasma glucose levels using a rat model of streptozocin-induced diabetes [61, 62].

A methanol-soluble extract of *C. zeylanicum* exhibited weak inhibition of α-glucosidase and α-amylase activity with IC_{50} values of 140.0 and 130.6 $\mu g/cm^3$, respectively. Also, it has been demonstrated that a *C. zeylanicum* bark aqueous extract had a mild effect on lowering blood glucose levels in vivo with alloxan-induced diabetic mice. Pharmacological studies have indicated that *C. zeylanicum* extract (as a traditional preparation) may enhance the hypoglycemic properties of insulin and could be used as an adjuvant remedy in diabetic patients [59, 309, 310].

A chloroform extract of *C. verum* bark stem exhibited an antidiabetic effect in vivo at 20 mg/kg using a rat model of streptozocin-induced diabetes. The aqueous extract of the bark of *C. verum* showed inhibition of α-glucosidase and α-amylase and, in an in vivo model, decreased glucose levels in streptozocin-diabetic rats [63]. Moreover, an extract of Cinnamon phenolic constituents improved lipid metabolism in an in vivo study through the modulation of the transcription factors SREBP-1c, and LXRs, suggesting that the effect of cinnamon phenols may have a potential use in the management of hyperlipidemia [68].

The purported antidiabetic mechanisms by *C. verum*, *C. cassia*, *C. tamala*, and *C. burmannii* include: (1) stimulation of insulin secretion by pancreatic β-cells, (2) increasing muscle and hepatic glycogen content, (3) inhibition of α-glucosidase and α-amylase enzymes, (4) stimulation of cellular glucose uptake by membrane translocation of GLUT-4, (5) reduction of gluconeogenesis, and (6) enhancement of insulin signaling [61, 62].

5.5 Citrus aurantium

Food supplements containing *C. aurantium*, commonly known as bitter orange, may increase energy expenditure, lipolysis, or reduce the appetite [311]. Limonoids from this species have been shown to exhibit several pharmacological activities. An in vivo study of male C57BL/6J mice fed a high-fat diet for nine weeks were further fed this type of diet either alone or supplemented with 0.2% w/w nomilin (**53**) for 77 days. The nomilin-treated mice had lower body weights, serum glucose, serum insulin and enhanced glucose tolerance. This result suggests a novel biological function of nomilin (**53**) as an agent having antiobesity and antihyperglycemic effects, likely mediated through the activation of the G protein-coupled receptor TGR5 [76, 312].

p-Synephrine (**54**) is an α-adrenergic agonist, but it also has weaker β-adrenergic properties [311]. Trials on human subjects have shown that *C. aurantium* naringin (**50**) and neohesperidin (**51**) from *C. aurantium* may increase resting metabolic rate and energy expenditure, as well as decrease weight gain when given for 6–12 weeks. The same study evaluated the thermogenic effects of *p*-synephrine (**54**) in conjunction with the flavonoids **50** and **51** (different doses) in a double-blinded, randomized, placebo-controlled protocol with ten subjects per treatment group. The results suggested a combination of the flavonoids **50** and **51** with *p*-synephrine (**54**) may assist in weight management [73, 313]. Another study demonstrated that feeding high fat diet-induced obese C57BL/6 mice with *C. aurantium* extract for eight weeks resulted in a considerable decrease in overall adipose tissue, body weight, and total serum cholesterol. The authors suggested the extract of *C. aurantium* and the two constituents **50** and **51** as potential anti-obesity agents that can inhibit white adipogenesis and induce brown adipocyte thermogenesis via activation of AMPKα [74, 75].

Haller et al. discussed the cardiovascular and weight loss properties of a combination of *p*-synephrine (**54**) (21 mg) and caffeine (304 mg) in subjects undergoing moderately intense exercise (30 min). The study revealed that the combination of **54** with caffeine increased the ergogenic properties of *p*-synephrine (**54**) [314]. In addition, the acute intake of *p*-synephrine (**54**) in healthy humans has not been associated with side effects, and the safety of this substance has been indicated by studies with animals with doses of up to 1000 mg/kg/day [315, 316]. Additionally, the intake of 49 mg of **54** for 15 days produced no significant changes in heart rate, electrocardiograms, or systolic or diastolic blood pressures in healthy individuals [317].

5.6 *Crataegus spp.*

Crataegus spp. (hawthorn) have been used in traditional medicines since ancient times for their cardiotonic and cardioprotective propensities [318]. The three medicinal species of the genus *Crataegus* officially recognized by the *European Pharmacopeia* include *Crataegus oxyacantha* L. ((Ph. Eur.), *Crataegi Folium cum Flore*), and *Crataegus monogyna* Jacq. [80]. Among these, *C. oxyacantha* has an extensive history for its cardiotonic properties and is considered an official plant in phytomedicine to treat cardiovascular conditions [80]. Other pharmacological activities of this botanical include antihyperlipidemic, antioxidant, and immunomodulatory activities that can be largely attributed to its content of phenolic constituents [80]. Various flavonoid derivatives are among the bioactive chemical constituents of the *Crataegus* species [80–83]. Recent research into the therapeutic benefits of *Crataegus* spp. has been in relation to its effects on cardiovascular disease, but this species is also known to have antihyperlipidemic, anti-inflammatory, and antioxidant capabilities [80, 319].

A study of the cardioprotective activity of this botanical showed that administration of the dried extract of *C. oxyacantha* (LT 132), standardized to contain 2.2% flavonoids, for three months to Wistar rats, demonstrated a significantly reduced average prevalence of malignant arrhythmias in the reperfusion period in the Langendorff heart model [320].

Another *Crataegus* extract, WS-1442, is a standardized aqueous alcoholic (45% ethanol) extract obtained from the dried leaves and flowers of *C. oxyacantha*, containing 18.75% oligomeric procyanidins. This extract is a widely studied preparation known for its beneficial indications for the treatment of cardiovascular disease [81, 318]. A study of the cardioprotective action of this extract demonstrated a significantly reduced mortality rate in male Sprague-Dawley rats subjected to reperfusion-induced arrhythmias after administration of a single dose of 100 mg/kg in comparison to the control group used [81]. Additionally, the occurrence and duration of severe ventricular fibrillation during the 15-minute reperfusion period was significantly lower in groups treated with repeated doses [81]. Also, a significantly lower duration of ventricular tachycardia was observed after administration of 100 mg/kg of WS-1442, for 4 and 8 h, before ischemia-inducing procedures [81]. The main properties of oligomeric procyanidins that are understood to contribute to their cardioprotective effects are free-radical scavenging and antioxidant activity, as well as vasorelaxant activity [318]. Their free-radical scavenging and antioxidative activities are attributed to their participation in the process of scavenging as donors of electrons of hydroxy groups to form, instead of free radicals, stable forms of radicals [318]. The vasorelaxant activity of oligomeric procyanidins is influenced by the release of nitric oxide (NO) from the vascular endothelium due to the activation of endothelial nitric oxide synthase (eNOS) in an isolated rat heart [318]. Further, the WS-1442 extract from *C. oxyacantha* induced an endothelium-dependent, NO-mediated vasorelaxation by eNOS phosphorylation at serine 1177 in rats and isolated human mammary arteries precontracted with 10 μM phenylephrine [321]. Mackenzie et al. reported the inhibition of the binding activity of NF-κB in T lymphoblast (Jurkat T) cells, using (−)-epicatechin (**59**), (+)-catechin (**60**), and procyanidin B2 (**61**), at concentrations of 1.7–17.2 μM, which may contribute to vasoprotection by their anti-inflammatory effects [322]. Dong et al. demonstrated that vitexin (**55**) (6 mg/kg, i.v.) exhibits cardioprotective effects and decreases the elevation of the ST segment (STEMI) of an electrocardiogram. It also reduces myocardial infarct size in myocardial ischemia-reperfusion in rats. Moreover, vitexin (**55**) decreased myocardial NF-κB, TNFα, phosphorylated c-Jun, and phosphorylated ERK expression in myocardial tissue [323, 324]. Many clinical studies have been reported for the *C. oxyacantha* aqueous alcoholic (45% ethanol) extract WS-1442®. In a study on 2681 patients suffering from congestive heart failure, the administration of WS-1442® extract (900 mg/day) for 620 days reduced the odds ratio of sudden cardiac death in patients with lower left ventricular function. In this study, WS-1442® had no significant effect on the primary endpoint and was safe to use in patients receiving optimal medication for heart failure. In addition, WS-1442® reduced the incidence or sudden cardiac death [325].

The antihyperlipidemic activity of *C. oxyacantha* fruits was demonstrated in male Wistar rats administered a hyperlipidemic diet, when a tincture was administered at a daily dose of 0.5 cm^3/100 g body weight for six weeks [80]. Notably, there was a significant reduction in total cholesterol synthesis in the liver (from 18.8 ± 1.1 mg/kg to 9.3 ± 1.4 mg/kg) and the cholesterol biosynthesis was significantly lower in the groups treated with the tincture (316.5 ± 66.6 mg/kg) compared to the control group (997.6 ± 109.2 mg/kg) [80]. Another study, involving New Zealand white rabbits, demonstrated the hypolipidemic activity of hawthorn fruits (*Crataegus pinnatifida*) after the rabbits were fed a high cholesterol diet [326]. Specifically, the rabbits were administered a high cholesterol diet supplemented with hawthorn fruit powder (2 g/100 g) and experienced a 23.4 and 22.2% reduction in serum total cholesterol and triacylglycerols, respectively, compared to rabbits fed an exclusively high cholesterol diet [326]. Further, supplementation of the hawthorn fruit resulted in significantly suppressed activity of intestinal acyl CoA: cholesterol acyltransferase (ACAT), suggesting that the hawthorn fruit decreases serum cholesterol levels by inhibition of cholesterol absorption mediated by down-regulation of intestinal ACAT activity [326]. In addition, (−)-epicatechin (**59**) showed an inhibitory effect on apolipoprotein B-100 (ApoB) secretion in human HepG2 hepatoma cells with an $IC_{50} \sim 50\ \mu M$ [327]. Furthermore, procyanidin C1 (**64**) inhibited triacylglycerol (TAG) and ApoB secretion in HepG2 cells at a dose of 25 mg/dm^3 [328]. Vitexin (**55**) at 100 μM reduced lipid droplet accumulation (37%) and inhibited triglyceride accumulation (39%). It also decreased C/EBPα and PPARγ protein expression levels in mouse adipocytes (3T3-L1 cells) [329].

Crataegus spp. have also been shown to have anti-inflammatory activity which might play a role in cardiovascular diseases [83]. Elango and Devaraj observed that 15-day treatment with a hawthorn extract at a dose of 100 mg/kg per day alleviated the pro-inflammatory response by reducing the levels of pro-inflammatory cytokines (interleukins (IL)-1β and -6, and tumor necrosis factor (TNFα) and intracellular adhesion molecule-1 (ICAM-1) gene expressions following ischemia/reperfusion induced injury [330]. Vitexin (**55**) (10 mg/kg, i.p., 30 min before stimulus with phenyl-*p*-benzoquinone, 1.89 g/kg) inhibited inflammation-associated pain and also inhibited 91% of the acetic acid-induced writhing response. Vitexin (**55**) exhibited an anti-hyperalgesic effect through reducing the pro-inflammatory cytokine (TNF-α, IL-1β, IL-6, and IL-33) and enhancing the anti-inflammatory cytokine (IL-10) production induced by carrageenan (300 μg/paw) in mice [331].

Overall, extensive studies on the flavonoid components of the *Crataegus* species have implicated their therapeutic application in many diseases, particularly cardiovascular disease, as a result of their antihyperlipidemic, antioxidant and anti-inflammatory activities.

5.7 Curcuma longa

Curcuminoid (diarylheptanoid) compounds from *C. longa* rhizomes (turmeric) are considered to have some potential in the supportive management of neurogenerative diseases, metabolic and cardiovascular diseases, and possibly in cancer based on the multitude of literature documenting their biological activity [97]. However, it is important to mention that despite the impressive spectrum of biological activities of curcuminoids using in vitro test systems, there has been a lack in the clinical applicability of these compounds because of their poor solubility, low absorption and bioavailability, and high metabolism rate [97].

First, the effects of curcuminoids on neurodegenerative diseases have been investigated in various in vivo models [97]. Curcumin (**65**), desmethoxycurcumin (**66**), and bisdemethoxycurcumin (**67**) have all been demonstrated to have potent activity in inhibiting microglial activation of primary microglia from the cortex of newborn Sprague-Dawley rats [332]. In the same study, all three compounds were observed to suppress nitric oxide and TNF-α production significantly in a concentration-dependent manner with a relative potency in order from greatest suppression to least being desmethoxycurcumin (**66**), bisdemethoxycurcumin (**67**), and then curcumin

(**65**) [332]. Further, curcumin (**65**) exhibited a neuroprotective effect through wnt3a and β-catenin expression that was increased after 500 n*M* treatment of isolated neural stem cells from the brains of fetal Sprague-Dawley rats [333]. In another study on compound **65**, involving streptozotocin-induced diabetic rats, beneficial neuroprotective effects were demonstrated by reducing the glutamate-mediated excitotoxicity in the rat cerebral cortex by modulating alterations in glutamatergic receptors, oxidative stress, and imbalanced glutamate metabolism [323].

The hypoglycemic, antidiabetic, hypolipidemic, atheroprotective, and cardioprotective activities of the curcuminoids in vivo support the use of turmeric for the alleviation of metabolic and cardiovascular disease [98]. An ethanol extract of turmeric suppressed significantly an increase in blood glucose in type 2 diabetic KK-A^y mice, stimulated human adipocyte differentiation dose dependently and showed human peroxisome proliferator-activated receptor (PPAR)-γ ligand binding activity [324]. The four curcuminoids curcumin (**65**), desmethoxycurcumin (**66**), bisdemethoxycurcumin (**67**), and Ar-turmerone (**70**) were identified from the extract, suggesting their potential role in the prevention of type 2 diabetes [324]. The hypolipidemic effect of curcumin (**65**) was demonstrated in LDL receptor knockout mice fed a high fat diet [334]. Treatment of the lowest dose of 500 mg of curcumin per kg of diet effectively lowered lipid accumulation in peritoneal macrophages and decreased mRNA expression of aP2 and ABCA1 that are otherwise increased by a high fat diet [334]. A further study using LDL receptor knockout mice administered 20 mg/kg curcumin (**65**) per day p.o. for four months led to ameliorated oxLDL-induced cholesterol accumulation in macrophages due to decreased oxLDL uptake and increased cholesterol efflux [335]. The main outcomes after curcumin treatment were a decrease in atherosclerotic lesion areas in aortic roots, a decrease in serum levels of IL-6, TNF-α, MCP-1, total cholesterol, triglyceride, and non-HDL-c, an increase in serum HDL-c and aortic ABCA1 levels, and a decrease in the aortic SR-A level suggesting an atherosclerotic potential for curcumin (**65**) [335]. The cardioprotective potential of curcumin was demonstrated in Sprague-Dawley rats administered 300 mg/kg curcumin daily one week before cardiac ischemia/reperfusion (I/R) injury [336]. The main outcomes were an increase in cardiac contractility and cardiac function parameters otherwise decreased by I/R injury, a decrease in cardiac fibrosis otherwise increased by I/R injury, a decrease in TLR2 mRNA and protein expression, and a decrease in macrophage infiltration [336].

A broad range of cytotoxic and antiproliferative activities in vitro have been demonstrated by curcuminoids [97]. For example, administration at dose of 5 μ*M* curcumin (**65**), desmethoxycurcumin (**66**), or bisdemethoxycurcumin (**67**) decreased urokinase plasminogen activator (uPA) and inhibited collagenase matrix metalloproteinases (MMPs) in HT1080 human fibrosarcoma cells [337]. The regulation of uPA and MMPS plays an important role in cancer cell invasion by cleavage of the extracellular matrix and the curcuminoids thus may have a role in preventing cancer metastasis [337]. Further, desmethoxycurcumin (**66**) has been demonstrated to have antiproliferative activity against the MDA-MB-231 breast cancer cell line with an IC_{50} value of 33.4 μ*M*, and desmethoxycurcumin (**66**), bisdemethoxycurcumin (**67**), and curcumin (**65**) were all shown to inhibit MCF-7 breast cancer cell line proliferation with IC_{50} values of 10, 15, and 20 μM, respectively [338, 339]. In a clinical study involving the breast carcinoma and normal breast tissues from 58 invasive ductal carcinoma patients, curcumin was demonstrated to modulate the expression of maspin, a serine protease inhibitor that can suppress tumor growth and metastasis [340]. This effect was corroborated in vitro in MCF-7 cells after treatment with 20 μM of curcumin, which inhibited cell growth, induced apoptosis and upregulated maspin gene expression [340].

Therefore, the curcuminoids found in abundance in turmeric are well known for their potential to treat or prevent a broad range of ailments. Despite all of the potential applications provided in the literature, an in-depth review of the main constituent of *Curcuma longa*, curcumin (**65**), suggests that clinical studies of this diarylheptanoid may not provide sufficient supportive evidence for the effective use of this botanical [341].

5.8 Echinacea spp.

Echinacea species are native to North America and have been used by Native Americans for multiple types of health problems. Echinacea is one of the most popular natural health products purchased worldwide, with the majority of commercially available products containing *E. purpurea* or *E. angustifolia* [99, 342, 343]. Of these two *Echinacea* species, *E. purpurea* is the more well-known [344]. *Echinacea* species have been traditionally utilized for toothache, bowel pain, snakebite, skin disorders, seizure, chronic arthritis, wound healing, and cancer. *Echinacea*-derived products are widely used as dietary supplements worldwide and are marketed and used as immunostimulants to alleviate the symptoms of the common cold, influenza, and upper respiratory tract infections [345, 346]. There is evidence that the constituents of *Echinacea* extracts have immunomodulatory properties in vitro and in vivo [100, 344, 346, 347]. Alkamide constituents of *Echinacea* species have been reported to have cannabinomimetic properties on both CB_1 ($K_i = 2.00$–48.77 μM) and CB_2 ($K_i = 1.86$–66.39 μM) receptors, which may be attributed to their structural similarity to the control anandamide (($K_i = 0.371$ μM) (an endogenous cannabinoid receptor ligand)) (Fig. 10) [348]. A study published by Raduner et al. demonstrated that the alkamides, dodeca-(2*E*,4*E*,8*Z*,10*Z*)-tetraenoic acid isobutylamide (**85**), and dodeca-(2*E*,4*E*)-dienoic acid isobutylamide (**89**) potently displaced the radioligand from the membrane recombinantly overexpressing CB_2 receptors, with *Ki* values of 57 ± 14 and 60 ± 14 n*M*, respectively (Fig. 10) [349]. These results suggest the potential mechanism of action of alkamides as immunomodulators. Moreover, echinalkamide (**90**) dose-dependently inhibited effects on receptor activator of nuclear factor kappa-B ligand (RANKL)-induced osteoclasts, and on proliferation of osteoclasts and efficiently attenuated osteoclastic bone resorption, without toxicity, and hence may contribute to the anti-inflammatory effect of *E. purpurea* [101].

Echinacoside (**95**) has shown relevant pharmacological properties such as neuroprotective and cardiovascular effects. Chen et al. reported that echinacoside (**95**) (7.0, 3.5 mg/kg for 7 days, i.p.) protects the striatal dopaminergic neurons from injury induced by 6-hydroxydopamine (6-OHDA) in rats. Accordingly, the authors concluded this compound may be useful in alleviating symptoms associated with Parkinson's disease (PD) [350]. Echinacoside (**95**) (concentration range 0.01–10 n*M*) has been reported to have antiosteoporotic activity, and to produce substantial increases in alkaline phosphatase activity, osteoblastic cell proliferation (MC3T3-E1), osteocalcin levels, collagen 1 secretion, and mineralization [351, 352]. In addition, echinacoside (**95**) has been found to inhibit the proliferation of pancreatic adenocarcinoma cells (SW1990) in a dose-dependent manner (20, 50, 100 μ*M*), by inducing the production of reactive oxygen species and the perturbation of mitochondrial membrane potential, thus triggering apoptosis [353].

Cichoric acid (**93**), a major antioxidant agent produced by *E. purpurea*, has exhibited multiple biological activities, including antioxidant and hypoglycemic effects [344]. In a study published by Zhu et al., cichoric acid (**93**) (60 mg/kg, p.o. for 4 weeks) inhibited hepatic injury and chronic inflammation in diabetic mice due to its antioxidant capacity and regulated the balance of gluconeogenesis and glycolysis. Also, **93** (100 μ*M*) regulated glucose metabolism and activated an antioxidant response in glucosamine-induced HepG2 cells [354]. In addition, cichoric acid (**93**) inhibited human immunodeficiency virus type 1 (HIV-1) integrase in vitro and prevented HIV replication in tissue culture [355]. Therefore, anti-inflammatory, antiosteoporotic, antioxidant, antiviral, immunomodulatory, and neuroprotective effects are all demostrated biological effects that have been shown for *E. purpurea.*

5.9 *Epimedium spp.*

Epimedium is a genus of the family Berberidaceae consisting of approximately 52 species of which the most well-known and phytochemically characterized are *Epimedium koreanum* Nakai, *Epimedium sagittatum* (Sieb & Zucc.), *Epimedium brevicornum* Maxim (EB) and *Epimedium grandiflorum* var. *thunbergianum* Miq. [105]. Traditionally, *Epimedium* (horny goat weed) has been used for a wide variety of disease states encouraging interest in its current use as a dietary supplement in the potential modulation of osteoporosis, climacteric period syndrome, breast lumps, hyperpiesia, and coronary heart disease [105]. Flavonoids from *Epimedium* spp. are the major bioactive constituents contributing to their pharmacological actions including documented osteoprotective, neuroprotective, immunomodulatory, and cancer-related activity [105].

Icariin (**96**) has been demonstrated to have an influence on many of these types of biological activities. In a 24-month randomized double-blind placebo-controlled clinical trial icariin was shown to be effective in preventing postmenopausal osteoporosis [356]. The role icariin (**96**) appears to have in bone formation can be attributed to its ability to stimulate the bone formation activity of osteoblasts, and the osteogenic differentiation of bone marrow-derived mesenchymal stem cells (BMSCs), while suppressing adipocytic trans-differentiation in primary osteoblasts [356]. Most recently, icariin has been demonstrated to promote bone formation by activating the sclerostin/Wnt/β-catenin signaling pathway in BMSCs extracted from twelve-week-old female Sprague Dawley rats [357].

Growing evidence of the possible therapeutic effect of icariin (**96**) in neurogenerative diseases has been recognized [358]. In animal models of Alzheimer's disease, this compound was shown to prevent the production of amyloid β (1–42) and to inhibit amyloid precursor protein (APP) expression and β-site APP cleaving enzyme 1 (BACE-1) [359]. Moreover, icariin has been shown to mitigate pro-inflammatory responses of microglia in culture in other animal models of cerebral ischemia, depression, Parkinson's disease, and multiple sclerosis [359, 360]. Learning and memory have also been demonstrated to be improved in both normal aging mice and in a ten-month old amyloid precursor protein (APP)/presenilin 1 (PS1) transgenic mice model after icariin treatment [359]. Icariin (**96**) and its main metabolite, icaritin (**104**), have been suggested to regulate the function and activation of immune cells, modulating the release of inflammatory factors, and restoring aberrant signaling pathways [360]. Both these compounds downregulated histiocytes and immune cell (M1-type Mφ and TH-17 cells)-secreted pro-inflammatory cytokines through a diverse array of signaling pathways and may improve injury and tissue stress in many disease states [360]. In a study with male APP/PS1 mice induced with significant depressive-like behaviors following restraint/isolation stress, elevated pro-inflammatory cytokines, decreased anti-inflammatory cytokines and increased M1 phenotype microglia were all observed after ten months; however, administration of icariin for six months attenuated all these changes [361]. Immune cells such as Mφ, NK cells, and T cells and their cytokines to influence immune surveillance, homeostasis, and defense in cancers, infectious diseases and immune thrombocytopenia have been demonstrated to be regulated by icariin (**96**) [360]. Icariin (**96**) has been shown to increase the cytotoxic T lymphocyte (CTL) response for P815AB peptide on DBA/2J mice bearing tumors suggesting that this compound could enhance the immunogenicity of P815AB and improve the ability of T cells and CTLs to recognize tumor cells [362].

Extracts of various *Epimedium* species have been determined to show effects on the cardiovascular system components including the vessels, blood and heart [105]. A water extract from the leaves of *Epimedium sagittatum* reduced ventricular fibrillation induced by chloroform in mice or induced by calcium chloride in rats [105]. Further, an extract of Herba Epimedii, a Chinese herbal medicine officially sourced from *Epimedium wushanense* T.S. Ying and *Epimedium koreanum* Nakai in the *Chinese Pharmacopeia*, slowed down the heart rate and reduced myocardial contraction of isolated rat hearts after treatment with doses of 0.25 and 0.50 mg/cm^3 [105]. The total flavones from this extract improved the abnormal electrocardiogram J-point of the acute myocardial ischemia model in mice and effectively prevented the increase of blood viscosity at a dose as low as 6 mg/kg [105].

Limited evidence of the cytotoxic capabilities of icariin (**96**) have been reported. Doses of 100–800 μg/cm^3 of the 95% EtOH extract of Herba Epimedii were shown to inhibit the growth of the MCF-7 breast cancer cell line, and icariin, at a dose of 100 μg/cm^3 inhibited the proliferation of the HL-60 hepatoma cell line [105]. In summary, the flavonoid components of *Epimedium* species, particularly icariin (**96**), have demonstrated a broad range of potential therapeutic benefits including osteoprotective, neuroprotective, immunomodulatory effects.

5.10 Euterpe oleracea

Euterpe oleracea Mart., also known by the popular name of açaí, is distributed widely in the Amazon region and its fruits are rich in phenolic constituents [363]. Interest in açaí has arisen due to its high in vitro antioxidant activity. Phenolic constituents are generally associated with health-promoting properties and the prevention of several degenerative diseases [364, 365]. Also, the açaí fruit is well-known for its potential antihypercholesterolemic activity [365].

The preventive potential of açaí phenols is also associated with lipidemia and obesity. A study published by Martino et al. investigated the antilipidemic and anti-inflammatory effects of açaí phenols in 3T3-L1 mouse adipocytes. Pre-adipocytes were differentiated with and without these compounds at concentrations of 2.5, 5, and 10 μg/cm^3. The açaí phenols reduced intracellular lipid accumulation in differentiated adipocytes in a dose-dependent manner and down-regulated PPARγ2. Also, gene expression of the adipogenic transcription factors C/ebpα, C/ebpβ, Klf5 and Srebp1c was decreased. In addition, these compounds protected cells against the production of ROS. Hence, açaí constituents may be useful in the prevention of adipogenesis and oxidative stress [366]. An in vivo study showed that the pulp of *E. oleracea* promoted hypercholesterolemic effect, through the enhancement of expression of transporter (ABCG5 and ABCG8), and low-density lipoprotein receptor (LDL-R), which are genes involved with the secretion and cholesterol biosynthesis in the rat, respectively [367]. A phenol-rich açaí extract reduced obesity, increased cholesterol excretion, and improved oxidative stress in the liver [363].

Tsuda et al. demonstrated that cyanidin 3-glucoside (**106**) at a dose of 100 μ*M*, significantly increased in human adipocytes the expression of genes involved in lipid metabolism (UCP2, ACOX1, PLN, adiponectin). The up-regulation of the genes may be performed to prevent excess lipid accumulation in adipocytes. These results provide a biochemical basis for the use of cyanidin 3-glucoside (**106**), which can also have important implications for preventing obesity and diabetes [368]. Cyanidin-3-rutinoside (**107**) is a mixed-type competitive inhibitor of pancreatic lipase with a IC_{50} value of 59.4 ± 1.41 μ*M*. Furthermore, compound **107** inhibited pancreatic cholesterol esterase by about 5–18% in a range of concentrations (0.125–1 m*M*). Also, cyanidin-3-rutinoside (**107**) (12.5–100 μ*M*) exhibited in Caco-2 cells a significant reduction in cholesterol uptake in both free cholesterol (17–41%) and mixed micelles (20–30%). Finally in Caco-2 cells, cyanidin-3-rutinoside (**107**) at 100 μ*M* was able to suppress mRNA expression of the Niemann-Pick C1-Like 1 [(NPC1L1) intracellular cholesterol transporter 1], after 24 h incubation [369].

A clinical trial on seven athletes (17.5 ± 1.2 years), who ingested once daily for six weeks 100 cm^3 of açaí juice blend, showed an increase in the antioxidant capacity of plasma, decreased exercise-induced muscle damage, and a substantial improvement of serum lipid profile. The authors attributed these observations to its high total phenolic content and the related high in vivo antioxidant and hypocholesterolemic activities of this supplement [365].

5.11 *Garcinia gummi-gutta*

Garcinia gummi-gutta (L.) N. Robson (Clusiaceae), commonly known by its previous scientific name *Garcinia cambogia* (Gaertn.) Desr., is native to Southeastern Asia and is distributed throughout the world [117, 370]. The fruit is used commonly as a food preservative and as a flavoring or food-bulking agent. This small fruit, reminiscent of a pumpkin in appearance, currently is most popularly used and widely advertised as a possible weight-loss supplement [371]. In many Asian countries, *G. gummi-gutta* is utilized as a traditional remedy to treat constipation, hemorrhoids, rheumatism, diarrhea, edema, irregular menstruation, and intestinal parasites [117].

Mazzio and Soliman demonstrated that a fruit extract of *G. cambogia* (IC_{50} = 0.235 mg/cm^3) exhibited cytotoxic activity against a murine neuroblastoma cell line (Neuro-2A) [372]. Garcinol (**129**), a polyisoprenylated benzophenone, has been reported as having cancer-related biological activities. Thus, garcinol (IC_{50} = 9.42 μM) (**129**) induced apoptosis of HL-60 cells, in a concentration and time-dependent manner [373]. In addition, garcinol (**129**) inhibited eukaryotic topoisomerase I and topoisomerase II, at concentrations comparable to etoposide (IC_{50} ~ 25–101 μM, respectively) [374]. Moreover, oxy-guttiferone M (**127**), oxy-guttiferone K (**128**) and guttiferone M (**131**) were found to be weak inhibitors of topoisomerase II (IC_{50} = 105.5, 109.0, and 73.0 μM, respectively) [374].

Hydroxycitric acid (HCA) (**120**) is a component present in the fruit rind of *Garcinia gummi-gutta*, which may be responsible for its weight-loss potential. HCA (**120**) reduced weight-gain by inhibiting adenosine triphosphate (ATP)-citrate lyase, the enzyme responsible of fatty acid synthesis [117]. Rao et al. administered HCA (**120**) and its lactone analogue (**124**) (1.1–5.5 mmol/kg, stomach tube, daily over an eight-week period) to Wistar albino rats. Feed intake was suppressed in both the HCA (**120**) and garcinia lactone (**124**) groups. The garcinia lactone (**124**) group was found to reduce weight gain in a dose-dependent manner [375]. Anton et al. investigated the effect of HCA (**120**) on food intake, satiety, weight-loss, and oxidative stress levels in humans. The authors used two doses of HCA (**120**) (2,800 mg/day and 5,600 mg/day) in obese individuals (body mass index; BMI range = 25.0–39.9) and found that both doses were safe for appetite suppression and weight management [376]. Therefore, further studies may provide more evidence to support the use of this botanical.

5.12 *Ginkgo biloba*

A diverse array of flavonoids isolated from *G. biloba* leaves has demonstrated several types of biological activity including antioxidative, antibacterial, antiviral, cancer-related, and neuroprotective effects [119]. The flavonol aglycone, isorhamnetin (**133**) has been shown to induce the expression of neurofilaments and potentiate NGF-induced neurite outgrowth and neurofilament expression, thus suggesting its potential in the treatment or prevention of neurogenerative disease [377]. Flavonol glycosides with antioxidant activity isolated from *G. biloba* include quercetin 3-*O*-β-D-glucopyranoside (**137**) and quercetin 3-*O*-α-L-[6‴-*p*-coumaroyl-(β-D)-glucopyranosyl-(1→2)-rhamnopyranoside]-7-*O*-β-D-glucopyranoside (**138**). These compounds were subjected to DPPH and cytochrome-c reduction assays in the HL-60 leukemia cell culture system and exhibited inhibitory activity against free-radical formation (IC_{50} 23.9 and 14.7 μM, respectively) and free-radical scavenging activity in the DPPH assay (IC_{50} values of 26.1 and 15.8 μM, respectively) [122].

The biflavones ginkgetin (**134**), isoginkgetin (**135**), and bilobetin (**136**) isolated from *G. biloba* exhibited inhibitory effects on human thrombin with IC_{50} values in the range of 8.05–17.83 μM [378]. Further, **134** has been demonstrated to have a role as a STAT5 inhibitor, blocking the differentiation of preadipocytes into adipocytes, suggesting its potential as an antiadipogenesis and antiobesity drug [379]. Compounds **134–136** have also shown lipid absorption reduction activities by inhibiting a pancreatic lipase enzyme with IC_{50} values ranging from 2.90 to 12.78 μM [380]. Additionally, these biflavones have been demonstrated by Song et al. to be highly specific and potent carboxylesterase 2 inhibitors (IC_{50} values ranging from 0.07–0.14 μM), suggesting their potential for development as novel agents to alleviate irinotecan-induced diarrhea [381].

Biginkgosides E (**139**) and H (**140**) have been demonstrated to inhibit nitric oxide production in lipopolysaccharide-activated BV-2 microglial cells with IC_{50} values of 2.91 and 17.23 μM, respectively, and biginkgoside F (**141**) demonstrated a 34.3% increase in cell viability at 1 μM in SH-SY5Y neuroblastoma cells [120].

The terpene lactones ginkgolides A (**143**) and B (**144**) from *G. biloba* leaves have been found to protect against synapse damage and the cognitive loss that occurs during the early stages of Alzheimer's disease as a result of their antagonistic activity at the platelet-activating factor (PAF) receptor [382]. Further, these compounds have been demonstrated as having neuroprotective activity by alleviating depressive and anxiety-like behaviors in rats [383]. Ginkgolides A (**143**), B (**144**), C (**145**), and J (**146**) have been determined to act potently as selective use-dependent blockers of glycine-activated chloride channels in the hippocampal neurons of rats with IC_{50} values of 1.97, 0.273, 0.267, and 2.0 μM, respectively [384]. Bilobalide, a further terpene lactone, has also been shown to have potential neuroprotective effects including reducing triethyltin-induced cerebral edema, reducing damage from cerebral ischemia, and decreasing cortical infarct volume in in vivo stroke models [384, 385].

The preparation EGb761 is a standard extract of *G. biloba* containing 24% (*w*/*w*) flavonoids and 6% (*w*/*w*) terpene lactones that has been used as a European phytomedicine for tinnitus treatment, neuron protection, and enhancements in cognitive function [118]. This extract has been observed to directly reduce tinnitus severity in patients with dementia and to indirectly improve secondary effects of tinnitus severity such as depression and anxiety symptoms and improved cognition [386]. Moreover, the protective potential of EGb761 has been studied extensively in animal models of neurogenerative disease [122]. Specifically, EGb761 protects against neuronal destruction in the hippocampus of gerbils and rats in models of cerebral ischemia and reduces cognitive decline in transgenic mice that overexpress growth hormone and consequently have elevated and increasing free radical processes that correlate strongly with their ability to survive [387, 388]. A human study demonstrated EGb761 to have beneficial effects on attention, memory and cognitive function in patients with multiple sclerosis [388]. Additionally, patients with mild to moderate cognitive impairment administered 40 mg of EGb761 three times daily experienced a significant improvement in their performance in the Kendrick Battery, a digit copying test, after 12 weeks and an improved median response time on a computerized version of a classification task after 24 weeks compared to placebo [389]. Therefore, the diverse and unique flavonoid and terpenoid components isolated from *G. biloba* contribute to the many biological activities of this botanical that support its use as a dietary supplement and in traditional medicine.

5.13 Hypericum perforatum

Traditionally, *Hypericum perforatum* L. (St. John's wort) (Hypericaceae) has been utilized in European countries for the treatment of excitability, neuralgia, fibrositis, sciatica, menopausal neurosis, anxiety, depression, as a nerve tonic, and in topical preparations for treating wounds [125]. An array of phytochemicals has been isolated from *H. perforatum* of which the biologically active constituents are the prenylated acylphloroglucinols (PAPs), xanthones, naphthodianthrones, bianthroquinones, and monoterpenes [125, 126]. Hyperforin (**149**), a prenylated phloroglucinol, and hypericin (**150**), a naphthodianthrone, are considered the major active constituents from St. John's wort and have displayed a range of biological activities including antidepressant, cytotoxic, antimicrobial and antioxidant activities [125].

In terms of the antidepressant activity of St. John's wort, hypericin acts by inhibiting the enzyme monoamine oxidase, as well as the neuronal uptake of serotonin, norepinephrine, and dopamine, and of γ-aminobutyric acid (GABA) and α-glutamate [130]. Hyperforin (**149**) has been demonstrated with similar inhibitory activities having IC_{50} values in the range of 0.05–0.10 $\mu g/cm^3$ against the uptake of all of serotonin, dopamine, noradrenaline, and GABA [390].

Several different constituents of St. John's wort have displayed cytotoxic activities against cancer cells [130]. Phloroglucinol derivatives have been shown to exhibit cytotoxic activity [130]. Hyperibone J (**151**) is an analog of hyperforin that has been demonstrated to have weak cytotoxicity activity against the MCF-7 breast cancer cell line with an IC_{50} value of 17.8 μg/cm^3 [391]. The prenylated acylphloroglucinol derivatives ialibinone B (**152**) and hyperpapuanone (**153**) have both shown cytotoxic activity against KB cells with IC_{50} values of 3.2 ± 1.1 and 6.6 ± 1.9 μg/cm^3, respectively [392]. Sampsoniones A (**154**) and I (**156**) are classified in a family of polyprenylated benzoylphloroglucinols that have displayed cytotoxic effects against the P388 cancer cell line with IC_{50} values of 13 and 6.6 μg/cm^3, respectively [130, 131]. Another phloroglucinol derivative, tomoenone F (**157**) was demonstrated to be cytotoxic against KB cells with an IC_{50} value of 6.2 μ*M* and to have more potent cytotoxicity activity against two MDR cancer cell lines (KB-C2 and K562/Adr) than doxorubicin [393]. Further, the spirocyclic acylphloroglucinols hyperbeanols B (**158**) and D (**159**) have exhibited cytotoxic activity against the K652 cancer cell line with IC_{50} values of 16.9 and 20.7 μ*M*, respectively [394]. The acylphloroglucinol derivatives hypercalyxones A (**160**) and B (**161**) were screened for their cytotoxicity against KB cells and were demonstrated to have inhibitory activities with IC_{50} values of 6.5 ± 0.78 and 7.0 ± 0.63 μg/cm^3, respectively [395]. Xanthone derivatives isolated from St. John's wort also have shown cytotoxic activity. Kielcorin (**162**) is a xanthonolignoid that has been determined to show cytotoxicity against KB cells with an IC_{50} value of 8.1 μg/cm^3 [130]. In addition, hyperxanthones C (**163**) and E (**164**) were active against the A549 and MCF-7 cancer cell lines with IC_{50} values in the range of 8.5–19.5 μg/cm^3 [130]. Finally, the monoterpene biyouyanagin A (**165**) displayed cytotoxic activity against KB-C2, MCF-7, COLO205, and K-562 cancer cell lines with IC_{50} values ranging from 16.6 to 38.8 μg/cm^3 [130].

Constituents of St. John's wort have shown evidence of antimicrobial activity including hyperforin, hypercalyxones A (**160**) and B (**161**), isohyperbrasilol B (**166**), and yojironin A (**167**). Hyperforin (**149**) has shown potent antibacterial activity against penicillin-resistant and methicillin-resistant *Staphylococcus aureus* with minimum inhibitory concentrations ranging from 0.1 to 1 μg/cm^3 [396]. Hypercalyxones A (**160**) and B (**161**) exhibited a range of antibacterial activities against *B. cereus*, *S. aureus*, *S. epidermidis*, and *M. luteus* [395]. In a bioautography assay on silica-gel glass-backed plates, isohyperbrasilol B (**166**) produced potent inhibitory activity at a minimum quantity of 0.2 μg level [397]. Furthermore, the monoterpenoid yojironin A (**167**), when screened for antimicrobial activity against *A. niger*, *C. albicans*, *C. neoformans*, *T. mentagrophytes*, *S. aureus*, and *B. subtilis*, displayed minimum inhibitory concentrations of 8, 2, 4, 2, 8, and 4 μg/cm^3, respectively [398].

The antioxidant activity of hyperjovinol A (**168**) and 3-geranyl-1-(2-methylpropanoyl)-phloroglucinol (**169**) was demonstrated in vitro using the 2,2-diphenyl-1-picrylhydrazyl (DPPH) and dichloro-dihydrofluorescein diacetate (DCFH-DA) assays [399]. Both had antioxidant potentials that were comparable to that of the known antioxidant Trolox, while hyperjovinol A (**168**) was able to prevent exogenous stimulation of reactive oxygen species [399]. Also, the benzophenone cariphenone A (**170**) showed antioxidant activity similar to quercetin (**108**) by exhibiting an inhibition of chemiluminescence in a total radical-trapping parameter assay [400].

Philipp et al. reported a randomized, double-blind, multicenter clinical trial in a primary care setting for 263 on patients with moderate depression who received St. John's wort extract at 350 mg three times daily [STEI 300, containing 0.2–0.3% hypericin (**150**) and 2–3% hyperforin (**149**); $n = 106$], imipramine 100 mg daily (divided 3 doses: 50 mg, 25 mg, and 25 mg; titrated from 50 mg on day 1, 75 mg on days 2–4; $n = 110$) or placebo ($n = 47$), for eight weeks. This *H. perforatum* extract was found to be more effective than placebo after six weeks of treatment, and to be as efficacious as imipramine after eight weeks of treatment [401]. In summary, St. John's wort contains a broad range of phytochemicals that support its traditional medicinal use as an antidepressant and other activities have been studied in the laboratory including cytotoxicity, antimicrobial and antioxidant effects.

5.14 *Lepidium meyenii*

Maca (*Lepidium meyenii* Walp.) is a plant native to Peru in the family Brassicaceae and is used traditionally by indigenous Andean populations to enhance fertility and to alleviate respiratory disorders and anemia [148]. The biological activities of maca constituents are largely attributed to their degradation products, isothiocyanates (ITCs), induced thermally or chemically, or by the action of the enzyme myrosinase [145, 148]. ITCs from *L. meyenii* have been demonstrated with cytotoxic and antitumor activity against lung, breast, colon, prostate, and ovarian cancers in preclinical studies in vitro and in certain animal models by suppressing cell proliferation, angiogenesis and metastasis [145]. Additionally, these ITCs have shown limited activity in vitro and in vivo against neurodegenerative diseases, although there have been only limited studies into the mechanisms of this type of activity [145]. Glucotropaeolin (**176**) significantly increased the endurance exercise capacity, decreased blood lactate concentrations, and significantly elevated non-esterified fatty acids in plasma during exercise after administration of 0.015 mg/kg by stomach intubation for six weeks to mice in a swimming endurance test [402].

Studies of the biological activities of macamides and macaenes are also limited. However, macamides exhibit moderate fatty acid amide hydrolase (FAAH) inhibitory activity. For example, *N*-(3-methoxybenzyl)-(9*Z*,12*Z*)-octadecadienamide (**180**) gave an IC_{50} value of 10.3 ± 1.3 μM in this assay [403]. Interestingly, *N*-benzyl-(9*Z*,12*Z*)-octadecadienamide (**179**) exerts selective binding affinity for the cannabinoid CB_1 receptor as well as inhibition of anandamide cellular uptake [404].

Compounds from maca containing a thiohydantoin moiety display a variety of biological effects including hypolipidemic, antimutagenic, antithyroidal, and cancer-related activities [142]. From a screen of seventeen thiohydantoins isolated, the compound (+)-meyeniin A (**181**) showed moderate cytotoxic activity against the HL-60, A549, and MCF-7 human cancer cell lines with IC_{50} values of 14.41, 32.22 and 33.14 μM, respectively [405]. Other thiohydantoins in this screen included meyeniin B (**182**) and meyeniin C (**183**), which did not show any perceived activity against the cell lines used [405]. The double-bonded sulfur atom possessed by these compounds offers the potential to form polar interactions with various biological targets [142].

Alkaloids isolated from *L. meyenii* have shown antiproliferative activity against several cancer cell lines [142]. The imidazole alkaloids, lepidiline A (**171**) and lepidiline B (**172**), demonstrated cytotoxic activity with ED_{50} values of $\leq$ 10 μg/cm^3 against the UMUC3, PACA2, MDA231, and FDIGROV cell lines, respectively [406]. Other alkaloids isolated from this botanical include the pyrrole alkaloids macapyrrolins A (**173**), B (**174**) and C (**175**), but these showed no discernible cancer cell line cytotoxic activity up to a concentration of 40 μ*M* [141].

Extracts of maca (*L. meyenii*) have given evidence of a range of other biological activities [143]. An effect on the reproductive system was shown in male rats induced with prostatic hyperplasia by testosterone enanthate, which experienced a reduction in prostate size after treatment with hydroalcoholic or aqueous extracts of maca [407]. A pentane extract of maca root exhibited neuroprotectant activity in vitro with an EC_{50} value of 2.8 μg/cm^3 using crayfish neurons with H_2O_2 induced oxidative stress and under stroke-inducing conditions in vivo using Sprague-Dawley rats at a dose of 3 mg/kg [408].

5.15 Linum usitatissimum

Flaxseed (*Linum usitatissimum* L.) is the seed from the flax plant [150]. The Latin name of flaxseed means "very useful", and it has brown and golden varieties (also known as golden linseeds). Flaxseed/flax oil has been a part of the human diet for thousands of years in Asia, Europe, Africa, and North America. Flaxseed is emerging as an important functional food ingredient. There are several ways to ingest flaxseed, namely, milled, in the form of an oil, or when added to bakery products. Currently, flaxseed has been the focus of increased interest in the field of diet and disease research due to its potential health benefits [150, 409]. Scientific evidence supports the consumption of flaxseed for its high content in ω-3 fatty acids (e.g., α-linolenic acid (**184**)) (Fig. 19), digestible proteins, phytoestrogenic lignans (e.g., secoisolariciresinol diglucoside (SDG) (**185**) (Fig. 19)), other phenols and flavonoids, sterols, as well as various soluble and insoluble fibers. Evidence showed that feeding with flaxseed (either as an oil or enriched product) may help prevent certain diseases such as respiratory, cardiovascular, and obesity disorders, and cancer [150–152]. SDG (**185**) has shown anti-inflammatory, antioxidant, antimicrobial, hypolipidemic, and neuroprotective effects [409, 410]. Rom et al. determined in vivo that SDG (**185**) (4 mg/mouse p.o.) diminished leukocyte adhesion to and migration across the blood-brain barrier (BBB) in a model of aseptic encephalitis (intracerebral, tumor necrosis factor-α (TNF-α) injection) and prevented enhanced BBB permeability during systemic inflammatory response (lipopolysaccharide (LPS)). SDG (**185**) can serve as an anti-inflammatory and barrier-protective agent in neuroinflammation [410]. Flaxseed oil (α-linolenic acid (**184**) (55%)) consumption exerts several effects on inflammatory mediators and markers depending on the dose used [402]. Some information on the antiulcer and anti-secretory properties of flaxseed oil has been obtained [150]. Thus, the oil (1.0, 2.0, and 3.0 cm^3/kg, i.p.) inhibited the formation of gastric lesions in a dose-dependent manner in an aspirin-induced gastric ulceration model in rats [411].

Additionally, the potential antineoplastic activity of SDG (**185**) was suggested to be associated with the inhibition of certain enzymes involved in carcinogenesis [409, 412]. In a clinical trial, 45 premenopausal women were selected with suspicious breast biopsies or who were former breast cancer survivors, and were administered SDG (**185**, 50 mg) daily for one year. Patients who received SDG (**185**) exhibited fewer breast precancerous changes, and 80% of them showed a decrease in the levels of a marker of proliferation (MKI67) [413]. Also, human studies have shown SDG (**185**) as a potential cardiovascular protector by mediating the mechanism of total cholesterol, LDL-cholesterol, HDL-cholesterol, triacylglycerides and glucose metabolism. Twenty patients with hypercholesterolemia and hypertriglyceridemia received SDG (**185**) (600 mg p.o. per day for 8 weeks) and it decreased the levels of cholesterol, LDL-cholesterol and glucose [414].

5.16 Lycium spp.

Among the most common species of *Lycium* are *L. barbarum* L., *L. chinense* Mill., and *L. ruthenicum* Murr. [155]. *Lycium barbarum* L. has been used in Chinese Traditional Medicine and has antioxidant, antidiabetic, cancer-related, immunomodulatory, and neuroprotective effects. This botanical has also been used as a functional and medicinal food [155]. Its biological activities can be attributed largely to the polysaccharides isolated from *L. barbarum* that comprise 5–8% of the dried fruits and consist of a complex mixture of highly branched and partly characterized polysaccharides and proteoglycans [155, 156]. The *L. barbarum* polysacharides (LBPs) have shown biological activities such as antioxidant, immune regulation, antitumor, and neuroprotective effects [155]. The antioxidant activities of LBPs have been studied in some in vivo models of disease [155]. Administration of 50, 100, or 200 mg/kg LBPs consisting of D-rhamnose, D-xylose, D-arabinose, D-fucose, D-glucose, and D-galactose units having β-glycosidic linkages in a molar ratio of 1:1.07:2.14:2.29:3.59:10.06 restored the abnormal oxidative capacity to nearly normal levels in streptozotocin-induced diabetic Wistar rats [415]. Further, the modulatory effects of LBPs on exercise-induced oxidative stress in the skeletal muscles of Wistar rats were investigated by Niu et al. [416]. These rats were subjected to a 30-day exhaustive exercise program and administered orally 100, 200, or 300 mg/kg body weight of LBPs or an isotonic saline solution [416]. The results suggested that administration of LBPs reduced oxidative stress induced by exhaustive exercise in a dose-dependent manner and increased superoxide dismutase and glutathione peroxidase activity and reduced MDA levels in skeletal muscles [416]. Zhang et al. observed the immunomodulating effects of the glycan part of *L. barbarum* polysaccharide-protein complex fraction 4 (LBPF4-OL) on mouse splenocytes, T cells, B cells, and macrophages [417]. The results showed that LBPF4-OL markedly stimulated splenocyte proliferation and activated macrophages in Balb/c mice that were injected intraperitoneally with 100 μg/cm^3 LBPF4-OL daily for 6 days, but could not induce T-and B-lymphocyte proliferation [417]. Moreover, LBPF4-OL has been demonstrated to act as an activator and inducer of the Toll-like receptor 4(TLR4)/MD2-MAPK signaling pathway using TLR4 knockout mice [418]. This glycan significantly induced the production of TNF-α and IL-1β in peritoneal macrophages isolated from C3H/HeN (wild type) mice but not in C3H/HeJ (TLR4-deficient) mice [418].

LBPs have been shown to have potent cytotoxic activity using cancer cell lines [155]. These polysaccharides (300 μg/cm^3) have been demonstrated to inhibit the growth of MCF-7 cells and the arrest of the cell cycle in the *S* phase as well as induce apoptosis via the ERK pathway [419]. Further, an increase of intracellular Ca^{2+} concentration and an overproduction of nitrite caused by LBP treatment of HeLa cells was demonstrated to change the cell cycle distribution and induce apoptosis through the mitochondrial pathway in this cervical cancer cell line [420]. Additionally, LBP (100–1000 μg/cm^3) exhibited in a dose-dependent manner effects on human prostate cancer cells in vitro by inhibiting PC-3 and DU-145 cell proliferation by damaging cellular DNA, inducing apoptosis, regulating gene expression of Bcl-2/Bax, and when tested in vivo, diminished PC-3 prostate tumor growth in BALB/c nude mice [421].

The neuroprotective effects of LBPs are one of their most studied therapeutic effects [157]. Chen et al. demonstrated the potential use of LBPs as a promotor of hippocampal neurogenesis and neuronal survival in neurodegenerative disorders [422]. LBPs were administered via gastric perfusion to Sprague-Dawley for 14 days followed by subcutaneous treatment of scopolamine, which is known to exert various toxic effects on the nervous system [422]. Groups that were treated with LBPs prior to scopolamine treatment had improved performance in novel object recognition and object location memory tasks when compared to groups treated only with scopolamine [422]. These results suggested that LBP administration can prevent the cognitive and spatial navigation and hippocampal neurogenesis from damage by scopolamine [422].

5.17 Marrubium vulgare

Marrubium vulgare L. is a widespread species of the genus that currently occurs in Central and Western Asia, North Africa, Europe, and South America [161, 162]. Traditionally, this herb has been used for the treatment of gastrointestinal and respiratory disorders as well as a remedy for women who were unable to conceive, appropriately leading to its common folk name in the Serbian language, očajnica, meaning "desperate woman" [161]. The constituents isolated from the *M. vulgare*

are largely responsible for a wide range of therapeutic properties observed including gastroprotective, antiedematogenic, analgesic, and antidiabetic activities [160, 161].

The gastroprotective potential of a methanol extract from the leaves of *M. vulgare* and the compound marrubiin (**187**) was demonstrated in both ethanol-induced and indomethacin/bethanecol-induced gastric ulcer model in Swiss mice. Ulcer-inducing treatments were administered 1 h after treatment with 25, 50, or 100 mg/kg of the methanol extract of *M. vulgare* or with either 25 mg/kg marrubiin (**187**), 30 mg/kg omeprazole, or vehicle control [423]. The results demonstrated that 50 and 100 mg/kg doses of the methanol extract of *M. vulgare* and 25 mg/kg marrubiin (**187**) reduced the lesion index, the total injured area, and the percentage injured area significantly [160]. Additionally, all doses of the methanol extract and marrubiin (**187**) were observed to inhibit ulcers by greater than 50% in the indomethacin/bethanechol-induced ulcer model [160].

The antiedemic properties of marrubiin (**187**) were observed in a model of microvascular leakage in the ears of mice [423]. Albino Swiss mice were treated with marrubiin (**187**) (1–100 mg/kg) simultaneously with the intradermal injection of 30 mg/kg Evans blue dye in the orbital venous plexus one hour before the experiment [423]. Carrageenan (300 μg/ear), histamine (3 μg/ear), and bradykinin (0.3 μg/ear) were injected intradermally to induce microvascular extravasation [423]. The ears were cut and submerged in 37°C formamide to extract the blue dye. The difference in volume of the extravasated plasma volume of the treated group and the volume of the control group was considered as showing activity. The results obtained from these studies with marrubiin (**187**) demonstrated ID_{50} values of the different pro-inflammatory agents to be 13.84 mg/kg after histamine treatment, 18.82 mg/kg after treatment with bradykinin, and 13.61 mg/kg after treatment with carrageenan. Therefore, marrubiin (**187**) attenuated the microvascular ear leakage of Evans blue dye induced by various pro-inflammatory agents [423].

Various reports of potential antidiabetic activity have appeared for *M. vulgare* extracts in vivo. In a study involving streptozotocin-induced diabetic rats administered 200 mg/kg of a methanolic extract of *M. vulgare* daily for five weeks, a significant reduction levels of all blood glucose, serum urea, uric acid, and creatinine was observed [424]. Additionally, a correction of the lipid profiles was observed compared to diabetic rats after treatment with this extract [424]. In another study involving albino male adult mice previously given cyclosporine and streptozotocin to induce autoimmune diabetes mellitus type 1, these were administered daily doses of 2 mg/cm^3 of a methanol extract, 2 mg/cm^3 of an aqueous extract, or 1 mg/ml of a butanol extract of *M. vulgare* for 28 days [425]. The results showed a significant decrease in total cholesterol, LDL cholesterol, VLDL cholesterol, and triglycerides after administration of the *M. vulgare* extracts [425]. Marrubiin (**187**) has been observed to have antidiabetic activity in an obese rat model, in which an increase in the respiratory rate and mitochondrial membrane potential under hyperglycemic conditions was observed [160]. Further studies in vitro on this diterpene lactone showed a stimulation of insulin secretion in INS-1 cells cultured under hyperglycemic conditions following their treatment with marrubiin (**187**) as well as a significant increase in insulin and glucose transporter-2 gene expression [160].

5.18 *Matricaria chamomilla*

Chamomile species (*Matricaria chamomilla*) are among the most popular traditional medicinal herbs. The name chamomile refers to several daisy-like plants in the family Asteraceae. The most popular species used in commercial chamomile preparations is *M. chamomilla* L. (syn. *M. recutita*), known as German chamomile [164, 165]. It is generally considered as a safe herbal remedy that may be used daily as a calming tea. In traditional medicine, chamomile has been used to treat multiple diseases and ailments, including fever, inflammation, muscle spasms, menstrual disorders, ulcers, wounds, gastrointestinal disorders, rheumatic pain, and hemorrhoids. An aqueous extract of *M. chamomilla* has been used frequently as a mild sedative to calm nerves and to reduce anxiety, to treat hysteria, nightmares, insomnia, and other sleep problems [165, 170, 426]. Chamomile is most often used as an herbal tea, but can also be taken orally as drops, capsules, or tablets, applied topically (poultice), and inhaled (essential oil) [164].

The aqueous extract (200, 400, and 600 mg/kg, p.o.) and essential oil (main compound in the preparation, chamazulene (**199**) (25.1%); (100, 200, and 300 mg/kg, p.o.)) of *M. chamomilla* were evaluated for antinociceptive activity in rodents. The analgesic effect was evaluated using the acetic acid-induced writhing and the tail immersion tests. In the acetic acid-induced writhing model, both preparations had an improved analgesic effect when compared to the control and a reference drug (aspirin). In the tail immersion test, the essential oil of chamomile exhibited significant analgesic activity at a low dose of 200 mg/kg, comparable to the aqueous extract at a higher dose of 600 mg/kg and morphine. The results of this study indicated that the aqueous extract and essential oil of *M. chamomilla* possesses analgesic properties, which are mediated through a central inhibitory mechanism [427].

Based on a study by Della Loggia et al. a dried flower extract of *M. chamomilla* was less effective (23.7% inhibition) than the fresh flower extract (31.6% inhibition), for which the activity was similar to that of benzydamine, an anti-inflammatory drug. The higher activity of the fresh flower was attributed to its higher concentration of matricin (**200**). Also, the anti-inflammatory activity of apigenin (**113**) was 10 times greater than that of matricin (**200**), which, in turn, was 10 times greater than that of chamazulene (**199**). Moreover, the flower essential oil was shown to lack anti-inflammatory activity (6.6% inhibition) [164, 428]. In addition, numerous studies have been performed to demonstrate the anti-inflammatory effects of α-bisabolol (**194**). In one such study, Rocha et al. reported that 100 and 200 mg/kg, p.o. of α-bisabolol (**194**) ameliorated an inflammatory response via leukocyte migration, protein extravasation, and the amount of TNF-α in mouse paw edemas induced by carrageenan and dextran [429]. Matricin (**200**) inhibits NF-κB transcriptional activity in endothelial cells and may contribute to its anti-inflammatory activity in vivo. Using ICAM-1 (Intercellular adhesion molecule 1) as a marker for NF-κB activation, matricin (**200**) was inhibitory in a concentration-dependent manner (10–75 μM) and did not show cytotoxic effects. At 75 μM, expression of ICAM-1 was reduced to 52.7 ± 3.3% and 20.4 ± 1.8% of the control in TNF-α and LPS-stimulated in endothelial cells (HMEC-1), respectively. Quantitative RT-PCR experiments revealed that TNF-α induced expression of the ICAM-1 gene was also reduced by compound **200** in a concentration-dependent manner, reaching 32.3 $\pm$ 6.2% of control at 100 μM matricin [430].

Maschi et al. demonstrated that an infusion, prepared from either the entire dried flowers or sifted flowers of *M. chamomilla* inhibited cyclic adenosine monophosphate (cAMP)-phosphodiesterase (PDE) activity (IC_{50} = 17.9–40.5 μM). Among the individual compounds occurring in chamomile, flavonoids (IC_{50} = 1.3–14.9 μM) contributed ~ 39% to the inhibitory effect of a chamomile infusion on cAMP-PDE. This is a likely mechanism for the spasmolytic activity of chamomile tea. The phenolic acids as well as (−)-α-bisabolol (**194**), and the coumarins were inactive [431].

Amsterdam et al. analyzed the antianxiety and antidepressant activity of an oral chamomile extract [220 mg with 1.2% apigenin (**113**)] in patients with symptoms of co-morbid anxiety and depression. The dose was increased weekly by 220 mg, reaching a maximum of 1,100 mg/day at week five, which continued through the end of the study (week eight). The study indicated that chamomile extract may reduce co-morbid depression in anxious patients [432]. In this regard, Han et al. evaluated the effect of apigenin (**113**) on monoamine oxidase (MAO) inhibition. Apigenin inhibited both MAO-A and MAO-B, with IC_{50} values of 1.7 and 12.8 μM, respectively [433].

5.19 Momordica charantia

M. charantia, commonly known as bitter melon, belongs to the family Cucurbitaceae and has been used traditionally for diabetes treatment in certain Asian countries [177]. Recent research has supported such antidiabetic activity that occurs largely by reducing blood sugar levels as well as other biological activities including antioxidant, antibacterial, antitumor, and enhancement of immune function effects [177]. The major active components responsible for these activities are the *M. charantia* polysaccharides (MCPs) found in various parts of the plant [176]. The heteropolysaccharide contents vary in their composition depending on environmental conditions, but in general the polysaccharide contents range from 5.91 to 10.62% of the dry powder and consist of arabinose, glucose, mannose, galactose, and rhamnose units [176]. Correspondingly, the biological activities of MCPs with different molecular weights and monosaccharide compositions can vary [176].

Three water-soluble MCPs (BP1, BP2, and BP3), isolated from the fruits of *Momordica charantia*, were evaluated for their antioxidant activity [434]. The results showed that all three had strong hydroxyl radical-scavenging activity and weak superoxide radical scavenging activity, but to varying degrees because of their different monosaccharide composition and structure [434]. Further, MCP has been demonstrated as a potential neuroprotective agent due to its antioxidant activities after results showed direct scavenging effects on superoxide, nitric oxide, and peroxynitrite as well as lipid peroxidation inhibition in vitro in oxygen glucose deprivation (OGD) neural cells [435].

The antidiabetic and antioxidant effects of MCPs from *M. charantia* were observed to correlate in an in vitro study in mice, as their antioxidant activity significantly increased with an improvement in blood glucose levels [436]. Further evidence of the antidiabetic activity of *M. charantia* has come from a study of an aqueous extract of the fruits in normal rats and in streptozotocin-induced diabetic rats administered a single dose of 250 mg/kg, which experienced a significantly diminished blood glucose level [437]. A separate study with the same extract demonstrated an improvement in all of fasting blood glucose, oral glucose tolerance, plasma insulin level, lipid profile, and hepatic and renal parameters in high fructose diet-fed rats and streptozotocin-induced diabetic rats, after consecutive 100 mg/kg doses every 24 h for 30 days [437].

MCPs demonstrated semi-lethal antibacterial activity against *Staphylococcus aureus*, *Escherichia coli*, *Bacillus subtilis*, and *Salmonella typhimurium* at doses of 0.15, 0.10, 0.13, and 0.13 mg/cm^3, respectively [438].

Two different polysaccharides isolated from *Momordica charantia* have shown antiproliferative activity against the human leukemia cell line K562 [439]. Also, sulfated modifications of MCPs exhibited the significant growth inhibition of HepG2 cells and Hela cells in vitro, suggesting the potential for sulfated MCPs to have an improved antiproliferative potential [440].

The immunomodulatory effects of MCPs have been demonstrated both in vitro and in vivo. The activity of the two polysaccharides, MCP composed of glucose and galactose, and MCP2 composed of glucose, mannose and galactose, led to the effective promotion of the proliferation of normal and ConA-induced splenic lymphocyte proliferation at doses in the range of 20–180 μg/cm^3 [441]. Further, in the same study, intragastric administration of a dose as low as 150 mg/kg per day of MCP to cyclophosphamide-induced immunosuppressed mice increased the carbolic particle clearance index, serum hemolysin production, spleen index, thymus index and NK cell cytotoxicity to normal levels, when compared with a control [441].

The major bioactivities of the cucurbitane-type compounds from *M. charantia* include cancer-related activities in multidrug resistance (MDR) reversion, in addition to antimalarial, antidiabetic, antiviral, and antibacterial effects.

Cucurbitacins B (**207**), E (**208**), and I (**209**) were able to induce apoptotic pathways, and block the cell cycle by inhibiting cyclins, increase autophagy, and inhibit the invasion and migration of cancer cells in in vitro and in vivo models [175]. Further, cucurbitane triterpenoid-type compounds have shown the ability to effect multidrug resistance reversal mechanisms in cancer [175]. For example, karavilagenin C (**212**), in particular, has demonstrated potent inhibitory activity (2 μM) against P-glycoprotein (P-gp) [175]. P-gp is an ATP-binding cassette (ABC) transporter protein that pumps out anticancer drugs from the cancer cells, preventing the drugs from reaching therapeutic levels. Thus, inhibiting the activity of this protein can be a valuable approach to combat MDR in cancer therapeutics [175]. Karavilagenin C (**212**) has also been demonstrated to have an IC_{50} value of 13.8 μM against

HT-29 colon cancer cells [175]. An antiproliferative screen of some cucurbitane-type compounds determined momordicoside I (**211**) to have antiproliferative activity against the MCF-7 ($IC_{50} = 12.36 \pm 0.13$ μg/cm^3), Doay (12.04 ± 0.61 μg/cm^3), WiDr (16.63 ± 1.54 μg/cm^3), and HEP-2 (13.27 ± 1.31 μg/cm^3) human cancer cell lines [181].

The antimalarial activity of cucurbitane-type compounds has been studied in vitro and in vivo [175]. Karavilagenin C (**212**) and kuguaglycoside A (**213**) were both active against chloroquine-sensitive (3D7) and resistant (Dd2) *P. falciparum* strains with IC_{50} values below 12 μ*M* [175]. Further, karavilagenin C (**212**) showed a 71.8% inhibition of parasitemia in hepatic cells at a dose of 5 μ*M* and 33.1% parasitemia inhibition in vivo on oral administration to mice of 33.1 mg/kg per day [175]. Karavilagenin C (**212**) exhibited antibacterial activity against methicillin-resistant *Staphylococcus aureus* (MRSA) COL_{OXA} [175]. At a dose of 3 μ*M* the intracellular accumulation of ethidium bromide was increased after a 30-minure period and the test compound modulated the bacterial efflux pump systems [175].

5.20 *Nigella sativa*

Nigella sativa, also known as black seed or black cumin in English, belongs to the family Ranunculaceae. This species grows in Eastern Europe, the Middle East, and Western Asia [442]. Traditionally, *N. sativa* has been used for a broad range of diseases including asthma, fever, bronchitis, cough, dizziness, chronic headache, back pain, and inflammation [181]. Growing evidence of the biological effectiveness of thymoquinone (**214**) from *N. sativa* has been demonstrated, including immunomodulatory, antidiabetic, and antioxidant effects, as well as hypotensive and hypolipidemic activities related to metabolic diseases [181, 442].

Among the most valuable and well-studied activities of *N. sativa* seeds are their effects on the immune system [181]. A comparative study of the immunomodulatory activities of *N. sativa* seeds and thymoquinone (**214**) showed diverse cellular immunity effects on the proliferation capabilities of T lymphocytes and splenocytes, inhibition of cytokine secretion, increased IL-3 secretion and IL-2 levels and increased peripheral lymphocyte and monocyte counts [181, 443, 444]. Thymoquinone (**214**) treatment significantly raised glutathione levels in red blood cells in addition to acting as a detoxifying agent due to its activity against oxidative stress linked with reactive oxygen species [181]. The antioxidant activities of thymoquinone (**214**) has been supported by multiple studies demonstrating the ability of this compound to preserve the activity of antioxidant enzymes such as glutathione, peroxidase, glutathione-*S*-transferase and catalase and its activity as a free-radical and superoxide scavenger [445]. Moreover, thymoquinone (**214**) has been shown to act as a nephroprotective agent by decreasing SSAT and CYP3A1 gene expression by various antioxidant mechanisms [446].

The antidiabetic properties of *N. sativa* were investigated in patients with diabetes mellitus type 2, who were administered 2 g of the seeds daily for three months

[447]. The results suggested that at this dose, the patients experienced a decrease in postprandial glucose levels, fasting blood glucose levels, insulin resistance, and a decrease in glycosylated hemoglobin [447]. Further, the effects of administration of 5% *N. sativa* seeds or 3 mg/cm^3 i.p. thymoquinone (**214**) for six days/week to streptozotocin-induced diabetic rats were monitored [448]. The results showed that administration of both of these reduced diabetes-induced increases in tissue malondialdehyde and serum glucose and increased serum insulin and tissue superoxide dismutase levels [448].

The hypotensive and hypolipidemic activities of thymoquinone support its potential role in metabolic syndrome, which is an important risk factor in cardiovascular diseases [442]. Administration of 0.5 cm^3 of a 10 mg/cm^3 thymoquinone preparation twice daily to rats fed an atherogenic suspension by gavage led to a reduction in HMG-CoA reductase activity and an increase in arylesterase activity [449]. Also, thymoquinone (**214**) was able to block a shift in buoyancy from the less atherogenic lb-LDL to the highly atherogenic sd-LDL, restoring the distribution of both LDL-C and apoB to their normal levels [449]. The hypotensive activity of *Nigella sativa* was shown in a study in rats intravenously administered the dethymoquinonated volatile oil, α-pinene (**216**), or *p*-cymene (**217**) in the dose range of 2–16 mm^3/kg, which decreased the arterial blood pressure and heart rate [450].

5.21 Oryza sativa

More than half of the world's population consumes rice as a staple food item owing to its high nutritional value [186]. *Oryza sativa* L. (Poaceae), also known as black rice, originates from Asia and is exported to many other countries worldwide [186]. The nutritional value of rice is attributed to the essential amino acids, dietary fiber, carotenoids, folate, lignin, minerals and many bioactive phytochemicals present [185]. Several *O. sativa* constituents have potent antioxidant activities [185–187]. Other health benefits of the phytochemicals in rice are concerned with its antioxidant effects, and include antidiabetic, antiinflammation, and antihyperlipidemic activities [185].

In a study with streptozotocin-induced diabetic rats, oral administration of 50 and 100 mg/kg γ-oryzanol (**228**) reduced the blood glucose level of normal and diabetic rats and reduced oxidative stress, lipid peroxidation, and increased levels of superoxide dismutase and glutathione in the liver [451].

The anthocyanidin *O. sativa* constituent, cyanidin-3-glucoside (**106**), and its metabolites expressed potent anti-inflammatory effects by regulating NF-κB and MAPK activation in vitro and in vivo [452]. Further, the phytochemicals ferulic acid (**39**), *p*-coumaric acid (**117**), γ-oryzanol (**228**), γ-tocotrienol (**227**), and GABA (**229**) have been shown to reduce liver inflammation in Sprague-Dawley rats in rats induced with CCl_4 hepatotoxicity and significantly decreased levels of TNF-α, IL-6, and IL-1β [453].

The effect of γ-oryzanol (**228**) on lipid metabolism was evident in an investigation in dyslipidemic patients that demonstrated its ability to lower oxidative stress via regulation of ROS levels, total antioxidant capacity, and inflammatory biomarkers, including TNF-α, IL-1β, and TXB2 [454]. These results have supported the notion that the antihyperlipidemic properties of γ-oryzanol are associated with its antioxidant properties. Thus, this compound has been demonstrated to decrease oxidative stress in SH-SY5Y cells treated with H_2O_2 and prevent neurotoxicity by upregulating genes that upregulate antioxidant activities and the antiapoptotic genes NF-κB and Bcl-2, while downregulating the proapoptotic genes TNF, BAX, and caspase-9 [455].

The antioxidant potential of γ-oryzanol (**228**) and ferulic acid (**39**) was demonstrated after administration of each compound at a dose of 5 g/kg for 30 days to c57BL mice after ethanol-induced liver injury [456]. The results showed a reduction in serum activities of plasma aspartate aminotransferase and alanine aminotransferase, along with decreases in hepatic lipid hyperoxide and thiobarbituric acid reactive substance (TBARS) levels [456].

Studies have confirmed the efficacy and tolerability of red yeast rice in patients that exhibit intolerance to conventional statin drugs. Stefanutti et al. studied selected 55 patients with familial hypercholesterolemia who discontinued statin use due to muscle pain. They received a cholesterol-lowering diet containing 300 mg of red yeast rice (containing 10 mg of monacolin K (**232**)) daily. LDL cholesterol levels in patients decreased significantly after six months of treatment (17% for men, 16% for women; $p < 0.005$). After 12 months, levels had decreased by 24% and 27% in men and women, respectively [457]. In addition, numerous reports suggest that statins can inhibit colon cancer cell growth. Hong et al. compared red yeast rice with purified monacolin K (**232**) in the inhibition of colon cancer. Monacolin K (**232**) at a concentration of 5.93 μM reduces the proliferation of HCT-116 and HT-29 human colon cancer cells and induces apoptosis in cancer cells. In turn, red yeast rice (50 $\mu g/cm^3$) inhibits HCT-116 tumor cell growth (41%) and increases cell apoptosis [458].

Rubropunctatin (**234**) has been reported with activity against HeLa cells with an IC_{50} value of 93.71 ± 1.96 μM (24 h in the dark), while the cytotoxic activity was enhanced under light irradiation (halogen lamp, 500 W, wavelength, 597–622 nm, and fluence rate, 15 mW cm^{-2}, for 30 min). Treatment of HeLa cells with rubropunctatin (**234**) in the dark or under light irradiation resulted in a dose-dependent apoptosis. This suggested that rubropunctatin (**234**) could be a promising natural component in red yeast rice with potential anticancer benefits [458, 459].

Monascin (**236**) has been shown to lower low-density lipoprotein cholesterol (LDL-C) and preserve high-density lipoprotein cholesterol contents. Also, monascin (**236**) inhibited acetyl-coenzyme A acetyltransferase, microsomal triglyceride transfer protein, and apolipoprotein (apo) B-100 expression, thereby preventing LDL assembly. In addition, enhanced LDL-receptor expression increased the transport of LDL-C to the liver for metabolism in Sprague-Dawley rats (0.55 mg of

236/day 400 g body weight) [460]. Furthermore, monascin (**236**) (1 μM treatment for five days) induced a decline in the Aβ-induced paralysis phenotype and Aβ deposits in the transgenic strains CL4176 and CL2006 of the nematode *Caenorhabditis elegans*, which expresses human muscle-specific $A\beta_{1-42}$ in the cytoplasm of body wall muscle cells [461]. Wang et al. reported that 10 mg/kg/day of monascin (**236**) improved neurological deficits, reduced the volume of hematoma in 1–7 days after intracerebral hemorrhage (ICH), and decreased blood brain barrier (BBB) permeability and edema formation in 1–3 days following ICH in male Sprague-Dawley (SD) rats [462].

5.22 Panax spp.

Korean Red (*Panax ginseng* C.A. Meyer) (Araliaceae) and American ginseng (*Panax quinquefolius* L.) (Araliaceae) are among the most popular *Panax* species ingested as dietary supplements [192]. Ginseng root has been a significant source of natural medicines and has been used for thousands of years in East Asia. *Panax ginseng* traditionally has been used to help prevent and treat a broad range of diseases including diabetes, inflammation, nervous system diseases, and cardiovascular disease [194]. The predominant bioactive constituents from ginseng are the ginsenosides. These compounds have been shown to have a potential beneficial therapeutic role in degenerative diseases such as cardiovascular disease and diabetes, and neurogenerative diseases such as Parkinson's disease and Alzheimer's disease [198].

Ginsenosides have had therapeutic indications in cardiovascular disease and diabetes [194]. Individuals with hypertension and diabetes treated with a combined American ginseng extract containing the major ginsenosides Rb1 (**240**), Rb2 (**241**), Rg1 (**247**), Rc (**242**), Rd (**243**), Re (**246**), and Rg3 (**244**) and Rg3-enriched Korean ginseng experienced a reduction in central systolic blood pressure compared to a control [192]. Further, ginsenoside Rb2 (**241**), in particular, has been shown to have antidiabetic properties [197]. Streptozotocin-induced diabetic rats treated with ginsenoside Rb2 (10 mg/rat/day, administered i.p. for six days) showed a significant decrease in triglycerides, non-esterified fatty acids, and total cholesterol in the serum, and a lowering action of the serum levels of 3-hydroxybutyrate and acetoacetate indicating this compound shows the potential for an improvement of diabetic ketoacidosis [463]. Also, ginsenoside Rb2 (**241**) (0.1 μM for 24 h) has a potential mechanism for antidiabetic activity and it has been suggested to be through inhibition of palmitate-induced gluconeogenesis via AMP-activated protein kinase (AMPK)-induced small heterodimer partner (SHP), by relieving estrogen receptor stress in H4IIE cells (rat hepatoma cells) [464].

Various ginsenosides have been shown with cytotoxic activity against cultured cancer cell lines [197]. Ginsenoside Rh2 (**245**) (0.2–120 mg/kg) showed antineoplastic activity in vivo by exhibiting antiproliferation, anti-invasion, and antimetastasis effects, and inducing cell cycle arrest, and promoting cancer cell differentiation in MDA-MB-231, PC-3, K562, and HepG2 human cancer cell xenografted nude mice [465]. The IC_{50} values of ginsenoside Rh2 ranged from 17 to 100 μM, against different tumor cells such as hepatocellular carcinoma (HepG2), hepatocellular carcinoma (SMMC-7721), human leukemia (HL-60, Jurkat), human erythroleukemia (K562), acute myeloid leukemia (KG-1a), human cervical carcinoma (HeLa), hepatocellular carcinoma (SK-HEP-1), colorectal cancer (SW480), human colorectal cancer (HCT116), prostate cancer (LNCaP, PC-3, DU-145), human breast cancer (MCF-7 and MDA-MB-231), human ovarian cancer (HRa, KK, KF and KFr), lung adenocarcinoma (A549), epidermoid carcinoma (A431), and nasopharyngeal carcinoma (HK-1) [465]. Further, ginsenoside Rg3 (**244**) exerted antitumor potential in lung, colon, breast, and colorectal cancers by affecting antioxidant, anti-inflammatory, and immune signaling pathways. Ginsenoside Rg3 has been reported with IC_{50} values that ranged from 30–150 μM in different cell lines, including MDA-MB-231, MCF-7, HT-29, and ovarian cancer cells (SKOV3) [466].

Ginsenoside Rg1 (**247**) has been demonstrated as having a beneficial therapeutic role in chronic neurogenerative diseases such as Alzheimer's disease (AD) and Parkinson's disease (PD) [198]. It has been shown to improve amyloid pathology, modulate the amyloid precursor protein process, improve cognition and activate hippocampal-dependent protein kinase/hippocampal-respond element-binding protein (PKA/CREB) to induce neuroprotection in an AD mouse model [198]. Ginsenoside Rg1 (**247**) administration via intraperitoneal injection for 15 days to PD models of rats dose-dependently reduced dopaminergic cell loss induced by 1-methyl-4-phenyl-1,2,3,6-tetrahydropyridine (MPTP) [198]. Zhou et al. confirmed that intraperitoneal injection of ginsenoside Rg1 (**247**) (20 mg/kg) for 15 days could protect anti-tyrosine hydroxylase positive cells in the mouse substantia nigra pars compacta (SNpc) region from MPTP toxicity [467]. Further, the CD3+ T cells accumulated in the SNpc, and reduced infiltration of T cells in this region, suggesting the immunoprotection of Rg1 (**247**) in the MPTP-induced PD mouse model used [467].

5.23 *Pausinystalia johimbe*

The stem bark of yohimbe, *Pausinystalia johimbe*, is sold popularly as an herbal supplement for the treatment of sexual disorders in men and is known to contain a number of structurally similar indole alkaloids [199, 468]. Yohimbine (**252**) has potent and selective antagonistic activity of the α_2-adrenoceptor and is a weaker antagonist of the α_1-adrenoreceptor, blocking the presynaptic feedback inhibition of noradrenaline release [468]. Accordingly, inhibition of the central α_2-adenoreceptor by yohimbine results in an increase in sympathetic tone and an increase in blood pressure [200]. The degree of erection or erectile dysfunction can be determined by the balance between nitric oxide stimulus originating from the noncholinergic nerve system and the counterbalancing effect of the sympathetic noradrenergic nerves [468]. Therefore, it has been postulated that the mechanism of yohimbine in inducing an erectile response includes antagonism of the presynaptic and postsynaptic α_2-adrenoreceptors [468]. This activity activates noradrenergic neurons to release norepinephrine, resulting in the activation of α-adrenoreceptors in the endothelium and the subsequent release of nitric oxide and prostanoids [468]. In turn, this results in the increases in intracellular cyclic GMP and cyclic CMP leading to the relaxation of the penile smooth muscles [468].

Several clinical trials have been conducted demonstrating the ability of yohimbine to stimulate an erectile response. First, in a 10-week placebo-controlled, double-blind, partial crossover trial of 48 patients with psychogenic impotence, 62% of patients who were administered 6 mg of oral yohimbine reported an improvement in sexual function compared to 16% who reported improvement in the placebo group [469]. In this study, the groups treated with yohimbine reported that the dosage was well tolerated with no serious adverse side effects [469]. Another similar 8-week, double-blind, placebo-controlled trial on 85 patients administered either yohimbine at a dose of 10 mg t.i.d. or placebo, resulted in a significantly better response in the subjective criteria determined (71% vs. 45%, respectively), including quality of erection during sexual contact or intercourse [469]. Again, no serious adverse events occurred, and the dosage was reported to be well tolerated [469]. However, other studies have shown responses that are not as consistent or as favorable as those mentioned above. For instance, in a blind study, twenty-two patients were administered a placebo for 30 days followed by administration of a single-high oral dose of 100 mg yohimbine [470]. The results demonstrated a trend toward increasing the erectile response, although the difference in comparison to the control group was not significant and, in addition, there was an increase in side effects reported after taking this high dose [470].

Other biological activities of the indole alkaloids from *Pausinystalia johimbe* include evidence for their immunosuppressive activity [203]. Both corynantheine (**256**) and corynanthine (**257**) exhibited inhibitory activity with IC_{50} values of 16.8 and 27.6 μM against ConA-induced T lymphocyte proliferation and 13.5 and 40.5 μM against LPS-induced B lymphocyte proliferation, respectively [203].

5.24 *Sambucus nigra and S. canadensis*

The genus *Sambucus* is commonly known as elderberry and is used both medicinally and as a food [221]. Due to the health-promoting properties of elderberry fruits, this species is widely used in herbal medicine and the food industry [207]. Consumption of products containing elderberries increased significantly throughout 2021 in response to the COVID-19 pandemic, caused by the virus severe acute respiratory syndrome-coronavirus 2 (SARS-CoV2) [346]. *Sambucus canadensis* and *S. nigra* refer, respectively, to American elderberry and black elderberry [471]. All parts of *S. canadensis* and *S. nigra* are used traditionally for medicinal purposes: elderberry leaves are used for cutaneous treatments, while elderberry fruits (dried or fresh) are used for the treatment of constipation, as a diuretic, and for respiratory tract and viral infections, colds, catarrh, and influenza. In turn, infusions of elderberry flowers are a traditional remedy for the treatment of inflammation, joint pain, skin disorders, colds, fever, and respiratory disturbances [206, 221, 223, 472].

Sambucus species are sources of several bioactive compounds reported to have pharmacological properties, such as antimicrobial, antiviral, antidiabetic, chronic stress, antioxidant, analgesic, anti-inflammation, and bone-protective activity [206, 221, 346, 473]. Anthocyanins, other flavonoids, and terpenoids have been suggested as being responsible for the anti-inflammatory and analgesic properties of *Sambucus* species [221]. Santin et al. demonstrated in an in vivo model (Swiss mice, inflammation induced by carrageenan) that an aqueous extract of *S. nigra* (30–600 mg/kg, p.o.), promoted a reduction in neutrophil migration as well as the decrease of TNF and cytokines, such as interleukin-1β (IL-1β), and IL-6 levels, in the inflamed exudate. In vitro, treatment with *S. nigra* decreased NO, TNF, IL-1β, IL-6, and promoted an increase of IL-10 in LPS-stimulated neutrophils [474].

The potential antiviral properties of extracts of *Sambucus* species, including against SARS-CoV2, HIV, influenza A (H1N1), and dengue virus serotype-2 (DENV-2), have been studied [221]. Castillo-Maldonado et al. reported on the antidengue activity of a methanol extract of *S. nigra* (400 $\mu g/cm^3$) using baby hamster kidney fibroblasts (BHK-21), and African green monkey kidney (Vero) cells. The extracts showed 60–80% cell viability, exhibiting protective effects against the DENV-2 virus with no apparent cytotoxicity [475]. Roschek et al. reported that a 1 mg/cm^3 dose of an extract of *S. nigra* produced 100% inhibition of H1N1 infection, with an IC_{50} of $252 \pm 34\ \mu g/cm^3$. A direct binding assay was used to establish that flavonoids from an elderberry extract binds to H1N1 virions, blocking the ability of the viruses to infect host cells [476]. Boroduske et al., using a competitive enzyme-linked immunosorbent assay (ELISA), assessed the activity of extracts prepared from *S. nigra* (fruits and flowers) on SARS-CoV2 receptor-binding domain-angiotensin converting enzyme 2 (SARS CoV2 RBD-ACE2). The results of the ELISA assay of *S. nigra* extracts on SARS CoV2 RBD-ACE2 binding revealed a significant concentration-dependent response (IC_{50} values of 1.66 mg/cm^3 fruits and 0.532 mg/cm^3 flowers) [477]. Rutin (**57**), isolated from the extract was presumed to be a principal compound linked to the anti-SARS-CoV2 effects of *Sambucus* species [221]. Recently, *S. nigra* preparations have been suggested as a potential COVID-19 supplementary treatment, in view of their effectiveness in the treatment of cold and influenza symptoms, based on randomized, double-blind, placebo-controlled studies and meta-analyses [478, 479].

Przybylska-Balcerek et al. showed the antibacterial activity of a *S. nigra* extract against selected bacterial strains within the concentration range of 0.5–0.05%. The following bacteria were the most sensitive to the extracts: *M. luteus*, *P. mirabilis*, *P. fragii*, and *E. coli*. Ferulic acid (**39**), *p*-coumaric acid (**117**), and apigenin (**113**) have been proposed as being responsible for the observed antibacterial effects [207].

Da Silva et al. investigated the cytotoxicity of an elderberry juice sample against MCF-7, HeLa, NCI-H460, and HepG2 cancer cells. HeLa and NCI-H460 cells were the most sensitive ($IC_{50} = 16 \pm 1\ \mu g/cm^3$ for both), while MCF-7 ($IC_{50} = 58 \pm 1\ \mu g/cm^3$) and HepG2 ($IC_{50} = 98 \pm 4\ \mu g/cm^3$) cells were less susceptible. The authors attributed the activity to the anthocyanin, cyanidin-3-*O*-sambubioside (**260**)

[480]. Another study reported the antiproliferative activity of an aqueous extract of *S. nigra*, which was evaluated against three human colorectal cancer cells lines (RKO, Caco-2, and HCT-116), but no discernible activity was shown in each case [223].

5.25 *Serenoa repens*

Serenoa repens (saw palmetto) is native to North America, specifically forming colonies in the sandy soils of Texas, Louisiana, South Carolina, and the south of Florida [226]. The commercial dietary supplements and biologically active constituents are derived from the fresh partially dried, or dried fruits of *Serenoa repens* [226]. This botanical is one of the most widely used herbal treatments for lower urinary tract symptoms and benign prostatic hyperplasia (BPH) in both preclinical and clinical studies [225]. The mechanism of action of saw palmetto extract is unknown in relation to their effects on lower urinary tract symptoms and BPH, although antiandrogenic, anti-inflammatory and proapoptotic activities have been suggested [225].

The most well-studied activity of SPE is its inhibitory activity on 5α-reductase. A saw palmetto berry lipid extract was reported to inhibit 5α-reductase, decrease the clinical symptoms of BPH, and showed a 34% decrease in the growth of prostatic carcinoma cells at a dose of 0.1 $\mu g/cm^3$ [481]. Additionally, the main fatty acid constituents of SPE, lauric acid (**286**), oleic acid (**226**), and myristic acid (**288**), effectively inhibit 5α-reductase [225]. These individual fatty acid components of SPEs were evaluated for their inhibitory activity against the two isoforms of 5α-reductase type 1 (R1) and type 2 (R2) [482]. The results demonstrated that lauric acid was active against both isoforms, while oleic acid and linoleic acid were selectively active against R1 and myristic acid had strong inhibitory activity against R2 [482]. Further, these fatty acid components exerted binding activities for α_1-adrenergic, muscarinic, and 1,4-DHP receptors and inhibited 5α-reductase activity with IC_{50} values of 42.1–67.6 $\mu g/cm^3$, and were more potent than SPE, which inhibited 5α-reductase activity with an IC_{50} value of 101 $\mu g/cm^3$ [483].

The anti-inflammatory effect of saw palmetto was demonstrated in a study involving the commercial standardized hexane extract, Permixon®, and this significantly reduced the expression of inflammatory factors IL-6, CCL-5, CCL-2, COX-2, and iNOS in LNCaP and PC3 human prostate cancer cells [484]. This effect contributed to the significant reduction in these cancer cell lines and normal cells after Permixon® treatment, suggesting its anticancer potential [484]. Additionally, a clinical study in patients administered an oral dose of Permixon® (320 mg) for three months showed reduced levels of inflammatory markers related to mRNA levels in the lumen of the prostate gland [485]. In this same study, the number of patients expressing the urinary pro-inflammatory cytokines CCL2 and CXCL10 was decreased, and the expression of urinary macrophage migration inhibitory factor was significantly reduced in patient groups treated with Permixon® [485].

The proapoptotic activity of SPE has been demonstrated in various cancer cell lines in vitro including an increase in the apoptotic index (PARP cleavage) in the P69 prostate epithelial cell line, and growth inhibition of PC3 and LNCaP prostate cancer cells through the intrinsic apoptotic pathway and in MCF-7 breast cancer and HCT116 colon cancer cells [225]. Also, the apoptotic index (Bax-to-Bcl2 ratio) was increased significantly in the transurethral prostate tissues from men with symptomatic BPH by surgery after treatment with 320 mg Permixon® for three months when compared to untreated controls [486].

The phytosterols from Saw Palmetto have a variety of biological activities ranging from anti-inflammatory and cholesterol-lowering effects, and inhibition of proliferation of human prostate cancer cells and growth of tumors derived from PC3 prostate cancer cells [225]. In particular, a dose of 130 mg of β-sitosterol (**280**) has been demonstrated to significantly improve the IPSS (international prostatic symptoms scale) and quality of life index in patients with symptoms of outlet obstruction caused by BPH [487]. These results were observed in a similar clinical study in patients with symptomatic BPH administered a mixture of phytosterols (60 mg of β-sitosterol) (**280**) [488]. Another randomized, double-blind clinical study of the phytosterols from saw palmetto administered for three months resulted in improved BPH symptoms and no reported side effects, suggesting the efficacy and safety of this botanical [489].

5.26 *Silybum marianum*

Silybum marianum (L.) or milk thistle is a botanical native to Northern Africa, Southern Europe, Southern Russia, and Turkey that has been used traditionally for liver and gallbladder disorders [490]. The bioactive constituents from *S. marianum* are represented by a range of flavonolignans that have a broad range of effects including hepatoprotective, antidiabetic, antimicrobial, antitumor, and neuroprotective activities [491]. Silymarin is a standardized extract of *Silybum marianum* seeds and is a widely used dietary supplement in the United States [491]. Silybins A (**294**) and B (**295**) make up approximately 70% of the total composition of silymarin and are considered to have the most important biological effects [490]. Other compounds reported in this extract include the flavonolignans isosilybin A (**296**), isosilybin B (**297**), silychristin A (**299**), isosilychristin (**301**), silydianin (**298**), and the flavonoid taxifolin (**302**) [491].

Several studies have investigated the hepatoprotective activity of *S. marianum* as a result of its known traditional uses [491]. For example, co-administration of 600 mg/kg p.o. of silymarin (as mentioned above) to rats with the hepatotoxic agent methotrexate significantly decreased levels of the serum liver enzymes alkaline phosphatase and bilirubin when compared to rats only administered methotrexate [492]. Also, silymarin (60 mg/kg orally for 12 days) was able to prevent body weight reduction observed in male Wistar rats after treatment with doxorubicin and to prevent increases in aspartate aminotransferase and creatine kinase serum levels in the liver tissues of rats subjected to doxorubicin-induced hepatotoxicity [493].

A silymarin seed extract has also been demonstrated as having antidiabetic potential in vitro and in animal models of the disease [491]. The levels of blood glucose, serum urea, serum creatinine and lipid peroxidation were all reduced in the liver tissue of streptozotocin-induced diabetic rats administered 100 mg/kg of silymarin [494]. Silymarin has also a broad range of antimicrobial activities, including methicillin-resistant *Staphylococcus aureus* (MRSA 43300) ($MIC = 60$ mg/dm^3) and *Candida albicans* ($MIC = 5$–30 mg/dm^3). It has also shown effectiveness in blocking hepatitis C virus (HCV) production at different doses (61.5–469 mg/kg mice); with the estimates of HCV-infected hepatocyte decline rates being higher in the 469 mg/kg treated mice [495]. Further, silymarin has been shown to inhibit Mayaro virus (MAYV) with an IC_{50} value of 25 mg/dm^3, by inhibiting MAYV replication and attenuating MAYV-induced oxidative stress [496]. Furthermore, it has reduced granulomatous peri-ovular reaction in the liver and decreased hepatic fibrosis in a murine model of schistosomiasis at doses of 10–25 mg/kg [495]. Additionally, the viability of the pathogenic human malaria parasite, *Plasmodium falciparum*, was reduced after administration of silymarin ($IC_{50} = 14 \pm 0.33\ \mu M$) by inducing membrane damage and apoptosis by heme interactions, with a high level of selectivity [495, 497]. Silybin (a 1:1 mixture of two diastereoisomers, silybin A (**294**) and silybin B (**295**)) has been reported with activity against human immunodeficiency virus (HIV-1) in a concentration range of 40–324 μM, through attenuating cellular functions involved in T-cell activation, proliferation, and HIV-1 infection [495].

The antitumor potential of *Silybum marianum* has been demonstrated in certain in vitro models [495]. The proliferation of various cancer cell lines was inhibited by silymarin by modulating signal transduction of transcription 3 (STAT3), caspase, and PI3K pathways. Silybin at a concentration range of 10–400 μM is a natural inhibitor of STAT3 activity, suggesting its preventive potential for STAT3-mediated cancer drug resistance [tested against HL-60, MCF-7, MDA-MB-231, breast cancer (T47D), HeLa, HT-29, esophageal squamous cell carcinoma (KYSE270), glioblastoma (LN229)] [498, 499]. Additionally, silybin (in a concentration range of 50–100 μM) has been demonstrated to inhibit the proliferation, migration, and adhesion of MDA-MB-231 human breast cancer cells by modulating the β1-integrin signaling pathway and Raf-1 function, while significantly inhibiting the expression of Cdc42 and D4-GDI mRNAs [499, 500].

There is growing evidence of the neuroprotective effects of *S. marianum* in animal models of neurogenerative diseases [495]. Anxiety/depression-like behaviors in male Sprague-Dawley rats induced by $A\beta_{1-42}$ were reduced after treatment with silibinin (silybin A (**294**)) and further evidenced by an increase in the BDNF/TrKB pathway and inhibition of autophagy in the hippocampus [501]. In a MPTP-induced animal model of Parkinson's disease a dose of 100 mg/kg silymarin showed neuroprotective effects and maintaining striatal dopamine levels by reducing apoptosis in the substantia nigra and protecting dopaminergic neurons [495]. Further, lipopolysaccharide (LPS)-induced neuroinflammatory impairment in rats treated with silybin A

(**294**) experienced a reduction in the impairment of learning and memory, possibly due to the activation of ROS-BDNF-TrKB pathway in the hippocampus [501].

5.27 *Trigonella foenum-graecum*

Trigonella foenum-graecum, commonly known as fenugreek, is used as a spice. It also has multiple documented medicinal properties, especially using its leaves and seeds [502]. Thus, *T. foenum-graecum* has been shown to possess hypoglycemic, antihypertensive, and hypolipidemic activities [237]. Fenugreek also has a long history of use in the treatment of reproductive disorders, in inducing labor, treating hormonal disorders, increasing milk production, and reducing menstrual pain. The plant extract has the ability to maintain blood glucose and cholesterol levels. In addition, *T. foenum-graecum* has been reported to have biological properties including antiviral, antimicrobial, hypotensive, antioxidant, anti-inflammatory, and cytotoxic activities [234, 235, 503].

Certain saponins possess many effects on tumor cells [235]. Protodioscin (**307**) induced cell death in the human promyelocytic leukemia cell line (HL-60) by apoptosis [504]. Dioscin (**308**) induced cytotoxicity on several tumor cell lines (BT-549, HepG2, SK-MEL, SK-OV-3) with IC_{50} values in the range of 1.9–6.8 $\mu g/cm^3$ [505]. Diosgenin (**309**) showed weak antiproliferative activity in a diverse panel of eleven cancer cell lines, with IC_{50} values between 20 to 80 μM. Moreover, studies have suggested that the cytotoxic mechanism of action of diosgenin (**309**) is associated with a modulation of multiple cell-signaling events involved in cell growth/proliferation, differentiation, and apoptosis, as well as oncogenesis and angiogenesis. Diosgenin (**309**) inhibits the proliferation of osteosarcoma cells by inducing apoptosis and cell cycle arrest in the G_1 phase, and also inhibits the proliferation of PC-3 (prostatic adenocarcinoma) in a dose-dependent manner, reducing cell migration and invasion by decreasing matrix metalloproteinase expression [503, 506].

Trigonelline (**304**) is a plant hormone (phytoestrogen) that has diverse regulatory functions with respect to plant cell cycle regulation, modulation and growth of the plant [238]. In addition, trigonelline (**304**) has been postulated to have therapeutic benefits, such as anticarcinogenic and hypocholesterolemic effects [238, 503]. Also, (2*S*,3*R*,4*S*)-4-hydroxyisoleucine (**312**) has been shown to cause an increase in glucose-dependent insulin secretion [238, 502]. Haeri et al. demonstrated that 4-hydroxy-isoleucine (**312**) (50 mg/kg, p.o.) per day given for four weeks to streptozotocin (STZ)-treated (type 1 diabetes model) rats lowered the blood glucose level from 500 to 330 mg/100 cm^3 [507].

A study published by Ahmadiani et al. evaluated the anti-inflammatory and antipyretic effects of *T. foenum-graecum* leaf extract in rats (1,000–2,000 mg/kg, i.p.) using a formalin-induced hind paw edema model for inflammation and intraperitoneal (i.p.) injection of an aqueous suspension of brewer's yeast as a model for hyperthermia. The results showed that the *T. foenum-graecum* extract and a positive control (sodium salicylate) had anti-inflammatory as well as antipyretic properties [508].

The polysaccharide galactomannan (**314**) component of *T. foenum-graecum* is high in galactose, which allows it to absorb water leading to the formation of highly viscous solutions in small concentrations and resulting in decreased glucose absorption in the digestive tract. This property helps in controlling type 2 diabetes [509]. Trask et al. successfully demonstrated the efficacy of this galactomannan (8–16 g) in 15 of 20 subjects with type 2 diabetes and being treated with oral medication and/or insulin, in controlling postprandial glucose [510]. Moreover, there have been anecdotal reports relating to the successful use of fenugreek as a galactogogue [238]. Reeder et al. evaluated the effect of fenugreek on milk production. In a double-blind placebo-controlled study involving 26 mothers of preterm infants born at less than 31 weeks of gestation, each mother took three fenugreek capsules (575 mg), three times/day for 21 days. However, the data obtained revealed no effect on milk production or on prolactin levels [511]. Further studies of this type are suggested to validate this particular use.

5.28 *Vaccinium macrocarpon*

Cranberry (*Vaccinium macrocarpon* L.) is an evergreen plant, native to wetlands in eastern parts of North America, and thus, it is also known commonly as American cranberry. Today, the plant grows in North America, some regions of South America, and in Eastern Europe [248, 249, 512, 513]. Recent research suggests that cranberry berry fruits are a source of numerous phytochemicals with a broad array of bioactivities that may have an impact on human health. Cranberry juice, fruit (fresh and dried), functional foods, and cranberry dietary supplements are promoted for prevention of urinary tract infections (UTI) and for maintenance of urinary tract health (UTH) [246]. Various phenols from *V. macrocarpon* have been associated in vitro with antibacterial, antiviral, antimutagenic, anticarcinogenic, antitumorigenic, antiangiogenic, antiinflammatory, and antioxidant properties [249].

In vivo animal models have revealed that cranberry extracts can reduce C-reactive protein (CRP) and proinflammatory interleukins, and suppress *Helicobacter pylori* infections, inhibit adherence of *Escherichia coli* to the urinary tract, and improve pancreatic β-cell glucose responsiveness. In turn, clinical studies have shown that cranberry products can lower LDL cholesterol (LDL-C) and total cholesterol, while increasing HDL cholesterol (HDL-C) [248, 249]. Cyanidin-3-*O*-arabinoside (**316**), cyanidin-3-*O*-galactoside (**315**), peonidin-3-*O*-arabinoside (**318**), and peonidin-3-*O*-galactoside (**317**), were identified as pancreatic lipase inhibitors with an IC_{50} value of 42.4, 48.9, 53.1, and 50.1 μM, respectively. These results could assist research on human lipase inhibitors from anthocyanin-rich natural sources. It has been suggested that cranberry anthocyanins could be developed as food supplements to facilitate the prevention and treatment of obesity [514].

American cranberry A-type procyanidins ((A-type PACs), (**319**)) exhibited in vitro antiadhesion activity at 60 μg/cm^3, and hence prevented *E. coli* bacteria from attaching to cells in the urinary tract and causing infections [248, 513, 515]. Lavigne et al. evaluated the antibacterial efficacy of cranberry capsules vs. placebo in the urine of healthy volunteers. A first double-blind, randomized, crossover trial involved eight volunteers who had followed three regimens, with or without cranberry. Twelve hours after consumption of cranberry or placebo hard capsules, the first urine of the morning was collected. Different *Escherichia coli* strains were cultured in the urine samples. Urinary antibacterial adhesion activity was measured in vitro using the human T-24 epithelial cell-line, and in vivo using the *Caenorhabditis elegans* (*C. elegans*) killing model. With the in vitro model, 108 mg of cranberry (commercial capsules) induced a significant reduction in bacterial adherence to T-24 epithelial cells as compared to placebo. Also, a dose-dependent decrease in bacterial adherence in vitro was noted after the consumption of 108 and 36 mg of cranberry (commercial capsules). The in vivo model confirmed that *E. coli* strains had a reduced ability to kill *C. elegans* after growth in the urine of patients who consumed cranberry capsules [516].

Various constituents in cranberry fruits may affect cancer-related processes. Flavonoids and A-type procyanidins ((A-type PAC), (**319**)) and (*Z*)-3-*O*-p-hydroxycinnamoyl ursolic acid (**320**) may limit processes involved in tumor invasion and metastasis. Cranberry extracts also inhibit the growth of breast, bladder, prostate, lung, and other human tumor cells. An 80% aqueous acetone extract of whole cranberry fruit increased the apoptosis of MCF-7 cells by 25% at a concentration of 50 g/dm^3 with a significant arrest in the G_1 phase. A cranberry PAC fraction also showed matrix metalloproteinases (MMP) inhibition in prostate cancer cells (DU-145), in a dose-dependent manner. A derivative of ursolic acid (**320**) inhibited the expression of MMP-2 and MMP-9 at micromolar concentrations [517]. In addition, Déziel et al. described the effects of a whole cranberry extract on the growth of DU-145 cancer cells in vitro at 10, 25 and 50 μg/cm^3 of extract. The cranberry extract was found to decrease the cellular viability of DU-145 cells. Also, the cranberry extract decreased the proportion of the cells in the G2-M phase of the cell cycle and increased the proportion of cells in the G1 phase of the cell cycle following treatment of DU-145 cells with 25 and 50 μg/cm^3 of extract for 6 h. The results demonstrated that extracts from the American cranberry (*Vaccinium macrocarpon*) can affect the behavior of human prostate cancer cells in vitro and further support their potential health benefits [518].

In a recent investigation by Skemiene et al. on rat brains, experiments involving the monitoring of caspase 3 activity and LDH release, as well as the expansion of infarct area in cerebral cortex and cerebellum in ischemia-damaged rat brains, indicated necrosis and apoptosis inhibiting ability of cyanidin-3-*O*-galactoside (**315**) in a concentration of 10 μ*M*, i.v. The authors suggested that reductive rather than antioxidant capacities of cyanidin-3-*O*-galactoside (**315**) may be an important component in providing protection against ischemic brain damage [519, 520]. Wen et al. further examined the coronal hippocampal sections of rats using Nissl staining and found cyanidin-3-*O*-galactoside (**315**) (50 mg/kg) to have a strong ability to ameliorate learning and memory impairment among amyloid-β-induced neurotoxicity rats. Such results have highlighted the potential of cyanidin-3-*O*-galactoside (**315**) as having a beneficial role in Alzheimer's disease [520, 521].

5.29 Valeriana officinalis

Valeriana officinalis (L.), commonly known as valerian, is an herbaceous plant native to Europe and Asia belonging to the Valerianaceae family [261, 522]. Traditionally, valerian has been used as a sedative [523]. Several biological activities inducing sedation, sleep promotion, antidepressant, antianxiety, and cardiovascular effects on *V. officinalis* extracts and its constituents have been reported [261, 522].

A number of studies have been conducted in a variety of patient populations to assess the effectiveness of valerian extracts to improve the quality of sleep and sleep problems. For example, a clinical study involving 100 postmenopausal women experiencing insomnia reported an improvement in quality of sleep after receiving 530 mg of a concentrated valerian extract twice a day for four weeks [524]. Valerian (530 mg, 1 h before sleep) was also able to improve the sleep quality and reduce anxiety in HIV-positive patients in the first four weeks of antiretroviral therapy in a double-blinded, placebo-controlled clinical trial [525]. The antianxiety activity of valerian was also demonstrated in patients with generalized anxiety disorder administered an average of 81.3 mg per day of valepotriates, indicating that this group of iridoids to be the active anxiolytic constituents [526]. An in vivo study showed an anxiolytic potential after a single injection of a valepotriate fraction into mice at a dose of 10 mg/kg in an elevated plus-maze test [527]. The results gave an increase in the percentage of time spent in the open arms in comparison to the closed arm areas, suggesting an antianxiety potential [527]. The antidepressant activity of a dichloromethane extract of valerian was demonstrated after a single dose administration of 20 mg/kg or 40 mg/kg to mice in a forced swim test [528]. The results demonstrated that this extract both reduced the immobility period and increased the levels of the neurotransmitters norepinephrine and dopamine in the forebrain of the mouse after two weeks of dosing [528].

The effects of valerian on the cardiovascular system can be attributed to an ability to reduce blood pressure and heart rate, and protection against myocardial ischemia reperfusion injury, and antiarrhythmic activity [522]. An essential oil sample of valerian was demonstrated to decrease the heart rate and diastolic blood pressure, and to prolong the duration of the ST segment and T wave in a dose-dependent manner [529]. These effects were attributed to its effects on relaxing vascular smooth muscle cells, enlarging vessel diameter, and decreasing blood resistance [529]. Furthermore, an ethanol extract from the roots of valerian has been demonstrated to be a potent smooth muscle dilator in the pulmonary vascular bed of male and female mongrel cats [530].

In an isolated rat ischemia reperfusion (I/R) heart model using a Langendorff-perfusion system, the effects of valerian extract (100 mg/dm^3) pretreatment on I/R injury and related biochemical factors, and cytosolic free calcium, were observed [531]. The results demonstrated that the valerian extract was able in a dose-dependent manner to prevent I/R injury and weakened vasospasm perfusion while sustaining heartbeat and reducing ventricular arrhythmias [531]. Further, significantly lower levels of lactate dehydrogenase, creatinine phosphokinase, and malondialdehyde and enhanced activities of superoxide dismutase, adenosine triphosphatase, and glutathione peroxidase were observed after administration of valerian essential oil [531]. The antiarrhythmic activity was shown after a valerian extract dose-dependently reduced the duration of the action potential and inhibited the sodium current (I_{Na}), L-type calcium current ($L_{Ca\text{-}L}$), and the transient outward potassium current (L_{to}) [522].

Bos et al. investigated several components of the volatile oil of *V. officinalis*. Test and control mouse groups were observed for various responses indicating sedative activity. Doses of 50 mg/kg for valerenic acid (**328**) showed significant effects including decreased motor activity, decreased rotarod performance and ataxia [252]. Khom et al. reported than valerenic acid (**328**) shifted the GABA concentration-effect curve towards lower GABA concentrations and elicited substantial currents through $GABA_A$ channels at $\geq$ 30 μM. Valerenic acid (**328**) expressed $GABA_A$ receptors with 13 different subunit compositions in *Xenopus oocytes* and measured I_{GABA} using the two-microelectrode voltage-clamp technique. At higher concentrations ($\geq$ 100 μM), valerenic acid (**328**) was found to inhibit $GABA_A$ receptors (I_{GABA}). Thus, valerenic acid (**328**) was identified as a subunit specific allosteric modulator of the $GABA_A$ receptor. This study suggests the proposed sedative, hypnotic and anxiolytic effects suggested for *Valerian* may be caused by interaction of valerenic acid (**328**) with $GABA_A$ channels [532]. Shi et al. reported the anxiolytic effects of valtrate (**321**) in rats. The animals were administered orally valtrate (**321**) (5, 10, and 20 g/kg daily) for 10 days and exposed to open field test (OFT) and elevated plus-maze (EPM) evaluations. Then, the corticosterone levels in the rat serum were measured by an enzyme-linked immunosorbent assay (ELISA). Valtrate (**321**) (10 mg/kg, p.o.) exhibited an anxiolytic effect in rats by increasing the time and entry percentage into the open arms in the EMP and the number of central entries in the OFT. Valtrate (**321**) (10 mg/kg, p.o.) significantly reduced the corticosterone level in the rat serum. These results suggest that valtrate (**321**) has anxiolytic activity in behavioral models that might be mediated via the function of hypothalamus-pituitary-adrenal axis [533]. Fernandez et al. reported that the flavone glycoside linarin (**351**) (4, 7, and 14 mg/kg, i.p.) has sedative and sleep-enhancing properties that are potentiated by simultaneous administration of valerenic acid (5 mg/kg) (**328**) in adult male Swiss mice. These results supported the presence in *Valeriana officinalis* of flavonoid glycosides with sedative and sleep-enhancing properties and demonstrated the potentiating effects in its extracts [534].

Zhang et al. determined the potential anticonvulsant activities of baldrinal (**323**) with a mouse model of pilocarpine induced epilepsy. The mice were treated with different doses of baldrinal (**323**) (50 and 100 mg/kg) prior to pilocarpine injection.

In addition, the brain tissues were analyzed for histological changes, and the in-situ levels of glutamic acid were also measured. Pilocarpine-treated mice showed a significant increase in glutamic acid levels, which was restored by baldrinal (**323**). Also, baldrinal (**323**) reduced the rate of seizures in the epileptic mice and ameliorated the increased levels of the *N*-methyl-D-aspartate receptor ($NMDAR_1$), brain-derived neurotrophic factor (BDNF), interleukin-1β (IL-1β) and tumor necrosis factor α (TNF-α). Baldrinal (**323**) has a potential antiepileptic effect, which may be mediated by reducing the inflammatory response in the pilocarpine-induced brain and restoring the balance of GABAergic and glutamatergic neurons [535].

Wang et al. reported the protective effect of pinoresinol-4,4′-di-*O*-*β*-D-glucoside (**348**) on the neurotoxicity of PC12 cells induced by amyloid-beta ($A\beta_{25-35}$). As a result, compound **348** (12–25 μ*M*) was found to afford protection against Aβ-induced toxicity in PC12 cells [536].

5.30 Withania somnifera

Withania somnifera, commonly known as Ashwagandha, has been used in Ayurvedic and other systems of traditional medicine for over 3,000 years [266, 269]. In this context, it has been claimed to have potent aphrodisiac, sedative, rejuvenating and life-prolonging properties. This species is used as dietary supplement, with a decoction of its roots employed as a nutrient and health restorative to pregnant and elderly people [266]. The pharmacological activities of *W. somnifera* extracts have been summarized in several reports. For example, methanol fractions of an extract showed high anti-inflammatory activity when compared to a 5 mg/kg dose of hydrocortisone sodium succinate. Oral administration of 1000 mg/kg of *W. somnifera* root powder decreased the synthesis of glycosaminoglycan in the granulation tissue of carrageenin-induced air pouch granuloma in rats, which was more potent than that of the positive control, hydrocortisone. The activity was attributed to the high content of biologically active withanolides in the plant, of which withaferin A (**355**) is a major component [268, 537]. Compound **355** (3 μ*M*) demonstrated anti-inflammatory properties in cystic fibrosis (CF), mediated by NF-κB, in an in vitro model of CF-related inflammation, wherein NF-κB is activated by filtrates of a clinically isolated strain of *Pseudomonas aeruginosa* [538]. Also, oral administration of an aqueous extract of *W. somnifera* (100 mg/kg, p.o.) prevented an increase in lipid peroxidation (LPO) in mice and rabbits, after receiving 0.2 μg/kg, i.v. of lipopolysaccharide or 100 μg/kg i.v. of peptidoglycan [539]. Wube et al. reported that withasomnine (**374**) (50 μM) inhibited leukotriene (25.6 ± 3.24%) metabolism in an in vitro bioassay using activated human neutrophile granulocytes. Therefore, withasomnine (**374**) exhibited anti-inflammatory activity [540].

Anxiolytic and antidepressant activities have been studied for *W. somnifera.* A mixture of withanolides (20 and 50 mg/kg) was administered to rats orally once daily for five days, and exhibited anxiolytic and antidepressant effects, when compared to lorazepam and imipramine [541]. A root extract (50, 100 and 200 mg/kg; p.o.)

was found to promote learning and memory in animal models [542]. Ghosal et al. reported that sitoindoside IX (**365**) and sitoindoside X (**366**), at a dose range of 50–200 mg/kg, p.o., produced anti-stress activity in albino mice and rats and augmented learning acquisition and memory retention in both young and old rats [543]. Also, the sitoindosides VII (**367**) and (VIII) (**368**) (25 and 50 mg/kg; p.o., for four days) exhibited in this manner anti-stress activity, which was potentiated by withaferin A (**355**) in Wistar rats [544].

In vitro studies have also indicated that an aqueous extract of *W. somnifera* (50 μg/cm^3) inhibited the formation of amyloid β (Aβ) fibrils [545]. Kuboyama et al. showed that withanoside IV (**361**) (10 μmol/kg/day, p.o.) improved memory deficits in $A\beta_{25-35}$ injected (25 nmol, i.c.v.) in mice and prevented the loss of axons, dendrites, and synapses. These data suggest that orally administrated withanoside IV (**361**) may ameliorate neuronal dysfunction in Alzheimer's disease [546]. In addition, withanolide A (**356**) and withanoside IV (**361**) showed neurite outgrowth activity at a concentration of 1 μM using a human neuroblastoma SH-SY5Y cell line. This investigation supports the potential neuritogenic role in traditional medicine [547]. Additionally, Pandey et al. reported that withanone (**359**) at doses (20 mg/kg/daily, p.o) to Wistar rats for a duration of 21 days showed a significant improvement in cognitive skill by inhibiting amyloid β-42 and attenuated the elevated levels of pro-inflammatory cytokines like TNF-α, IL-1β, IL-6, MCP-1, nitric oxide and lipid peroxidation [548].

Withania somnifera was evaluated for its antitumor effect in urethane-induced lung adenomas using adult male albino mice. Simultaneous administration of an extract (200 mg/kg; p.o.) and urethane (125 mg/kg) was found to reduce tumor incidence [549]. In turn, withaferin A (**355**) inhibited human umbilical vein endothelial cell (HUVEC) sprouting in a three-dimensional collagen-I matrix, with an IC_{50} value of 12 nM, thus demonstrating a potent antiangiogenic effect [550]. Compound **355** has been tested for its antiproliferative activity against the NCI-H460 (lung), HCT-116 (colon), SF-268 [central nervous system (CNS)] and MCF-7 (breast) cancer cell lines, leading to IC_{50} values ranging from 0.24 ± 0.01 to 0.36 ± 0.04 μg/cm^3 [551]. Withanolide D (**358**) exhibited in vitro cytotoxic activity against the HT-29 (colon), MCF-7, MKN-45 (gastric), HeLa (cervical), and U-937 (histiocytic lymphoma) cancer cell lines in the micromolar range ($IC_{50} = 1.0 \pm 0.02$ to 1.69 ± 0.09 μM) [552].

5.31 *Zingiber officinale*

Ginger rhizomes (*Zingiber officinale* Roscoe) (Zingiberaceae) have been consumed as a spice and an herbal medicine for many years [282, 285]. Today, different pharmacopoeias list ginger extract for various afflictions. Ginger root is used to attenuate and treat several common diseases, such as headaches, colds, fever, hypertension, nausea, and emesis. Currently, ginger is recognized as possessing different bioactive metabolites such as antiemetic, anti-inflammatory, and cancer-related properties [285, 553]. Traditionally, ginger is probably most highly-recognized for alleviating symptoms of nausea and vomiting, and several controlled studies have indicated that it may be effective as an antiemetic [282]. The activation of vagal afferent nerves mediated by serotonin (5-HT) is crucial in the mechanism of emesis [285]. In an in vitro study, 6-shogaol (**379**), 6-gingerol (**376**), and zingerone (**384**) from ginger blocked 5-HT-evoked currents in nodose neurons, exhibiting IC_{50} values of 9.27×10^{-6}, 3.03×10^{-5}, and $1.19 \times 10^{-3} M$, respectively [554].

Much of the biological activity attributed to ginger is due to its phenolic compounds such as 4-gingerol (**375**), 6-gingerol (**376**), 8-gingerol (**377**), 10-gingerol (**378**), and also shogaols and paradols [555]. An extract of *Z. officinale* and two phenolic compounds (6-shogaol (**379**) and 6-gingerol (**376**)) showed antiproliferative effects against several tumor cell lines. The plant extract displayed cytotoxic activity against pancreatic cancer cells by inducing ROS mediated apoptosis [556]. Qi et al. observed that 6-shogaol (**379**) (15 mg/kg) inhibited colorectal tumor growth in a xenograft mouse model. Also, this same compound inhibited the growth of human colon cancer cells (HCT-116 and SW-480) with IC_{50} values of 7.5 and 10 μM, respectively [557]. In addition, 6-shogaol (**379**) reduced the viability of human pancreatic cancer cells (Panc-1) with an IC_{50} value of 5.2 $\mu g/cm^3$ [556]. Sang et al. demonstrated that shogaols (6-shogaol (**379**), 8-shogaol (**380**), 10-shogaol (**381**)) were more potent than 6-gingerol (**376**), 8-gingerol (**377**), and 10-gingerol (**378**) in suppressing the growth of human lung cancer cells (H-1299) and HCT-116 cells [558]. 6-Gingerol (**376**) (200 μM) induced cell cycle arrest in the G_1 phase using several colorectal cancer cell lines (HCT-116, SW480, HT-29, LoVo). In the case of the LoVo cells, compound **376** induced cell cycle arrest by reducing cyclin D1 and also induced non-steroidal anti-inflammatory drug-activated gene 1 (NAG-1) [559]. Also, 10-gingerol (**378**) exhibited the highest leukotriene A4 hydrolase (LTA4H) and epoxidase hydrolase inhibitory activities with IC_{50} values of 21.59 ± 1.02 and 15.24 ± 0.84 μM, respectively [560]. HeLa cells were sensitive to 10-gingerol (**378**) with an IC_{50} value of 29.19 μM and IC_{80} of 50.87 μM, and underwent altered cell morphology and cell cycle arrest in the G_0/G_1 phase [561]. El-Naggar et al. reported the cytotoxic activity of 4-gingerol (**375**), 6-gingerol (**376**), 8-gingerol (**377**), 10-gingerol (**378**) against HCT-116 cells. Of these, 10-gingerol (**378**) was shown to be the most potent compound against HCT-116 cells ($IC_{50} = 33.36 \pm 0.27$ μM). Sakulnarmrat et al. reported that a phenolic constituent-containing fraction of commercially available dried ginger rhizome powder, which consisted of 4-gingerol (**375**) (3.6%), 6-gingerol (**376**) (41.9%), 8-gingerol (**377**) (7.2%), 10-gingerol (**378**) (5.1%), 6-shogaol (**379**) (24.3%), and 6-paradol (**382**) (5.9%) inhibited the proliferation of HT-29 ($IC_{50} = 1.06 \pm 0.02$ mg/cm^3) and AGS cells ($IC_{50} = 1.29 \pm 0.03$ mg/cm^3) [562]. In another study, 6-gingerol (**376**), 8-gingerol (**377**), and 10-gingerol (**378**) found in this long-used spice exhibited IC_{50} values of 666.2 ± 134.6, 135.6 ± 22.6, and 12.1 ± 0.3 μM, respectively, against breast cancer cells (MDA-MB-231). 10-Gingerol (**378**), with a longer unbranched alkyl side chain, was observed to be more potent than 6-gingerol (**376**) and 8-gingerol (**377**), in suppressing the growth of human and mouse mammary carcinoma cells and MDA-MB-231, as well as human lung and colon cancer cells [558]. Gingerol derivatives having longer unbranched alkyl side chains may be effective inhibitors of cancer cell growth because of their increased lipophilicity, resulting in enhanced permeation of the cancer cell membrane [555].

Anti-inflammatory effects of ginger and its constituents have been reported widely. Organic solvent extracts of ginger containing both gingerols (6-gingerol (**376**), 8-gingerol (**377**), 10-gingerol (**378**)) and shogaols (6-shogaol (**379**), 8-shogaol (**380**), 10-shogaol (**381**)) inhibited LPS-induced PGE2 production (IC_{50} < 0.1 μg/cm^3) [563]. In another study in vitro, 6-gingerol (**376**), 8-gingerol (**377**), and 10-gingerol (**378**) inhibited LPS-induced PGE2 and TNF-α production [555]. Also, 6-gingerol (**376**) (50–100 mg/kg, i.p.) inhibited paw edema induced by carrageenin. This same compound (25–50 mg/kg, i.p.) produced inhibition of both the acetic acid-induced writhing response and formalin-induced licking time in the late phase. These results suggested that 6-gingerol (**376**) possesses analgesic and anti-inflammatory activities [564]. In addition, 6-shogaol (**379**) has shown to inhibit the arachidonic acid cascade and nitric oxide (NO) synthesis [558]. Ginger rhizome constituents have demonstrated analgesic action mediated by cyclooxygenase-1 (COX-1) inhibition. Thus, 8-paradol (**383**) ($IC_{50} = 4 \pm 1\ \mu M$), was shown to be a potent COX-1 inhibitor and antiplatelet aggregation agent, more active in this regard than its analogs, 8-gingerol (**377**) and 8-shogaol (**380**) [565].

6 Future Perspectives

Botanicals used for human wellness and resilience are designated in the U.S. as dietary supplements and are regulated as foods and not drugs. Due to their long history of use, dietary supplements are of great interest for further scientific studies. In the United States, these botanical products are free of prescription requirements leading to them to be widely available over the counter and the internet. An increased use of these products has been reported in recent years, particularly in 2020 with a 17.3% increase in sales [12]. However, most of these products lack sufficient clinical evidence to validate their purported applications. Despite many studies that have been reported on many of these products, some of the studies are either incomplete or the data have been obtained with inadequate models, using inappropriate statistical methods for analysis of data, and unsuitable study designs that do not correlate with the traditional use of the botanical [566]. The increase in use and popularity of botanical products has elicited a greater need to assess and ensure their safety and overall quality to protect consumers, particularly for botanicals with limited prior studies [11]. Hence, there is a definite need for more comprehensive studies on these botanical dietary supplements.

Some of the areas of development in this field include detailed studies on mechanisms of action, secondary effects, pharmacokinetics, toxicity (at all levels such as children, pregnant women, and adults), and clinical trials in order to validate the efficacy and safety of the dietary supplement under study [2, 221]. After the molecular mechanism is defined using in vitro and in vivo models, the therapeutic properties should be evaluated in advanced larger animal models and in human clinical trials, which should correlate with the therapeutic properties of the plant [7]. It is also important to point out that more detailed investigations should be carried out, which would provide valuable information on the interaction of the biocomponents of the preparation with drugs or food [2]. Moreover, stability studies on products available in the market can solve key questions regarding effectiveness and toxicity of the products that are stored under unsuitable storage conditions. These challenges further underscore the importance of developing robust and suitable analytical methods to ensure the safety and quality of botanical supplements. Therefore, new methods, models, and experimental designs in the analysis of understudied botanical products will play an important role in the advancement of research in the dietary supplement field.

Other key elements to consider in the evaluation of botanical dietary supplements include quality control and quality assurance of the products available in the market as discussed elsewhere in this volume. A critical initial step in developing validated analytical methods for downstream quality control of in-process or finished botanical products is authentication and correct characterization of the raw plant parts, extracts, or powders. Verification of the plant material facilitates the avoidance of species misidentification, cross-contamination from different plant species, and deters adulteration by manufacturers of botanical dietary supplements [567]. Complex chemical composition in botanical products intrinsically varies from batch to batch generating lack of uniformity. Other factors involved in the lack of uniformity are differences in geographic origin, growth conditions, harvesting periods, and processing methods [567]. Hence, proper methods for adequate analysis should be selected based on the composition of matter in the individual products. Accordingly, comprehensive examination of the components in products, product formulations, methods and processes used in the evaluation of botanical dietary supplements is greatly needed to provide high quality, safe, and effective products to the consumer [568]. Thus, it will be a significant contribution to the field when additional translational work is completed on those presently incompletely studied botanicals.

7 Conclusions

The use of botanical dietary supplements has increased in recent years in the United States, particularly during the recent 2020 pandemic caused by COVID-19. Botanicals have been used worldwide for centuries as part of traditional medicine systems. In the U.S., these products are regulated by the Dietary Supplement Health and Education Act (DSHEA), with dietary supplements playing an increasing prominent role in the comprehensive healthcare plan for many Americans. However, to date, safety and efficacy studies of botanical dietary supplements are still somewhat limited. This chapter summarizes information on marker compounds present in extracts of botanicals dietary supplements and herbal medicine, and on in vitro and in vivo studies of selected bioactive compounds as well as the evaluation in humans of a selection of widely used dietary supplements in the U.S.

Although many studies have focused on biological and phytochemical studies, there is limited evidence about the scientific implications related to the medicinal applications of these traditionally used plants. There are key elements to consider in the analysis of these botanicals to provide the required evidence. It is important to establish internationally recognized guidelines to evaluate the quality and guarantee the efficacy and safety of herbal products. It is also recommended that clinical and preclinical studies to standardize dosage, homogenize herbal products, identify marker compounds, identify active compounds, and identify toxic metabolites or adverse effects of these products are expanded in scope.

Several pharmacological studies reported on these botanicals have focused mainly on their organic-solvent soluble extracts, but the actual constituents related to the biological activities, for which the plant is known, are still incompletely understood. Therefore, it is important to continue with phytochemical analysis and pharmacological studies of these botanicals with a long history of use.

The same rigorous preclinical and clinical approaches used to discover and develop drugs may be applied to the investigation of botanical dietary supplements using a stepwise approach. Also, the requirement that good manufacturing practice (GMP) be used in the preparation of these products has contributed significantly to the safety of these products. In addition, chemical and biological standardization would help improve the reproducibility of botanical dietary supplement preparations, while further absorption, distribution, metabolism and excretion (ADME) studies as well as clinical investigations of drug-botanical interactions would enhance the safe use of these products.

References

1. WHO Organización Mundial de la Salud (2013) Estrategia de la oms sobre medicina tradicional 2014–2023. https://apps.who.int/iris/handle/10665/95008. Accessed 17 Oct 2022
2. Ullah H, De Filippis A, Baldi A, Dacrema M, Esposito C, Garzarella EU, Santarcangelo C, Tantipongpiradet A, Daglia M (2021) Beneficial effects of plant extracts and bioactive food components in childhood supplementation. Nutrients 13:3157
3. Efferth T, Greten H (2012) Quality control for medicinal plants. Med Aromat Plants 1:7
4. Lordan R, Rando HM, Greene CS (2021) Dietary supplements and nutraceuticals under investigation for COVID-19 prevention and treatment. mSystems 6:e0122
5. Wang Y-K, Li WQ, Xia S, Guo L, Miao Y, Zhang B-K (2021) Metabolic activation of the toxic natural products from herbal and dietary supplements leading to toxicities. Front Pharmacol 12:758468
6. Vovk I, Glavnik V (2015) Analysis of dietary supplements. In: Poole CF (ed) Instrumental thin-layer chromatography. Elsevier, Boston, MA, p 589
7. van Breemen RB, Fong HHS, Farnsworth NR (2008) Ensuring the safety of botanical dietary supplements. Am J Clin Nutr 87:509s
8. Kunle OF, Egharevba HO, Ahmadu PO (2012) Standardization of herbal medicines—a review. Int J Biodivers Conserv 4:101
9. Bandaranayake WM (2006) Quality control, screening, toxicity, and regulation of herbal drugs. In: Ahmad I, Aqil F, Owais M (eds) Modern phytomedicine: turning medicinal plants into drugs. Wiley, Weinheim, p 25
10. Sahoo N, Manchikanti P, Dey S (2010) Herbal drugs: standards and regulation. Fitoterapia 81:462
11. Smillie TJ, Khan IA (2010) A comprehensive approach to identifying and authenticating botanical products. Clin Pharmacol Ther 87:175

12. Smith T, Majid F, Eckl V, Reynolds CM (2021) Herbal supplement sales in US increase by record-breaking 17.3% in 2020. HerbalGram 131:52
13. U.S. Food & Drug Administration (2021) FDA regulation of cannabis and cannabis-derived products, including cannabidiol (CBD). https://www.fda.gov/news-events/public-health-focus/fda-regulation-cannabis-and-cannabis-derived-products-including-cannabidiol-cbd. Accessed 13 Sept 2022
14. Cogan PS (2019) On healthcare by popular appeal: critical assessment of benefit and risk in cannabidiol based dietary supplements. Expert Rev Clin Pharmacol 12:501
15. Koturbash I, MacKay D (2020) Cannabidiol and other cannabinoids: from toxicology and pharmacology to the development of a regulatory pathway. J Diet Suppl 17:487
16. Wink M, Botschen F, Gosmann C, Schäfer H, Waterman PG (2010) Chemotaxonomy seen from a phylogenetic perspective and evolution of secondary metabolism. In: Wink M (ed) Annual plant reviews, biochemistry of plant secondary metabolism, vol 40. Wiley, Chichester, UK, p 364
17. Zidorn C (2019) Plant chemophenetics—a new term for plant chemosystematics/plant chemotaxonomy in the macro-molecular era. Phytochemistry 163:147
18. Kellogg JJ, Paine MF, McCune JS, Oberlies NH, Cech NB (2019) Selection and characterization of botanical natural products for research studies: a NaPDI center recommended approach. Nat Prod Rep 36:1196
19. Skiba MB, Luis PB, Alfafara C, Billheimer D, Schneider C, Funk JL (2018) Curcuminoid content and safety-related markers of quality of turmeric dietary supplements sold in an urban retail marketplace in the United States. Mol Nutr Food Res 62:1800143
20. Cerrato A, Citti C, Cannazza G, Capriotti AL, Cavaliere C, Grassi G, Marini F, Montone CM, Paris R, Piovesana S, Laganà A (2021) Phytocannabinomics: untargeted metabolomics as a tool for *Cannabis* chemovar differentiation. Talanta 230:122313
21. Guo Y, Yin T, Wang X, Zhang F, Pan G, Lv H, Owoicho Orgah J, Zhu Y, Wu H (2017) Traditional uses, phytochemistry, pharmacology and toxicology of the genus *Cimicifuga*: a review. J Ethnopharmacol 209:264
22. Gafner S (2016) Adulteration of *Actaea racemosa*. In: Botanical adulterants bulletin. Available via American Botanical Council. http://www.herbalgram.org/resources/botanical-adulterants-prevention-program/adulterants-bulleteins/black-cohosh-bulletin-june-2016/. Accessed 17 Oct 2022
23. Ali Z, Khan SI, Fronczek FR, Khan IA (2007) 9,10-seco-9,19-Cyclolanostane arabinosides from the roots of *Actaea podocarpa*. Phytochemistry 68:373
24. Qiu F, McAlpine JB, Krause EC, Chen S-N, Pauli GF (2014) Pharmacognosy of black cohosh: the phytochemical and biological profile of a major botanical dietary supplement. In: Kinghorn AD, Falk H, Kobayashi J (eds) Progress in the chemistry of organic natural products, vol 99. Springer, Cham, Switzerland, p 1
25. Hao DC, Gu XJ, Xiao PG, Liang ZG, Xu LJ, Yong P (2013) Recent advance in chemical and biological studies on Cimicifugeae pharmaceutical resources. Chin Herbal Med 5:81
26. Sun LR, Qing C, Zhang YL, Jia SY, Li ZR, Pei SJ, Qiu MH, Gross ML, Qiu SX (2007) Cimicifoetisides A and B, two cytotoxic cycloartane triterpenoid glycosides from the rhizomes of *Cimicifuga foetida*, inhibit proliferation of cancer cells. Beilstein J Org Chem 3:3
27. Geng P, Chen P, Sun J, McCoy JH, Harnly JM (2019) Authentication of black cohosh (*Actaea racemosa*) dietary supplements based on chemometric evaluation of hydroxycinnamic acid esters and hydroxycinnamic acid amides. Anal Bioanal Chem 411:7147
28. Salari S, Amiri MS, Ramezani M, Moghadam AT, Elyasi S, Sahebkar A, Emami SA (2021) Ethnobotany, phytochemistry, traditional and modern uses of *Actaea racemosa* L. (black cohosh): a review. In: Barreto GE, Sahebkar A (eds) Pharmacological properties of plant-derived natural products and implications for human health. Springer, Cham, Switzerland, p 403
29. Jöhrer K, Stuppner H, Greil R, Çiçek SS (2020) Structure-guided identification of black cohosh (*Actaea racemosa*) triterpenoids with in vitro activity against multiple myeloma. Molecules 25:766

30. Çiçek SS, Girreser U, Zidorn C (2018) Quantification of the total amount of black cohosh cycloartanoids by integration of one specific ^{1}H NMR signal. J Pharm Biomed Anal 155:109
31. Alam A, Al Arif Jahan A, Bari MS, Khandokar L, Mahmud MH, Junaid M, Chowdhury MS, Khan MF, Seidel V, Haque MA (2022) *Allium* vegetables: traditional uses, phytoconstituents, and beneficial effects in inflammation and cancer. Crit Rev Food Sci Nutr 16:1
32. Kuete V (2017) *Allium sativum*. In: Kuete V (ed) Medicinal spices and vegetables from Africa: therapeutic potential against metabolic, inflammatory, infectious and systemic diseases. Academic Press, London, p 363
33. Shang A, Cao SY, Xu XY, Gan RY, Tang GY, Corke H, Mavumengwana V, Li HB (2019) Bioactive compounds and biological functions of garlic (*Allium sativum* L.). Foods 8:246
34. El-Saber Batiha G, Beshbishy AM, Wasef LG, Elewa YHA, Al-Sagan AA, Abd El-Hack ME, Taha AE, Abd-Elhakim YM, Devkota HP (2020) Chemical constituents and pharmacological activities of garlic (*Allium sativum* L.): a review. Nutrients 12:872
35. Putnik P, Gabrić D, Roohinejad S, Barba FJ, Granato D, Mallikarjunan K, Lorenzo JM, Kovačević DB (2019) An overview of organosulfur compounds from *Allium* spp.: from processing and preservation to evaluation of their bioavailability, antimicrobial, and anti-inflammatory properties. Food Chem 276:680
36. Kothari D, Lee W-D, Niu K-M, Kim S-K (2019) The genus *Allium* as poultry feed additive: a review. Animals 9:1032
37. Amagase H (2006) Clarifying the real bioactive constituents of garlic. J Nutr 136:716S
38. Diretto G, Rubio-Moraga A, Argandoña J, Castillo P, Gómez-Gómez L, Ahrazem O (2017) Tissue-specific accumulation of sulfur compounds and saponins in different parts of garlic cloves from purple and white ecotypes. Molecules 22:1359
39. Lanzotti V, Scala F, Bonanomi G (2014) Compounds from *Allium* species with cytotoxic and antimicrobial activity. Phytochem Rev 13:769
40. Reynolds T (ed) (2004) Aloes: the genus *Aloe*. CRC Press, Boca Raton, FL
41. Kumar S, Kumar R (2019) Role of acemannan *O*-acetyl group in murine radioprotection. Carbohydr Polym 207:460
42. Sierra-García GD, Castro-Ríos R, González-Horta A, Lara-Arias J, Chávez-Montes A (2014) Acemannan, an extracted polysaccharide from *Aloe vera*: a literature review. Nat Prod Commun 9:1217
43. Ahluwalia B, Magnusson MK, Isaksson S, Larsson F, Öhman L (2016) Effects of *Aloe barbadensis* Mill. extract (AVH200®) on human blood T cell activity in vitro. J Ethnopharmacol 179:301
44. Christaki E, Florou-Paneri P (2010) *Aloe vera*: a plant for many uses. J Food Agric Environ 8:245
45. Puia A, Puia C, MoiȘ E, Graur F, Fetti A, Florea M (2021) The phytochemical constituents and therapeutic uses of genus *Aloe*: a review. Not Bot Horti Agrobot Cluj-Napoca 49:12332
46. Liu C, Cui Y, Pi F, Cheng Y, Guo Y, Qian H (2019) Extraction, purification, structural characteristics, biological activities and pharmacological applications of acemannan, a polysaccharide from *Aloe vera*: a review. Molecules 24:1554
47. Reynolds T, Dweck AC (1999) *Aloe vera* leaf gel: a review update. J Ethnopharmacol 68:3
48. Cock IE (2015) The genus *Aloe*: phytochemistry and therapeutic uses including treatments for gastrointestinal conditions and chronic inflammation. In: Rainsford KD, Powanda MC, Whitehouse MW (eds) Novel natural products: therapeutic effects in pain, arthritis and gastro-intestinal diseases. Springer, Heidelberg, p 179
49. Sibhat G, Kahsay G, Van Schepdael A, Adams E (2021) Fast and easily applicable LC-UV method for analysis of bioactive anthrones from aloe leaf latex. J Pharm Biomed Anal 195:113834
50. Cardarelli M, Rouphael Y, Pellizoni M, Colla G, Lucini L (2017) Profile of bioactive secondary metabolites and antioxidant capacity of leaf exudates from eighteen *Aloe* species. Ind Crops Prod 108:44
51. Yang F, Cao Y, Yu H, Guo Y, Cheng Y, Qian H, Yao W, Xie Y (2021) Transformation and degradation of barbaloin in aqueous solutions and aloe powder under different processing conditions. Food Biosci 43:101279

52. Salehi B, Albayrak S, Antolak H, Kręgiel D, Pawlikowska E, Sharifi-Rad M, Uprety Y, Tsouh Fokou PV, Yousef Z, Zakaria ZA, Varoni EM, Sharopov F, Martins N, Iriti M, Sharifi-Rad J (2018) *Aloe* genus plants: from farm to food applications and phytopharmacotherapy. Int J Mol Sci 19:2843
53. Viljoen AM, Van Wyk B (2000) The chemotaxonomic significance of the phenyl pyrone aloenin in the genus *Aloe*. Biochem Syst Ecol 28:1009
54. Hęś M, Dziedzic K, Górecka D, Jędrusek-Golińska A, Gujska E (2019) *Aloe vera* (L.) Webb.: natural sources of antioxidants—a review. Plant Foods Hum Nutr 74:255
55. Nalimu F, Oloro J, Kahwa I, Ogwang PE (2021) Review on the phytochemistry and toxicological profiles of *Aloe vera* and *Aloe ferox*. Future J Pharm Sci 7:145
56. Dey P, Dutta S, Chowdhury A, Das AP, Chaudhuri TK (2017) Variation in phytochemical composition reveals distinct divergence of *Aloe vera* (L.) Burm.f. from other *Aloe* species: rationale behind selective preference of *Aloe vera* in nutritional and therapeutic use. J Evid Based Complement Altern Med 22:624
57. Bera TK, Kar SK (2018) Phytochemical constituents of *Aloe vera* and their multifunctional properties: a comprehensive review. Int J Pharm Sci Res 9:1416
58. Banik S, Sharangi AB (2019) Phytochemistry, health benefits and toxicological profile of aloe. J Pharmacogn Phytochem 8:4499
59. Wang J, Su B, Jiang H, Cui N, Yu Z, Yang Y, Sun Y (2020) Traditional uses, phytochemistry and pharmacological activities of the genus *Cinnamomum* (Lauraceae): a review. Fitoterapia 146:104675
60. Oketch-Rabah HA, Marles RJ, Brinckmann JA (2018) Cinnamon and Cassia nomenclature confusion: a challenge to the applicability of clinical data. Clin Pharmacol Ther 104:435
61. Sharifi-Rad J, Dey A, Koirala N, Shaheen S, El Omari N, Salehi B, Goloshvili T, Cirone Silva NC, Bouyahya A, Vitalini S, Varoni EM, Martorell M, Abdolshahi A, Docea AO, Iriti M, Calina D, Les F, López V, Caruntu C (2021) *Cinnamomum* species: bridging phytochemistry knowledge, pharmacological properties and toxicological safety for health benefits. Front Pharmacol 12:600139
62. Błaszczyk N, Rosiak A, Kałużna-Czaplińska J (2021) The potential role of cinnamon in human health. Forests 12:648
63. Wang Y-H, Avula B, Nanayakkara NPD, Zhao J, Khan IA (2013) Cassia cinnamon as a source of coumarin in cinnamon-flavored food and food supplements in the United States. J Agric Food Chem 61:4470
64. Monteiro Nogueria I, dos Santos MO, Costa-Junior LM, da Silva LA, de Aguiar Andrade EH, Soares Maia JG, Mouchrek Filho VE (2017) Chemical composition and acaricide activity of an essential oil from a rare chemotype of *Cinnamomum verum* Presl on *Rhipicephalus microplus* (Acari: Ixodidae). Vet Parasitol 238:54
65. Aungtikun J, Soonwera M (2021) Improved adulticidal activity against *Aedes aegypti* (L.) and *Aedes albopictus* (Skuse) from synergy between *Cinnamomum* spp. essential oils. Sci Rep 11:4685
66. Wang Y, Harrington PDB, Chen P (2020) Metabolomic profiling and comparison of major cinnamon species using UHPLC–HRMS. Anal Bioanal Chem 412:7669
67. Barreca D, Mandalari G, Calderaro A, Smeriglio A, Trombetta D, Felice MR, Gattuso G (2020) *Citrus* flavones: an update on sources, biological functions, and health promoting properties. Plants 9:288
68. Yi L, Ma S, Ren D (2017) Phytochemistry and bioactivity of *Citrus* flavonoids: a focus on antioxidant, anti-inflammatory, anticancer and cardiovascular protection activities. Phytochem Rev 16:479
69. Dugrand-Judek A, Olry A, Hehn A, Costantino G, Ollitrault P, Froelicher Y, Bourgaud F (2015) The distribution of coumarins and furanocoumarins in *Citrus* species closely matches *Citrus* phylogeny and reflects the organization of biosynthetic pathways. PLoS One 10:e0142757
70. Shi Y-S, Zhang Y, Li H-T, Wu C-H, El-Seedi HR, Ye W-K, Wang Z-W, Li C-B, Zhang X-F, Kai G-Y (2020) Limonoids from *Citrus*: chemistry, anti-tumor potential, and other bioactivities. J Funct Foods 75:104213

71. Liu N, Li X, Zhao P, Zhang X, Qiao O, Huang L, Guo L, Gao W (2021) A review of chemical constituents and health-promoting effects of citrus peels. Food Chem 365:130585
72. Stohs SJ, Shara M, Ray SD (2020) *p*-Synephrine, ephedrine, *p*-octopamine and *m*-synephrine: comparative mechanistic, physiological and pharmacological properties. Phytother Res 34:1838
73. Končić MZ (2017) Role of selected medicinal plants in sports nutrition and energy homeostasis. In: Bagchi D (ed) Sustained energy for enhanced human functions and activity. Academic Press, London, p 119
74. Maksoud S, Abdel-Massih RM, Rajha HN, Louka N, Chemat F, Barba FJ, Debs E (2021) *Citrus aurantium* L. active constituents, biological effects and extraction methods. An updated review. Molecules 26:5832
75. Park J, Kim HL, Jung Y, Ahn KS, Kwak HJ, Um JY (2019) Bitter orange (*Citrus aurantium* Linné) improves obesity by regulating adipogenesis and thermogenesis through AMPK activation. Nutrients 11:1988
76. Sato R (2013) Nomilin as an anti-obesity and anti-hyperglycemic agent. Vitam Horm 91:425
77. Suntar I, Khan H, Patel S, Celano R, Rastrelli L (2018) An overview on *Citrus aurantium* L.: its functions as food ingredient and therapeutic agent. Oxid Med Cell Longev 2018:7864269
78. Edwards JE, Brown PN, Talent N, Dickinson TA, Shipley PR (2012) A review of the chemistry of the genus *Crataegus*. Phytochemistry 79:5
79. Djordjević S, Nikolić NĆ (2021) Hawthorn (*Crataegus* spp.) from botanical source to phytopreparations. Lekovite Sirovine 41:63
80. Orhan IE (2018) Phytochemical and pharmacological activity profile of *Crataegus oxycantha* L. (Hawthorn)—a cardiotonic herb. Curr Med Chem 25:4854
81. Zorniak M, Porc MP, Krzeminski TF (2019) Hawthorn revisited: time- and dose-dependent cardioprotective action of WS-1442 special extract in the reperfusion-induced arrhythmia model in rats in vivo. J Physiol Pharmacol 70:307
82. Fong HHS, Bauman JL (2002) Hawthorn. J Cardiovasc Nurs 16:1
83. Wang J, Xiong X, Feng B (2013) Effect of *Crataegus* usage in cardiovascular disease prevention: an evidence-based approach. Evid Based Complement Altern Med 2013:149363
84. Nahar L, Sarker S (2007) Phytochemistry of the genus *Curcuma*. In: Ravindran P, Babu KN, Sivaraman K (eds) Turmeric: the genus *Curcuma*. CRC Press, Boca Raton, FL, p 71
85. Booker A, Frommenwiler D, Johnston D, Umealajekwu C, Reich E, Heinrich M (2014) Chemical variability along the value chains of turmeric (*Curcuma longa*): a comparison of nuclear magnetic resonance spectroscopy and high performance thin layer chromatography. J Ethnopharmacol 152:292
86. Bejar E (2018) Turmeric (*Curcuma longa*) root and rhizome, and root and rhizome extracts. In: Botanical adulterants bulletin. Available via American Botanical Council. http://www.herbalgram.org/resources/botanical-adulterants-prevention-program/adulterants-bulletin-may-2018/. Accessed 17 Oct 2022
87. Tayyem RF, Heath DD, Al-Delaimy WK, Rock CL (2006) Curcumin content of turmeric and curry powders. Nutr Cancer 55:126
88. Kulyal P, Acharya S, Ankari AB, Kokkiripati PK, Tetali SD, Raghavendra AS (2021) Variable secondary metabolite profiles across cultivars of *Curcuma longa* L. and *C. aromatica* Salisb. Front Pharmacol 12:659546
89. Syu WJ, Shen CC, Don MJ, Ou JC, Lee GH, Sun CM (1998) Cytotoxicity of curcuminoids and some novel compounds from *Curcuma zedoaria*. J Nat Prod 61:1531
90. Avula B, Wang YH, Khan IA (2012) Quantitative determination of curcuminoids from the roots of *Curcuma longa*, *Curcuma* species and dietary supplements using an UPLC-UV-MS method. J Chromatogr Sep Tech 3(1):1000120
91. Zhang J, Jinnai S, Ikeda R, Wada M, Hayashida S, Nakashima K (2009) A simple HPLC-fluorescence method for quantitation of curcuminoids and its application to turmeric products. Anal Sci 25:385
92. Dosoky NS, Satyal P, Setzer WN (2019) Variations in the volatile compositions of *Curcuma* species. Foods 8:53

93. Braga ME, Leal PF, Carvalho JE, Meireles MA (2003) Comparison of yield, composition, and antioxidant activity of turmeric (*Curcuma longa* L.) extracts obtained using various techniques. J Agric Food Chem 51:6604
94. Jantan I, Saputri FC, Qaisar MN, Buang F (2012) Correlation between chemical composition of *Curcuma domestica* and *Curcuma xanthorrhiza* and their antioxidant effect on human low-density lipoprotein oxidation. Evid Based Complement Altern Med 2012:438356
95. Ib J, Ahmad AS, Ali NAM, Ahmad AR, Ibrahim H (1999) Chemical composition of the rhizome oils of four *Curcuma* species from Malaysia. J Essent Oil Res 11:719
96. Septama AW, Tasfiyati AN, Kristiana R, Jaisi A (2022) Chemical profiles of essential oil from Javanese turmeric (*Curcuma xanthorrhiza* Roxb.), evaluation of its antibacterial and antibiofilm activities against selected clinical isolates. S Afr J Bot 146:728
97. Amalraj A, Pius A, Gopi S, Gopi S (2016) Biological activities of curcuminoids, other biomolecules from turmeric and their derivatives—a review. J Tradit Complement Med 7:205
98. Zhang HA, Kitts DD (2021) Turmeric and its bioactive constituents trigger cell signaling mechanisms that protect against diabetes and cardiovascular diseases. Mol Cell Biochem 476:3785
99. Kindscher K (2016) The uses of *Echinacea angustifolia* and other *Echinacea* species by Native Americans. In: Kindscher K (ed) Echinacea: herbal medicine with a wild history. Springer, Cham, Switzerland, p 9
100. Cao C, Kindscher K (2016) The medicinal chemistry of *Echinacea* species. In: Kindscher K (ed) Echinacea: herbal medicine with a wild history. Springer, Cham, Swizerland, p 127
101. Chang BY, Lee SK, Kim DE, Bae JH, Ho TT, Park S-Y, Lee MK, Kim SY (2020) Effect of echinalkamide identified from *Echinacea purpurea* (L.) Moench on the inhibition of osteoclastogenesis and bone resorption. Sci Rep 10:10914
102. Bruni R, Brighenti V, Caesar LK, Bertelli D, Cech NB, Pellati F (2018) Analytical methods for the study of bioactive compounds from medicinally used *Echinacea* species. J Pharm Biomed Anal 160:443
103. Handy SM, Pawar RS, Ottesen AR, Ramachandran P, Sagi S, Zhang N, Hsu E, Erickson DL (2021) HPLC-UV, metabarcoding and genome skims of botanical dietary supplements: a case study in *Echinacea*. Planta Med 87:314
104. Ren L, Guo M-Y, Pang X-H (2018) Identification and classification of medicinal plants in *Epimedium*. Chin Herb Med 10:249
105. Ma H, He X, Yang Y, Li M, Hao D, Jia Z (2011) The genus *Epimedium*: an ethnopharmacological and phytochemical review. J Ethnopharmacol 134:519
106. Schulz M, Campelo Borges GdS, Valdemiro Gonzaga L, Oliveira Costa AC, Fett R (2016) Juçara fruit (*Euterpe edulis* Mart.): sustainable exploitation of a source of bioactive compounds. Food Res Int 89:14
107. Schulz M, Tischer Seraglio S, Valdemiro Gonzaga L, Oliveira Costa AC, Fett R (2021) Phenolic compounds in *Euterpe* fruits: composition, digestibility, and stability—a review. Food Rev Int. https://doi.org/10.1080/87559129.2021.1909060
108. de Souza C, Pereira D, dos Santos GF, Valeriano Tonon R, Beres C, Maria Corrêa Cabral L (2022) Towards chemical characterization and possible applications of juçara fruit: an approach to remove *Euterpe edulis* Martius from the extinction list. J Food Sci Technol. https://doi.org/10.1007/s13197-021-05342-8
109. de Lima Yamaguchi KK, Ravazi Pereira LF, Lamarão CV, Silva Lima E, da Veiga-Junior VF (2015) Amazon açaí: chemistry and biological activities: a review. Food Chem 179:137
110. de Almeida Magalhães TSS, de Oliveira Macedo PC, Converti A, Neves de Lima ÁA (2020) The use of *Euterpe oleracea* Mart. as a new perspective for disease treatment and prevention. Biomolecules 10:813
111. Spontoni do Espirito Santo BL, Figueiredo Santana L, Kato Junior WH, de Oliveira de Araújo F, Bogo D, de Cássia Freitas K, de Cássia Avellaneda Guimarães R, Aiko Hiane P, Pott A, de Oliveira Filiú WF (2020) Medicinal potential of *Garcinia* species and their compounds. Molecules 25:4513
112. The World Flora Online (2022) *Garcinia* L. http://www.worldfloraonline.org/taxon/wfo-4000015318. Accessed 14 Sept 2022

113. Seethapathy GS, Tadesse M, Urumarudappa SKJ, Gunaga SV, Vasudeva R, Malterud KE, Shaanker RU, de Boer HJ, Ravikanth G, Wangensteen H (2018) Authentication of *Garcinia* fruits and food supplements using DNA barcoding and NMR spectroscopy. Sci Rep 8:10561
114. Lim TK (2012) Introduction. In: Lim TK (ed) Edible medicinal and non-medicinal plants, vol 2. Springer, Dordrecht, The Nederlands, p 1
115. Pandey R, Chandra P, Kumar B, Srivastva M, Aravind AA, Shameer P, Rameshkumar K (2015) Simultaneous determination of multi-class bioactive constituents for quality assessment of *Garcinia* species using UHPLC–QqQ$_{\mathrm{LIT}}$–MS/MS. Ind Crops Prod 77:861
116. Sripradha R, Sridhar MG, Maithilikarpagaselvi N (2016) Antihyperlipidemic and antioxidant activities of the ethanolic extract of *Garcinia cambogia* on high fat diet-fed rats. J Complement Integr Med 13:9
117. Semwal RB, Semwal DK, Vermaak I, Viljoen A (2015) A comprehensive scientific overview of *Garcinia cambogia*. Fitoterapia 102:134
118. Shaito A, Thi Bich Thuan D, Thi Phu H, Hie Nguyen T, Hasan H, Halabi S, Abdelhady SK, Nasrallah GH, Eid A, Pintus G (2020) Herbal medicine for cardiovascular diseases: efficacy, mechanisms, and safety. Front Pharmacol 11:422
119. Liu L, Wang Y, Zhang J, Wang S (2020) Advances in the chemical constituents and chemical analysis of *Ginkgo biloba* leaf, extract, and phytopharmaceuticals. J Pharm Biomed Anal 193:113704
120. Ma G-L, Xiong J, Yang G-X, Pan L-L, Hu C-L, Wang W, Fan H, Zhao Q-H, Zhang H-Y, Hu J-F (2016) Biginkgosides A-I, unexpected minor dimeric flavonol diglycosidic truxinate and truxillate esters from *Ginkgo biloba* leaves and their antineuroinflammatory and neuroprotective activities. J Nat Prod 79:1354
121. Ude C, Schubert-Zsilavecz M, Wurglics M (2013) *Ginkgo biloba* extracts: a review of the pharmacokinetics of the active ingredients. Clin Pharmacokinet 52:727
122. Tang Y, Lou F, Wang J, Li Y, Zhuang S (2001) Coumaroyl flavonol glycosides from the leaves of *Ginkgo biloba*. Phytochemistry 58:1251
123. Gafner S (2018) Adulteration of *Ginkgo biloba* leaf extract. In: Botanical adulterants bulletin. Available via American Botanical Council. https://www.herbalgram.org/resources/botanical-adulterants-prevention-program/adulterants-bulletins/ginkgo-bulletin-january-2018/. Accessed 25 Oct 2022
124. Collins BJ, Kerns SP, Aillon K, Mueller G, Rider CV, DeRose EF, London RE, Harnly JM, Waidyanatha S (2020) Comparison of phytochemical composition of *Ginkgo biloba* extracts using a combination of non-targeted and targeted analytical approaches. Anal Bioanal Chem 412:6789
125. Barnes J, Anderson LA, Phillipson JD (2001) St John's wort (*Hypericum perforatum* L.): a review of its chemistry, pharmacology and clinical perspectives. J Pharm Pharmacol 53:583
126. Xiao C-Y, Mu Q, Gibbons S (2020) The phytochemistry and pharmacology of *Hypericum*. Progr Chem Org Nat Prod 112:85
127. Orhan N (2021) St. John's wort (*Hypericum perforatum*) laboratory guidance document. In: Laboratory guidance documents. Available via American Botanical Council. https://www.herbalgram.org/resources/botanical-adulterants-prevention-program/laboratory-guidance-documents/st-johns-wort-laboratory-guidance-document-december-2021/. Accessed 25 Oct 2022
128. Silva AR, Taofiq O, Ferreira IC, Barros L (2021) *Hypericum* genus cosmeceutical application—a decade comprehensive review on its multifunctional biological properties. Ind Crops Prod 159:113053
129. Raclariu AC, Paltinean R, Vlase L, Labarre A, Manzanilla V, Ichim MC, Crisan G, Brysting AK, de Boer H (2017) Comparative authentication of *Hypericum perforatum* herbal products using DNA metabarcoding, TLC and HPLC-MS. Sci Rep 7:1291
130. Zhao J, Liu W, Wang J-C (2015) Recent advances regarding constituents and bioactivities of plants from the genus *Hypericum*. Chem Biodivers 12:309
131. Hu L-H, Sim K-Y (1998) Complex caged polyisoprenylated benzophenone derivatives, sampsoniones A and B, from *Hypericum sampsonii*. Tetrahedron Lett 39:7999

132. Al-Shehbaz IA (2012) A generic and tribal synopsis of the Brassicaceae (Cruciferae). Taxon 61:931
133. Wu X, Huang H, Childs H, Wu Y, Yu L, Pehrsson PR (2021) Glucosinolates in *Brassica* vegetables: characterization and factors that influence distribution, content, and intake. Annu Rev Food Sci Technol 12:485
134. Đulović A, Burčul F, Čulić VČ, Ruščić M, Brzović P, Montaut S, Rollin P, Blažević I (2021) *Lepidium graminifolium* L.: glucosinolate profile and antiproliferative potential of volatile isolates. Molecules 26:5183
135. Blažević I, Montaut S, Burčul F, Rollin P (2015) Glucosinolates: novel sources and biological potential. In: Mérillon JM, Ramawat K (eds) Glucosinolates. Springer, Cham, Switzerland, p 3
136. Pagnotta E, Agerbirk N, Olsen CE, Ugolini L, Cinti S, Lazzeri L (2017) Hydroxyl and methoxyl derivatives of benzylglucosinolate in *Lepidium densiflorum* with hydrolysis to isothiocyanates and non-isothiocyanate products: substitution governs product type and mass spectral fragmentation. J Agric Food Chem 65:3167
137. Kwapong AA, Stapleton P, Gibbons S (2018) A new dimeric imidazole alkaloid plasmid conjugation inhibitor from *Lepidium sativum*. Tetrahedron Lett 59:1952
138. Maier UH, Gundlach H, Zenk MH (1998) Seven imidazole alkaloids from *Lepidium sativum*. Phytochemistry 49:1791
139. Jin W, Chen X, Dai P, Yu L (2016) Lepidiline C and D: two new imidazole alkaloids from *Lepidium meyenii* Walpers (Brassicaceae) roots. Phytochem Lett 17:158
140. Le HTN, Van Roy E, Dendooven E, Peeters L, Theunis M, Foubert K, Pieters L, Tuenter E (2021) Alkaloids from *Lepidium meyenii* (Maca), structural revision of macaridine and UPLC-MS/MS feature-based molecular networking. Phytochemistry 190:112863
141. Zhou M, Zhang R-Q, Chen Y-J, Liao L-M, Sun Y-Q, Ma Z-H, Yang Q-F, Li P, Ye Y-Q, Hu Q-F (2018) Three new pyrrole alkaloids from the roots of *Lepidium meyenii*. Phytochem Lett 23:137
142. Huang Y-J, Peng X-R, Qiu M-H (2018) Progress on the chemical constituents derived from glucosinolates in maca (*Lepidium meyenii*). Nat Prod Bioprospect 8:405
143. Wang S, Zhu F (2019) Chemical composition and health effects of maca (*Lepidium meyenii*). Food Chem 288:422
144. Zhang S-Z, Yang F, Shao J-L, Pu H-M, Ruan Z-Y, Yang W-L, Li H (2020) The metabolic formation profiles of macamides accompanied by the conversion of glucosinolates in maca (*Lepidium meyenii*) during natural air drying. Int J Food Sci Technol 55:2428
145. McCollom MM, Villinski JR, McPhail KL, Craker LE, Gafner S (2005) Analysis of macamides in samples of maca (*Lepidium meyenii*) by HPLC-UV-MS/MS. Phytochem Anal 16:463
146. Meisner HO, Mscisz A, Piatkowska E, Baraniak M, Mielcarek S, Kedzia B, Holderna-Kedzia E, Pisulewski P (2016) Peruvian maca (*Lepidium peruvianum*): (ii) phytochemical profiles of four prime maca phenotypes grown in two geographically-distant locations. Int J Biomed Sci 12:9
147. Geng P, Sun J, Chen P, Brand E, Frame J, Meissner H, Stewart J, Gafner S, Clark S, Miller J, Harnly J (2020) Characterization of maca (*Lepidium meyenii/Lepidium peruvianum*) using a mass spectral fingerprinting, metabolomic analysis, and genetic sequencing approach. Planta Med 86:674
148. Cullis C (2011) *Linum*. In: Kole C (ed) Wild crop relatives: genomic and breeding resources: oilseeds. Springer, Berlin, Heidelberg, p 177
149. Westcott ND, Muir AD (2003) Chemical studies on the constituents of *Linum* spp. In: Muir AD, Westcott ND (eds), Flax: the genus *Linum*. CRC Press, London, p 67
150. Bernacchia R, Preti R, Vinci G (2014) Chemical composition and health benefits of flaxseed. Austin J Nutr Food Sci 2:1045
151. Ansari R, Zarshenas MM, Dadbakhsh AH (2019) A review on pharmacological and clinical aspects of *Linum usitatissimum* L. Curr Drug Discov Technol 16:148

152. Akter Y, Junaid M, Afrose SS, Nahrin A, Alam SM, Sharmin T, Akter R, Hosen ZSM (2021) A comprehensive review on *Linum usitatissimum* medicinal plant: its phytochemistry, pharmacology, and ethnomedicinal uses. Mini Rev Med Chem 21:2801
153. Engels G, Brinckmann J (2017) *Lycium* (goji berry). HerbalGram 113:8
154. Qian D, Zhao Y, Yang G, Huang L (2017) Systematic review of chemical constituents in the genus *Lycium* (Solanaceae). Molecules 22:911
155. Tian X, Liang T, Liu Y, Ding G, Zhang F, Ma Z (2019) Extraction, structural characterization, and biological functions of *Lycium barbarum* polysaccharides: a review. Biomolecules 9:389
156. Cheng J, Zhou Z-W, Sheng H-P, He L-J, Fan X-W, He Z-X, Sun T, Zhang X, Zhao RJ, Gu L, Cao C, Zhou S-F (2015) An evidence-based update on the pharmacological activities and possible molecular targets of *Lycium barbarum* polysaccharides. Drug Des Devel Ther 9:33
157. Kwok SS, Bu Y, Cheuk-Yin Lo A, Chung-Yan Chan T, So KF, Shiu-Ming Lai J, Co Shih K (2019) A systematic review of potential therapeutic use of *Lycium barbarum* polysaccharides in disease. BioMed Res Int 2019:4615745
158. Talebi SM, Sheidai M, Arianejad F, Mahdieh M (2022) Biosystematics relationships among *Marrubium* L. (Lamiaceae) species in Iran. Genet Resour Crop Evol 69:2833
159. Argyropoulou C, Karioti A, Skaltsa H (2009) Labdane diterpenes from *Marrubium thessalum*. Phytochemistry 70:635
160. Popoola OK, Elbagory AM, Ameer F, Hussein AA (2013) Marrubiin. Molecules 18:9049
161. Aćimović M, Jeremić K, Salaj N, Gavarić N, Kiprovski B, Sikora V, Zeremski T (2020) *Marrubium vulgare* L.: a phytochemical and pharmacological overview. Molecules 25:2898
162. Villanueva JR, Esteban JM, Villanueva LR (2017) A reassessment of the *Marrubium vulgare* L. herb's potential role in diabetes mellitus type 2: first results guide the investigation toward new horizons. Medicines 4:57
163. Brouillet L (2022) *Matricaria*. Flora of North America. http://floranorthamerica.org/Matricaria. Accessed 22 Jun 2022
164. Kolanos R, Stice SA (2021) German chamomile. In: Gupta RC, Lall R, Srivastava A (eds) Nutraceuticals: efficacy, safety, and toxicity, 2nd edn. Academic Press, London, p 757
165. Ghizlane H, Aziz B (2016) Pharmacological properties of some medicinal plants, its components and using fields. In: Watson RR, Preedy VR (eds) Fruits, vegetables, and herbs: bioactive foods in health promotion. Academic Press, London, San Diego, Cambridge, Oxford, p 41
166. El Mihyaoui A, Esteves da Silva JCG, Charfi S, Candela Castillo ME, Lamarti A, Arnao MB (2022) Chamomile (*Matricaria chamomilla* L.): a review of ethnomedicinal use, phytochemistry and pharmacological uses. Life 12:479
167. Höferl M, Wanner J, Tabanca N, Ali A, Gochev V, Schmidt E, Kaul VK, Singh V, Jirovetz L (2020) Biological activity of *Matricaria chamomilla* essential oils of various chemotypes. Planta Med Int Open 7:e114
168. Tsivelika N, Sarrou E, Gusheva K, Pankou C, Koutsos T, Chatzopoulou P, Mavromatis A (2018) Phenotypic variation of wild chamomile (*Matricaria chamomilla* L.) populations and their evaluation for medicinally important essential oil. Biochem Syst Ecol 80:21
169. Sharifi-Rad M, Nazaruk J, Polito L, Morais-Braga MFB, Rocha JE, Coutinho HDM, Salehi B, Tabanelli G, Montanari C, del Mar CM (2018) *Matricaria* genus as a source of antimicrobial agents: from farm to pharmacy and food applications. Microbiol Res 215:76
170. Kapalka GM (2010) Anxiety disorders. In: Kapalka GM (ed) Nutritional and herbal therapies for children and adolescents: a handbook for mental health clinicians. Academic Press, London, p 219
171. Petruľová-Poracká V, Repčák M, Vilková M, Imrich J (2013) Coumarins of *Matricaria chamomilla* L.: aglycones and glycosides. Food Chem 141:54
172. Bharathi LK, John KJ (2013) Introduction. In: *Momordica* genus in Asia—an overview. Springer, New Delhi, p 1
173. Nagarani G, Abirami A, Siddhuraju P (2014) Food prospects and nutraceutical attributes of *Momordica* species: a potential tropical bioresources—a review. Food Sci Hum Wellness 3:117

174. Laczkó-Zöld E, Bacsadi B, Horváth A, Csupor D (2021) Development and validation of a RP-HPLC-DAD method for quantification of charantin in *Momordica charantia* products. J Food Compos Anal 104:104161
175. Ramalhete C, Gonçalves BM, Barbosa F, Duarte N, Ferreira M-JU (2022) *Momordica balsamina*: phytochemistry and pharmacological potential of a gifted species. Phytochem Rev 21:617
176. Jia S, Shen M, Zhang F, Xie J (2017) Recent advances in *Momordica charantia*: functional components and biological activities. Int J Mol Sci 18:2555
177. Zhang F, Lin L, Xie J (2016) A mini-review of chemical and biological properties of polysaccharides from *Momordica charantia*. Int J Biol Macromol 92:246
178. Hsiao P-C, Liaw C-C, Hwang S-Y, Cheng H-L, Zhang L-J, Shen C-C, Hsu F-L, Kuo Y-H (2013) Antiproliferative and hypoglycemic cucurbitane-type glycosides from the fruits of *Momordica charantia*. J Agric Food Chem 61:2979
179. Badalamenti N, Modica A, Bazan G, Marino P, Bruno M (2022) The ethnobotany, phytochemistry, and biological properties of *Nigella damascene*—a review. Phytochemistry 198:113165
180. Salehi B, Quispe C, Imran M, Ul-Haq I, Živković J, Abu-Reidah IM, Sen S, Taheri Y, Acharya K, Azadi H (2021) *Nigella* plants—traditional uses, bioactive phytoconstituents, preclinical and clinical studies. Front Pharmacol 12:625386
181. Ahmad MF, Ahmad FA, Ashraf SA, Saad HH, Wahab S, Khan MI, Ali M, Mohan S, Hakeem KR, Athar MT (2021) An updated knowledge of black seed (*Nigella sativa* Linn.): review of phytochemical constituents and pharmacological properties. J Herb Med 25:100404
182. Biswas SK, Kim D-E, Keum Y-S, Saini RK (2018) Metabolite profiling and antioxidant activities of white, red, and black rice (*Oryza sativa* L.) grains. J Food Meas Charact 12:2484
183. Pang Y, Ahmed S, Xu Y, Beta T, Zhu Z, Shao Y, Bao J (2018) Bound phenolic compounds and antioxidant properties of whole grain and bran of white, red and black rice. Food Chem 240:212
184. Samyor D, Das AB, Deka SC (2017) Pigmented rice a potential source of bioactive compounds: a review. Int J Food Sci Technol 52:1073
185. Rungratanawanich W, Memo M, Uberti D (2018) Redox homeostasis and natural dietary compounds: focusing on antioxidants of rice (*Oryza sativa* L.). Nutrients 10:1605
186. Ito VC, Lacerda LG (2019) Black rice (*Oryza sativa* L.): a review of its historical aspects, chemical composition, nutritional and functional properties, and applications and processing technologies. Food Chem 301:125304
187. Goufo P, Trindade H (2014) Rice antioxidants: phenolic acids, flavonoids, anthocyanins, proanthocyanidins, tocopherols, tocotrienols, γ-oryzanol, and phytic acid. Food Sci Nutr 2:75
188. Song J, Luo J, Ma Z, Sun Q, Wu C, Li X (2019) Quality and authenticity control of functional red yeast rice—a review. Molecules 24:1944
189. Zhu B, Qi F, Wu J, Yin G, Hua J, Zhang Q, Qin L (2019) Red yeast rice: a systematic review of the traditional uses, chemistry, pharmacology, and quality control of an important Chinese folk medicine. Front Pharmacol 10:1449
190. Avula B, Cohen PA, Wang Y-H, Sagi S, Feng W, Wang M, Zweigenbaum J, Shuangcheng M, Khan IA (2014) Chemical profiling and quantification of monacolins and citrinin in red yeast rice commercial raw materials and dietary supplements using liquid chromatography-accurate QToF mass spectrometry: chemometrics application. J Pharm Biomed Anal 100:243
191. Zuo Y-J, Wen J, Zhou S-L (2017) Intercontinental and intracontinental biogeography of the eastern Asian–eastern North American disjunct *Panax* (the ginseng genus, Araliaceae), emphasizing its diversification processes in eastern Asia. Mol Phylogenet Evol 117:60
192. Jovanovski E, Lea-Duvnjak-Smircic KA, Au-Yeung F, Zurbau A, Jenkins AL, Sung M-K, Josse R, Vuksan V (2020) Vascular effects of combined enriched Korean red ginseng (*Panax ginseng*) and American ginseng (*Panax quinquefolius*) administration in individuals with hypertension and type 2 diabetes: a randomized controlled trial. Complement Ther Med 49:102338

193. Chen W, Balan P, Popovich DG (2019) Comparison of the ginsenoside composition of Asian ginseng (*Panax ginseng*) and American ginseng (*Panax quinquefolius* L.) and their transformation pathways. In: Atta-ur-Rahman (ed) Studies in natural products chemistry, vol 63. Elsevier, Amsterdam, p 161
194. Park SK, Hyun SH, In G, Park C-K, Kwak Y-S, Jang Y-J, Kim B, Kim J-H, Han C-K (2021) The antioxidant activities of Korean red ginseng (*Panax ginseng*) and ginsenosides: a systemic review through in vivo and clinical trials. J Ginseng Res 45:41
195. Ichim MC, de Boer HJ (2021) A review of authenticity and authentication of commercial ginseng herbal medicines and food supplements. Front Pharmacol 11:612071
196. Yang Y, Ju Z, Yang Y, Zhang Y, Yang L, Wang Z (2021) Phytochemical analysis of *Panax* species: a review. J Ginseng Res 45:1
197. Ru W, Wang D, Xu Y, He X, Sun Y-E, Qian L, Zhou X, Qin Y (2015) Chemical constituents and bioactivities of *Panax ginseng* (C. A. Mey.). Drug Discov Ther 9:23
198. Huang X, Li N, Pu Y, Zhang T, Wang B (2019) Neuroprotective effects of ginseng phytochemicals: recent perspectives. Molecules 24:2939
199. Raman V, Avula B, Galal AM, Wang Y-H, Khan IA (2013) Microscopic and UPLC-UV-MS analyses of authentic and commercial yohimbe (*Pausinystalia johimbe*) bark samples. J Nat Med 67:42
200. Tam SW, Worcel M, Wyllie M (2001) Yohimbine: a clinical review. Pharmacol Ther 91:215
201. Lucas D, Neal-Kababick J, Zweigenbaum J (2015) Characterization and quantitation of yohimbine and its analogs in botanicals and dietary supplements using LC/QTOF-MS and LC/QQQ-MS for determination of the presence of bark extract and yohimbine adulteration. J AOAC Int 98:330
202. Cohen PA, Wang YH, Maller G, DeSouza R, Khan IA (2016) Pharmaceutical quantities of yohimbine found in dietary supplements in the USA. Drug Test Anal 8:357
203. Liu Y, Yu H-Y, Xu H-Z, Liu J-J, Meng X-G, Zhou M, Ruan H-L (2018) Alkaloids with immunosuppressive activity from the bark of *Pausinystalia yohimbe*. J Nat Prod 81:1841
204. Applequist WL (2015) A brief review of recent controversies in the taxonomy and nomenclature of *Sambucus nigra* sensu lato. Acta Hortic 1061:25
205. Thomas AL, Byers PL, Avery JD Jr, Kaps M, Gu S, Johnson HY, Millican M (2015) 'Marge': a European elderberry for North American producers. Acta Hortic 1061:191
206. Sidor A, Gramza-Michałowska A (2015) Advanced research on the antioxidant and health benefit of elderberry (*Sambucus nigra*) in food—a review. J Funct Foods 18:941
207. Przybylska-Balcerek A, Szablewski T, Szwajkowska-Michałek L, Świerk D, Cegielska-Radziejewska R, Krejpcio Z, Suchowilska E, Tomczyk Ł, Stuper-Szablewska K (2021) *Sambucus nigra* extracts—natural antioxidants and antimicrobial compounds. Molecules 26:2910
208. Mikulic-Petkovsek M, Schmitzer V, Slatnar A, Todorovic B, Veberic R, Stampar F, Ivancic A (2014) Investigation of anthocyanin profile of four elderberry species and interspecific hybrids. J Agric Food Chem 62:5573
209. Vlachojannis C, Zimmermann BF, Chrubasik-Hausmann S (2015) Quantification of anthocyanins in elderberry and chokeberry dietary supplements. Phytother Res 29:561
210. Lee J, Finn CE (2007) Anthocyanins and other polyphenolics in American elderberry (*Sambucus canadensis*) and European elderberry (*S. nigra*) cultivars. J Sci Food Agric 87:2665
211. Tasinov O, Dincheva I, Badjakov I, Kiselova-Kaneva Y, Galunska B, Nogueiras R, Ivanova D (2021) Phytochemical composition, anti-inflammatory and ER stress-reducing potential of *Sambucus ebulus* L. fruit extract. Plants 10:2446
212. Mikulic-Petkovsek M, Ivancic A, Todorovic B, Veberic R, Stampar F (2015) Fruit phenolic composition of different elderberry species and hybrids. J Food Sci 80:C2180
213. Christensen LP, Kaack K, Fretté XC (2008) Selection of elderberry (*Sambucus nigra* L.) genotypes best suited for the preparation of elderflower extracts rich in flavonoids and phenolic acids. Eur Food Res Technol 227:293
214. Kaltsa O, Lakka A, Grigorakis S, Karageorgou I, Batra G, Bozinou E, Lalas S, Makris DP (2020) A green extraction process for polyphenols from elderberry (*Sambucus nigra*) flowers using deep eutectic solvent and ultrasound-assisted pretreatment. Molecules 25:921

215. Uzlasir T, Kadiroglu P, Selli S, Kelebek H (2021) LC-DAD-ESI-MS/MS characterization of elderberry flower (*Sambucus nigra*) phenolic compounds in ethanol, methanol, and aqueous extracts. J Food Process Preserv 45:e14478
216. Buhrmester RA, Ebinger JE, Seigler DS (2000) Sambunigrin and cyanogenic variability in populations of *Sambucus canadensis* L. (Caprifoliaceae). Biochem Syst Ecol 28:689
217. Senica M, Stampar F, Mikulic-Petkovsek M (2019) Harmful (cyanogenic glycoside) and beneficial (phenolic) compounds in different *Sambucus* species. J Berry Res 9:395
218. Jiménez P, Cabrero P, Basterrechea JE, Tejero J, Cordoba-Diaz D, Cordoba-Diaz M, Girbes T (2014) Effects of short-term heating on total polyphenols, anthocyanins, antioxidant activity and lectins of different parts of dwarf elder (*Sambucus ebulus* L.). Plant Foods Hum Nutr 69:168
219. Jiménez P, Cabrero P, Cordoba-Diaz D, Cordoba-Diaz M, Garrosa M, Girbés T (2017) Lectin digestibility and stability of elderberry antioxidants to heat treatment in vitro. Molecules 22:95
220. Senica M, Stampar F, Veberic R, Mikulic-Petkovsek M (2016) Processed elderberry (*Sambucus nigra* L.) products: a beneficial or harmful food alternative? Food Sci Technol 72:182
221. Waswa EN, Li J, Mkala EM, Wanga VO, Mutinda ES, Nanjala C, Odago WO, Katumo DM, Gichua MK, Gituru RW, Hu G-W, Wang Q-F (2022) Ethnobotany, phytochemistry, pharmacology, and toxicology of the genus *Sambucus* L. (Viburnaceae). J Ethnopharmacol 292:115102
222. Schmitzer V, Veberic R, Stampar F (2012) European elderberry (*Sambucus nigra* L.) and American elderberry (*Sambucus canadensis* L.): botanical, chemical and health properties of flowers, berries and their products. In: Tuberoso CIG (ed) Berries: properties, consumption and nutrition. Nova Science Publishers, Hauppauge, NY, p 127
223. Ferreira SS, Silva AM, Nunes FM (2020) *Sambucus nigra* L. fruits and flowers: chemical composition and related bioactivities. Food Rev Int 38:1237
224. The World Flora Online (2022) *Serenoa* Hook.f. http://www.worldfloraonline.org/taxon/wfo-4000035109. Accessed 14 Sept 2022
225. Kwon Y (2019) Use of saw palmetto (*Serenoa repens*) extract for benign prostatic hyperplasia. Food Sci Biotechnol 28:1599
226. Saw palmetto (2000) USP 24-NF 19. United States Pharmacopeial Convention, Rockville, MD
227. Agbabiaka TB, Pittler MH, Wider B, Ernst E (2009) *Serenoa repens* (saw palmetto): a systematic review of adverse events. Drug Saf 32:637
228. Gresta F, Avola G, Guarnaccia P (2007) Agronomic characterization of some spontaneous genotypes of milk thistle (*Silybum marianum* L. Gaertn.) in Mediterranean environment. J Herbs Spices Med Plants 12:51
229. Marceddu R, Dinolfo L, Carrubba A, Sarno M, Di Miceli G (2022) Milk thistle (*Silybum marianum* L.) as a novel multipurpose crop for agriculture in marginal environments: a review. Agronomy 12:729
230. McCutcheon A (2020) Adulteration of milk thistle (*Silybum marianum*). Available via American Botanical Council. https://www.herbalgram.org/resources/botanical-adulterants-prevention-program/adulterants-bulletins/milk-thistle-bulletin-october-2020/ Accessed 19 Oct 2022
231. Křen V, Valentová K (2022) Silybin and its congeners: from traditional medicine to molecular effects. Nat Prod Rep 39:1264
232. Basu SK, Zandi P, Cetzal-Ix W (2019) Fenugreek (*Trigonella foenum-graecum* L.): distribution, genetic diversity, and potential to serve as an industrial crop for the global pharmaceutical, nutraceutical, and functional food industries. In: Singh RB, Watson RR, Takahashi T (eds) The role of functional food security in global health. Academic Press, London, p 471
233. The World Flora Online (2022) *Trigonella* L. http://www.worldfloraonline.org/taxon/wfo-4000039145. Accessed 14 Sept 2022
234. Mandal S, DebMandal M (2016) Fenugreek (*Trigonella foenum-graecum* L.) oils. In: Preedy VR (ed) Essential oils in food preservation, flavor and safety. Academic Press, London, p 421
235. Nagulapalli Venkata KC, Swaroop A, Bagchi D, Bishayee A (2017) A small plant with big benefits: fenugreek (*Trigonella foenum-graecum* Linn.) for disease prevention and health promotion. Mol Nutr Food Res 61:1600950

236. Dini I (2018) Spices and herbs as therapeutic foods. In: Holban AM, Grumezescu AM (eds) Food quality: balancing health and disease. Academic Press, London, p 433
237. Pal D, Mukherjee S (2020) Fenugreek (*Trigonella foenum*) seeds in health and nutrition. In: Preedy VR, Watson RR (eds) Nuts and seeds in health and disease prevention, 2nd edn. Academic Press, London, p 161
238. Gupta RC, Doss RB, Garg RC, Srivastava A, Lall R, Sinha A (2021) Fenugreek: multiple health benefits. In: Gupta RC, Lall R, Srivastava A (eds) Nutraceuticals, 2nd edn. Academic Press, London, p 585
239. Shawky E, Sobhy AA, Ghareeb DA, Eldin SMS, Selim DA (2022) Comparative metabolomics analysis of bioactive constituents of the leaves of different *Trigonella* species: correlation study to α-amylase and α-glycosidase inhibitory effects. Ind Crops Prod 182:114947
240. Song G-Q, Hancock JF (2011) Vaccinium. In: Kole C (ed) Wild crop relatives: genomic and breeding resources: temperate fruits. Springer, Berlin, p 197
241. Brendler T, Howell A (2020) American cranberry (*Vaccinium macrocarpon* Ait.) and the maintenance of urinary tract health. In: Máthé Á (ed) Medicinal and aromatic plants of North America. Springer, Cham, Switzerland, p 81
242. Hurkova K, Uttl L, Rubert J, Navratilova K, Kocourek V, Stranska-Zachariasova M, Paprstein F, Hajslova J (2019) Cranberries versus lingonberries: a challenging authentication of similar *Vaccinium* fruit. Food Chem 284:162
243. Cesonienė L, Daubaras R, Jasutienė I, Vencloviene J, Miliauskienė I (2011) Evaluation of the biochemical components and chromatic properties of the juice of *Vaccinium macrocarpon* Aiton and *Vaccinium oxycoccos* L. Plant Foods Hum Nutr 66:238
244. Brown PN, Turi CE, Shipley PR, Murch SJ (2012) Comparisons of large (*Vaccinium macrocarpon* Ait.) and small (*Vaccinium oxycoccos* L., *Vaccinium vitis-idaea* L.) cranberry in British Columbia by phytochemical determination, antioxidant potential, and metabolomic profiling with chemometric analysis. Planta Med 78:630
245. Lee J, Finn CE (2012) Lingonberry (*Vaccinium vitis-idaea* L.) grown in the Pacific Northwest of North America: anthocyanin and free amino acid composition. J Funct Foods 4:213
246. Krueger CG, Reed JD, Feliciano RP, Howell AB (2013) Quantifying and characterizing proanthocyanidins in cranberries in relation to urinary tract health. Anal Bioanal Chem 405:4385
247. Jungfer E, Zimmermann BF, Ruttkat A, Galensa R (2012) Comparing procyanidins in selected *Vaccinium* species by UHPLC-MS^2 with regard to authenticity and health effects. J Agric Food Chem 60:9688
248. Nemzer BV, Al-Taher F, Yashin A, Revelsky I, Yashin Y (2022) Cranberry: chemical composition, antioxidant activity and impact on human health: overview. Molecules 27:1503
249. Blumberg JB, Camesano TA, Cassidy A, Kris-Etherton P, Howell A, Manach C, Ostertag LM, Sies H, Skulas-Ray A, Vita JA (2013) Cranberries and their bioactive constituents in human health. Adv Nutr 4:618
250. Houghton PJ (1988) The biological activity of Valerian and related plants. J Ethnopharmacol 22:121
251. Houghton PJ (1997) The chemistry of *Valeriana*. In: Houghton PJ (ed) Valerian: the genus *Valeriana*. CRC Press, London, p 21
252. Bos R, Hendriks H, Scheffer JJC, Woerdenbag HJ (1998) Cytotoxic potential of valerian constituents and valerian tinctures. Phytomedicine 5:219
253. Hobbs C (1989) Valerian (*Valeriana officinalis*): a literature review. HerbalGram 21:19
254. Estrada-Soto S, Rivera-Leyva J, Ramírez-Espinosa JJ, Castillo-España P, Aguirre-Crespo F, Hernández-Abreu O (2010) Vasorelaxant effect of *Valeriana edulis* ssp. *procera* (Valerianaceae) and its mode of action as calcium channel blocker. J Pharm Pharmacol 62:1167
255. Gehlot A, Chaudhary N, Devi J, Joshi R, Kumar D, Bhushan S (2022) Induction and submerged cultivation of *Valeriana jatamansi* adventitious root cultures for production of valerenic acids and its derivatives. Plant Cell Tissue Organ Cult 148:347
256. Singh N, Gupta AP, Singh B, Kaul VK (2006) Quantification of valerenic acid in *Valeriana jatamansi* and *Valeriana officinalis* by HPTLC. Chromatographia 63:209

257. Raal A, Arak E, Orav A, Kailas T, Müürisepp M (2007) Variation in the composition of the essential oil of commercial *Valeriana officinalis* L. roots from different countries. J Essent Oil Res 20:524
258. Bhatt ID, Dauthal P, Rawat S, Gaira KS, Jugran A, Rawal RS, Dhar U (2012) Characterization of essential oil composition, phenolic content, and antioxidant properties in wild and planted individuals of *Valeriana jatamansi* Jones. Sci Hortic 136:61
259. Alfaro-Romero A, Balderas-López JL, Duarte-Lisci G, Navarrete A (2016) Root scent composition in *Valeriana officinalis* and *Valeriana edulis* ssp. *procera* analyzed by HS-SPME-GC-MS. J Essent Oil-Bear Plants 19:1821
260. Navarrete A, Avula B, Choi Y-W, Khan IA (2006) Chemical fingerprinting of *Valeriana* species: simultaneous determination of valerenic acids, flavonoids, and phenylpropanoids using liquid chromatography with ultraviolet detection. J AOAC Int 89:8
261. Patočka J, Jakl J (2010) Biomedically relevant chemical constituents of *Valeriana officinalis*. J Appl Biomed 8:11
262. Atta-ur-Rahman, Dur-e-Shahwar, Naz A, Choudhary MI (2003) Withanolides from *Withania coagulans*. Phytochemistry 63:387
263. Girme A, Saste G, Pawar S, Balasubramaniam AK, Musande K, Darji B, Satti NK, Verma MK, Anand R, Singh R, Vishwakarma RA, Hingorani L (2020) Investigating 11 withanosides and withanolides by UHPLC-PDA and mass fragmentation studies from ashwagandha (*Withania somnifera*). ACS Omega 5:27933
264. White PT, Subramanian C, Motiwala HF, Cohen MS (2016) Natural withanolides in the treatment of chronic diseases. In: Gupta SC, Prasad S, Aggarwal BB (eds) Anti-inflammatory nutraceuticals and chronic diseases. Springer, Cham, Switzerland, p 329
265. Xia G-Y, Cao S-J, Chen L-X, Qiu F (2022) Natural withanolides, an update. Nat Prod Rep 39:784
266. Mirjalili MH, Moyano E, Bonfill M, Cusido RM, Palazón J (2009) Steroidal lactones from *Withania somnifera*, an ancient plant for novel medicine. Molecules 14:2373
267. Dhanani T, Shah S, Gajbhiye NA, Kumar S (2017) Effect of extraction methods on yield, phytochemical constituents and antioxidant activity of *Withania somnifera*. Arab J Chem 10:S1193
268. Jain R, Kachhwaha S, Kothari SL (2012) Phytochemistry, pharmacology, and biotechnology of *Withania somnifera* and *Withania coagulans*: a review. J Med Plant Res. 6:5388
269. Dutta R, Khalil R, Green R, Mohapatra SS, Mohapatra S (2019) *Withania somnifera* (ashwagandha) and withaferin A: potential in integrative oncology. Int J Mol Sci 20:5310
270. Ali A, Maher S, Khan SA, Chaudhary MI, Musharraf SG (2015) Sensitive quantification of six steroidal lactones in *Withania coagulans* extract by UHPLC electrospray tandem mass spectrometry. Steroids 104:176
271. Modi SJ, Tiwari A, Ghule C, Pawar S, Saste G, Jagtap S, Singh R, Deshmukh A, Girme A, Hingorani L (2022) Pharmacokinetic study of withanosides and withanolides from *Withania somnifera* using ultra-high performance liquid chromatography-tandem mass spectrometry (UHPLC-MS/MS). Molecules 27:1476
272. Subbaraju GV, Vanisree M, Rao CV, Sivaramakrishna C, Sridhar P, Jayaprakasam B, Nair MG (2006) Ashwagandhanolide, a bioactive dimeric thiowithanolide isolated from the roots of *Withania somnifera*. J Nat Prod 69:1790
273. Khan MI, Maqsood M, Saeed RA, Alam A, Sahar A, Kieliszek M, Miecznikowski A, Muzammil HS, Aadil RM (2021) Phytochemistry, food application, and therapeutic potential of the medicinal plant *Withania coagulans*: a review. Molecules 26:6881
274. Alam N, Hossain M, Khalil MI, Moniruzzaman M, Sulaiman SA, Gan SH (2011) High catechin concentrations detected in *Withania somnifera* (ashwagandha) by high performance liquid chromatography analysis. BMC Complement Altern Med 11:65
275. Filipiak-Szok A, Kurzawa M, Szłyk E, Twarużek M, Błajet-Kosicka A, Grajewski J (2017) Determination of mycotoxins, alkaloids, phytochemicals, antioxidants and cytotoxicity in Asiatic ginseng (Ashwagandha, Dong quai, *Panax ginseng*). Chem Zvesti 71:1073

276. Maqsood M, Qureshi R, Ikram M, Ahmad MS, Jabeen B, Asi MR, Khan JA, Ali S, Lilge L (2018) In vitro anticancer activities of *Withania coagulans* against HeLa, MCF-7, RD, RG2, and INS-1 cancer cells and phytochemical analysis. Integr Med Res 7:184
277. Mundkinajeddu D, Sawant LP, Koshy R, Akunuri P, Singh VK, Mayachari A, Sharaf MH, Balasubramanian M, Agarwal A (2014) Development and validation of high performance liquid chromatography method for simultaneous estimation of flavonoid glycosides in *Withania somnifera* aerial parts. Int Sch Res Notices 2014:351547
278. Singh P, Guleri R, Singh V, Kaur G, Kataria H, Singh B, Kaur G, Kaul SC, Wadhwa R, Pati PK (2015) Biotechnological interventions in *Withania somnifera* (L.) Dunal. Biotechnol Genet Eng Rev 31:1
279. Shukla K, Dikshit P, Shukla R, Gambhir JK (2012) The aqueous extract of *Withania coagulans* fruit partially reverses nicotinamide/streptozotocin-induced diabetes mellitus in rats. J Med Food 15:718
280. Jayaprakasam B, Strasburg GA, Nair MG (2004) Potent lipid peroxidation inhibitors from *Withania somnifera* fruits. Tetrahedron 60:3109
281. Patil S, Shinde RD, Leong-Škorničková J, Chaudhari R (2021) Fixing stray traditions in gingers: the identity and nomenclatural history of *Zingiber neesanum* and other entwined names. Taxon 70:1339
282. Khan S, Pandotra P, Qazi AK, Lone SA, Muzafar M, Gupta AP, Gupta S (2016) Medicinal and nutritional qualities of *Zingiber officinale*. In: Watson RR, Preedy VR (eds) Fruits, vegetables, and herbs. Academic Press, London, p 525
283. Mbaveng AT, Kuete V (2017) *Zingiber officinale*. In: Kuete V (ed) Medicinal spices and vegetables from Africa. Academic Press, London, p 627
284. Semwal RB, Semwal DK, Combrinck S, Viljoen AM (2015) Gingerols and shogaols: Important nutraceutical principles from ginger. Phytochemistry 117:554
285. Mao Q-Q, Xu X-Y, Cao S-Y, Gan R-Y, Corke H, Beta T, Li H-B (2019) Bioactive compounds and bioactivities of ginger (*Zingiber officinale* Roscoe). Foods 8:185
286. Baliga MS, Shivashankara AR, Haniadka R, Palatty PL, Arora R, Fayad R (2013) Ginger (*Zingiber officinale* Roscoe): an ancient remedy and modern drug in gastrointestinal disorders. In: Watson RR, Preedy VR (eds) Bioactive food as dietary interventions for liver and gastrointestinal disease. Academic Press, San Diego, p 187
287. Wuttke W, Jarry H, Haunschild J, Stecher G, Schuh M, Seidlova-Wuttke D (2014) The non-estrogenic alternative for the treatment of climacteric complaints: black cohosh (*Cimicifuga* or *Actaea racemosa*). J Steroid Biochem Mol Biol 139:302
288. Einbond LS, Shimizu M, Xiao D, Nuntanakorn P, Lim JTE, Suzui M, Seter C, Pertel T, Kennelly EJ, Kronenberg F, Weinstein IB (2004) Growth inhibitory activity of extracts and purified components of black cohosh on human breast cancer cells. Breast Cancer Res Treat 83:221
289. Yue GG-L, Xie S, Lee JK-M, Kwok H-F, Gao S, Nian Y, Wu X-X, Wong C-K, Qiu M-H, Lau CB-S (2016) New potential beneficial effects of actein, a triterpene glycoside isolated from *Cimicifuga* species, in breast cancer treatment. Sci Rep 6:35263
290. Düker E-M, Kopanski L, Jarry H, Wuttke W (1991) Effects of extracts from *Cimicifuga racemosa* on gonadotropin release in menopausal women and ovariectomized rats. Planta Med 57:420
291. Mathew B, Biju RS (2008) Neuroprotective effects of garlic: a review. Libyan J Med 3:23
292. Gull I, Saeed M, Shaukat H, Aslam SM, Samra ZQ, Athar AM (2012) Inhibitory effect of *Allium sativum* and *Zingiber officinale* extracts on clinically important drug resistant pathogenic bacteria. Ann Clin Microbiol Antimicrob 11:8
293. Matsuura H, Ushiroguchi T, Itakura Y, Hayashi N, Fuwa T (1988) A furostanol glycoside from garlic, bulbs of *Allium sativum* L. Chem Pharm Bull 36:3659
294. Lanzotti V, Barile E, Antignani V, Bonanomi G, Scala F (2012) Antifungal saponins from bulbs of garlic, *Allium sativum* L. var. *voghiera*. Phytochemistry 78:126
295. Gebhardt R, Beck H (1996) Differential inhibitory effects of garlic-derived organosulfur compounds on cholesterol biosynthesis in primary rat hepatocyte cultures. Lipids 31:1269

296. Sobenin IA, Nedosugova LV, Filatova LV, Balabolkin MI, Gorchakova TV, Orekhov AN (2008) Metabolic effects of time-released garlic powder tablets in type 2 diabetes mellitus: the results of double-blinded placebo-controlled study. Acta Diabetol 45:1
297. Sobenin IA, Pryanishnikov VV, Kunnova LM, Rabinovich YA, Martirosyan DM, Orekhov AN (2010) The effects of time-released garlic powder tablets on multifunctional cardiovascular risk in patients with coronary artery disease. Lipids Health Dis 9:119
298. Ţigu AB, Moldovan CS, Toma V-A, Farcaş AD, Moţ AC, Jurj A, Fischer-Fodor E, Mircea C, Pârvu M (2021) Phytochemical analysis and in vitro effects of *Allium fistulosum* L. and *Allium sativum* L. extracts on human normal and tumor cell lines: a comparative study. Molecules 26:574
299. Jung Y, Park H, Zhao HY, Jeon R, Ryu JH, Kim WY (2014) Systemic approaches identify a garlic-derived chemical, *Z*-ajoene, as a glioblastoma multiforme cancer stem cell-specific targeting agent. Mol Cells 37:547
300. Womble D, Helderman JH (1988) Enhancement of allo-resposiveness of human lymphocytes by acemannan (Carrisyn^{TM}). Int J Immunopharmacol 10:967
301. Im S-A, Oh S-T, Song S, Kim M-R, Kim D-S, Woo S-S, Jo TH, Park YI, Lee C-K (2005) Identification of optimal molecular size of modified *Aloe* polysaccharides with maximum immunomodulatory activity. Int Immunopharmacol 5:271
302. Quezada MP, Salinas C, Gotteland M, Cardemil L (2017) Acemannan and fructans from *Aloe vera* (*Aloe barbadensis* Miller) plants as novel prebiotics. J Agric Food Chem 65:10029
303. Purwar R (2019) Antimicrobial textiles. In: Shahid-ul I, Butola BS (eds) The impact and prospects of green chemistry for textile technology. Woodhead Publishing, Cambridge, MA, p 281
304. Salah F, Ghoul YE, Mahdhi A, Majdoub H, Jarroux N, Sakli F (2017) Effect of the deacetylation degree on the antibacterial and antibiofilm activity of acemannan from *Aloe vera*. Ind Crops Prod 103:13
305. Ali SW, Purwar R, Joshi M, Rajendran S (2014) Antibacterial properties of *Aloe vera* gel-finished cotton fabric. Cellulose 21:2063
306. Yates KM, Rosenberg LJ, Harris CK, Bronstad DC, King GK, Biehle GA, Walker B, Ford CR, Hall JE, Tizard IR (1992) Pilot study of the effect of acemannan in cats infected with feline immunodeficiency virus. Vet Immunol Immunopathol 35:177
307. Huseini HF, Kianbakht S, Hajiaghaee R, Dabaghian FH (2012) Anti-hyperglycemic and anti-hypercholesterolemic effects of *Aloe vera* leaf gel in hyperlipidemic type 2 diabetic patients: a randomized double-blind placebo-controlled clinical trial. Planta Med 78:311
308. Bunyapraphatsara N, Yongchaiyudha S, Rungpitarangsi V, Chokechaijaroenporn O (1996) Antidiabetic activity of *Aloe vera* L. juice II. Clinical trial in diabetes mellitus patients in combination with glibenclamide. Phytomedicine 3:245
309. Mesripour A, Moghimi F, Rafieian-Kopaie M (2016) The effect of *Cinnamomum zeylanicum* bark water extract on memory performance in alloxan-induced diabetic mice. Res Pharm Sci 11:318
310. Wariyapperuma WANM, Kannangara S, Wijayasinghe YS, Subramanium S, Jayawardena B (2020) In vitro anti-diabetic effects and phytochemical profiling of novel varieties of *Cinnamomum zeylanicum* (L.) extracts. PeerJ 8:e10070
311. Witkamp R (2010) Biologically active compounds in food products and their effects on obesity and diabetes. In: Liu HW, Mander L (eds) Comprehensive natural products II: chemistry and biology. Elsevier Science, London, p 509
312. Ono E, Inoue J, Hashidume T, Shimizu M, Sato R (2011) Anti-obesity and anti-hyperglycemic effects of the dietary citrus limonoid nomilin in mice fed a high-fat diet. Biochem Biophys Res Commun 410:677
313. Stohs SJ, Preuss HG, Keith SC, Keith PL, Miller H, Kaats GR (2011) Effects of *p*-synephrine alone and in combination with selected bioflavonoids on resting metabolism, blood pressure, heart rate and self-reported mood changes. Int J Med Sci 8:295
314. Haller C, Duan M, Jacob P, Benowitz N (2008) Human pharmacology of a performance-enhancing dietary supplement under resting and exercise conditions. Br J Clin Pharmacol 65:833

315. Ruiz-Moreno C, Del Coso J, Giráldez-Costas V, González-García J, Gutiérrez-Hellín J (2021) Effects of *p*-synephrine during exercise: a brief narrative review. Nutrients 13:233
316. Deshmukh N, Stohs S, Magar C, Kale A, Sowmya B (2017) Bitter orange (*Citrus aurantium* L.) extract subchronic 90-day safety study in rats. Toxicol Rep 4:598
317. Shara M, Stohs SJ, Smadi MM (2018) Safety evaluation of *p*-synephrine following 15 days of oral administration to healthy subjects: a clinical study. Phytother Res 32:125
318. Zorniak M, Szydlo B, Krzeminski TF (2017) *Crataegus* special extract WS 1442: up-to-date review of experimental and clinical experiences. J Physiol Pharmacol 68:521
319. Tassell MC, Kingston R, Gilroy D, Lehane M, Furey A (2010) Hawthorn (*Crataegus* spp.) in the treatment of cardiovascular disease. Pharmacogn Rev 4:32
320. Al Makdessi S, Sweidan H, Dietz K, Jacob R (1999) Protective effect of *Crataegus oxyacantha* against reperfusion arrhythmias after global no-flow ischemia in the rat heart. Basic Res Cardiol 94:71
321. Brixius K, Willms S, Napp A, Tossios P, Ladage D, Bloch W, Mehlhorn U, Schwinger RHG (2006) *Crataegus* special extract WS®1442 induces an endothelium-dependent, no-mediated vasorelaxation via enos-phosphorylation at serine 1177. Cardiovasc Drugs Ther 20:177
322. Mackenzie GG, Carrasquedo F, Delfino JM, Keen CL, Fraga CG, Oteiza PI (2003) Epicatechin, catechin, and dimeric procyanidins inhibit PMA-induced NF-kappaB activation at multiple steps in Jurkat T cells. FASEB J 18:167
323. Jayanarayanan S, Smijin S, Peeyush KT, Anju TR, Paulose CS (2013) NMDA and AMPA receptor mediated excitotoxicity in cerebral cortex of streptozotocin induced diabetic rat: ameliorating effects of curcumin. Chem Biol Interact 201:39
324. Kuroda M, Mimaki Y, Nishiyama T, Mae T, Kishida H, Tsukagawa M, Takahashi K, Kawada T, Nakagawa K, Kitahara M (2005) Hypoglycemic effects of turmeric (*Curcuma longa* L. rhizomes) on genetically diabetic KK-Ay mice. Biol Pharm Bull 28:937
325. Holubarsch CJ, Colucci WS, Meinertz T, Gaus W, Tendera M (2008) The efficacy and safety of *Crataegus* extract WS® 1442 in patients with heart failure: the SPICE trial. Eur J Heart Fail 10:1255
326. Zhang Z, Ho WKK, Huang Y, James AE, Wang Lam L, Chen Z-Y (2002) Hawthorn fruit is hypolipidemic in rabbits fed a high cholesterol diet. J Nutr 132:5
327. Yee WL, Wang Q, Agdinaoay T, Ko D, Chang H, Grandinetti A, Franke AA, Theriault A (2002) Green tea catechins decrease apolipoprotein B-100 secretion from HepG2 cells. Mol Cell Biochem 229:85
328. Montagut G, Baiges I, Valls J, Terra X, del Bas JM, Vitrac X, Richard T, Mérillon J-M, Arola L, Blay M, Bladé C, Fernández-Larrea J, Pujadas G, Salvadó J, Ardévol A (2009) A trimer plus a dimer-gallate reproduce the bioactivity described for an extract of grape seed procyanidins. Food Chem 116:265
329. Kim J, Lee I, Seo J, Jung M, Kim Y, Yim N, Bae K (2010) Vitexin, orientin and other flavonoids from *Spirodela polyrhiza* inhibit adipogenesis in 3T3-L1 cells. Phytother Res 24:1543
330. Elango C, Devaraj SN (2010) Immunomodulatory effect of hawthorn extract in an experimental stroke model. J Neuroinflammation 7:97
331. Borghi SM, Carvalho TT, Staurengo-Ferrari L, Hohmann MS, Pinge-Filho P, Casagrande R, Verri WA Jr (2013) Vitexin inhibits inflammatory pain in mice by targeting TRPV1, oxidative stress, and cytokines. J Nat Prod 76:1141
332. Zhang LJ, Wu CF, Meng XL, Yuan D, Cai XD, Wang QL, Yang JY (2008) Comparison of inhibitory potency of three different curcuminoid pigments on nitric oxide and tumor necrosis factor production of rat primary microglia induced by lipopolysaccharide. Neurosci Lett 447:48
333. Chen F, Wang H, Xiang X, Yuan J, Chu W, Xue X, Zhu H, Ge H, Zou M, Feng H, Lin J (2014) Curcumin increased the differentiation rate of neurons in neural stem cells via wnt signaling in vitro study. J Surg Res 192:298
334. Zingg J-M, Hasan ST, Cowan D, Ricciarelli R, Azzi A, Meydani M (2012) Regulatory effects of curcumin on lipid accumulation in monocytes/macrophages. J Cell Biochem 113:833

335. Zhao J-F, Ching L-C, Huang Y-C, Chen C-Y, Chiang A-N, Kou YR, Shyue S-K, Lee T-S (2012) Molecular mechanism of curcumin on the suppression of cholesterol accumulation in macrophage foam cells and atherosclerosis. Mol Nutr Food Res 56:691
336. Kim YS, Kwon JS, Cho YK, Jeong MH, Cho JG, Park JC, Kang JC, Ahn Y (2012) Curcumin reduces the cardiac ischemia-reperfusion injury: involvement of the toll-like receptor 2 in cardiomyocytes. J Nutr Biochem 23:1514
337. Yodkeeree S, Chaiwangyen W, Garbisa S, Limtrakul P (2009) Curcumin, demethoxycurcumin, and bisdemethoxycurcumin differentially inhibit cancer cell invasion through the down-regulation of MMPs and uPA. J Nutr Biochem 20:87
338. Simon A, Allais DP, Duroux JP, Basly JP, Durand-Fontanier S, Delage C (1998) Inhibitory effect of curcuminoids on MCF-7 cell proliferation and structure-activity relationships. Cancer Lett 129:111
339. Lin H-Y, Lin J-N, Ma J-W, Yang N-S, Ho C-T, Kuo S-C, Way T-D (2015) Demethoxycurcumin induces autophagic and apoptotic responses on breast cancer cells in photodynamic therapy. J Funct Foods 12:439
340. Prasad CP, Rath G, Mathur S, Bhatnagar D, Ralhan R (2010) Expression analysis of maspin in invasive ductal carcinoma of breast and modulation of its expression by curcumin in breast cancer cell lines. Chem Biol Interact 183:455
341. Nelson KM, Dahlin JL, Bisson J, Graham J, Pauli GF, Walters MA (2017) The essential medicinal chemistry of curcumin. J Med Chem 60:1620
342. Aucoin M, Cooley K, Saunders PR, Carè J, Anheyer D, Medina DN, Cardozo V, Remy D, Hannan N, Garber A (2020) The effect of *Echinacea* spp. on the prevention or treatment of COVID-19 and other respiratory tract infections in humans: a rapid review. Adv Integr Med 7:203
343. Woelkart K, Bauer R (2007) The role of alkamides as an active principle of *Echinacea.* Planta Med 73:615
344. Sharifi-Rad M, Mnayer D, Morais-Braga MFB, Carneiro JNP, Bezerra CF, Coutinho HDM, Salehi B, Martorell M, Del Mar CM, Soltani-Nejad A, Uribe YAH, Yousaf Z, Iriti M, Sharifi-Rad J (2018) *Echinacea* plants as antioxidant and antibacterial agents: from traditional medicine to biotechnological applications. Phytother Res 32:1653
345. Fu R, Zhang P, Deng Z, Jin G, Guo Y, Zhang Y (2021) Diversity of antioxidant ingredients among *Echinacea* species. Ind Crops Prod 170:113699
346. Kim H, Calderón AI (2022) Rational and safe use of the top two botanical dietary supplements to enhance the immune system. Comb Chem High Throughput Screen 25:1129
347. Manayi A, Vazirian M, Saeidnia S (2015) *Echinacea purpurea*: pharmacology, phytochemistry and analysis methods. Pharmacogn Rev 9:63
348. Woelkart K, Xu W, Pei Y, Makriyannis A, Picone RP, Bauer R (2005) The endocannabinoid system as a target for alkamides from *Echinacea angustifolia* roots. Planta Med 71:701
349. Raduner S, Majewska A, Chen JZ, Xie XQ, Hamon J, Faller B, Altmann KH, Gertsch J (2006) Alkylamides from *Echinacea* are a new class of cannabinomimetics. Cannabinoid type 2 receptor-dependent and -independent immunomodulatory effects. J Biol Chem 281:14192
350. Chen H, Jing FC, Li CL, Tu PF, Zheng QS, Wang ZH (2007) Echinacoside prevents the striatal extracellular levels of monoamine neurotransmitters from diminution in 6-hydroxydopamine lesion rats. J Ethnopharmacol 114:285
351. Burlou-Nagy C, Bănică F, Jurca T, Vicaș LG, Marian E, Muresan ME, Bácskay I, Kiss R, Fehér P, Pallag A (2022) *Echinacea purpurea* (L.) Moench: biological and pharmacological properties. A review. Plants 11:1244
352. Li F, Yang Y, Zhu P, Chen W, Qi D, Shi X, Zhang C, Yang Z, Li P (2012) Echinacoside promotes bone regeneration by increasing OPG/RANKL ratio in MC3T3-E1 cells. Fitoterapia 83:1443
353. Wang W, Luo J, Liang Y, Li X (2016) Echinacoside suppresses pancreatic adenocarcinoma cell growth by inducing apoptosis via the mitogen-activated protein kinase pathway. Mol Med Rep 13:2613
354. Zhu D, Zhang N, Zhou X, Zhang M, Liu Z, Liu X (2017) Cichoric acid regulates the hepatic glucose homeostasis via AMPK pathway and activates the antioxidant response in high glucose-induced hepatocyte injury. RSC Adv 7:1363

355. McDougall B, King PJ, Wu BW, Hostomsky Z, Reinecke MG, Robinson WE (1998) Dicaffeoylquinic and dicaffeoyltartaric acids are selective inhibitors of human immunodeficiency virus type 1 integrase. Antimicrob Agents Chemother 42:140
356. Wang Z, Wang D, Yang D, Zhen W, Zhang J, Peng S (2018) The effect of icariin on bone metabolism and its potential clinical application. Osteoporos Int 29:535
357. Gao J, Xiang S, Wei X, Yadav RI, Han M, Zheng W, Zhao L, Shi Y, Cao Y (2021) Icariin promotes the osteogenesis of bone marrow mesenchymal stem cells through regulating sclerostin and activating the wnt/β-catenin signaling pathway. Biomed Res Int 2021:6666836
358. Li C, Li Q, Mei Q, Lu T (2015) Pharmacological effects and pharmacokinetic properties of icariin, the major bioactive component in Herba Epimedii. Life Sci 126:57
359. Jin J, Wang H, Hua X, Chen D, Huang C, Chen Z (2019) An outline for the pharmacological effect of icariin in the nervous system. Eur J Pharmacol 842:20
360. Bi Z, Zhang W, Yan X (2022) Anti-inflammatory and immunoregulatory effects of icariin and icaritin. Biomed Pharmacother 151:113180
361. Wang Y, Zhu T, Wang M, Zhang F, Zhang G, Zhao J, Zhang Y, Wu E, Li X (2019) Icariin attenuates M1 activation of microglia and Aβ plaque accumulation in the hippocampus and prefrontal cortex by up-regulating PPARγ in restraint/isolation-stressed APP/PS1 mice. Front Neurosci 13:291
362. Zhang X, Kang Z, Li Q, Zhang J, Cheng S, Chang H, Wang S, Cao S, Li T, Li J, Wang Y, Song Y, Yu H (2018) Antigen-adjuvant effects of icariin in enhancing tumor-specific immunity in mastocytoma-bearing DBA/2J mice. Biomed Pharmacother 99:810
363. de Oliveira PRB, da Costa CA, de Bem GF, Cordeiro VS, Santos IB, de Carvalho LC, da Conceição EP, Lisboa PC, Ognibene DT, Sousa PJ, Martins GR, da Silva AJ, de Moura RS, Resende AC (2015) *Euterpe oleracea* Mart.-derived polyphenols protect mice from diet-induced obesity and fatty liver by regulating hepatic lipogenesis and cholesterol excretion. PLoS One 10:e0143721
364. Gordon A, Cruz APG, Cabral LMC, de Freitas SC, Taxi CMAD, Donangelo CM, de Andrade Mattietto R, Friedrich M, da Matta VM, Marx F (2012) Chemical characterization and evaluation of antioxidant properties of açaí fruits (*Euterpe oleracea* Mart.) during ripening. Food Chem 133:256
365. Sadowska-Krępa E, Kłapcińska B, Podgórski T, Szade B, Tyl K, Hadzik A (2015) Effects of supplementation with açaí (*Euterpe oleracea* Mart.) berry-based juice blend on the blood antioxidant defence capacity and lipid profile in junior hurdlers. A pilot study. Biol Sport 32:161
366. Martino HSD, dos Santos Dias MM, Noratto G, Talcott S, Mertens-Talcott SU (2016) Anti-lipidaemic and anti-inflammatory effect of açaí (*Euterpe oleracea* Martius) polyphenols on 3T3-L1 adipocytes. J Funct Foods 23:432
367. Oliveira de Souza M, Souza E Silva L, Lopes de Brito Magalhães C, Baroos de Figueiredo B, Costa DC, Silva ME, Pedrosa ML (2012) The hypocholesterolemic activity of açaí (*Euterpe oleracea* Mart.) is mediated by the enhanced expression of the ATP-binding cassette, subfamily G transporters 5 and 8 and low-density lipoprotein receptor genes in the rat. Nutr Res 32:976
368. Tsuda T, Ueno Y, Yoshikawa T, Kojo H, Osawa T (2006) Microarray profiling of gene expression in human adipocytes in response to anthocyanins. Biochem Pharmacol 71:1184
369. Thilavech T, Adisakwattana S (2019) Cyanidin-3-rutinoside acts as a natural inhibitor of intestinal lipid digestion and absorption. BMC Complement Altern Med 19:242
370. Choppa T, Selvaraj CI, Zachariah A (2015) Evaluation and characterization of Malabar tamarind [*Garcinia cambogia* (Gaertn.) Desr.] seed oil. J Food Sci Technol 52:5906
371. Roy S, Rink C, Khanna S, Phillips C, Bagchi D, Bagchi M, Sen CK (2004) Body weight and abdominal fat gene expression profile in response to a novel hydroxycitric acid-based dietary supplement. Gene Expr 11:251
372. Mazzio EA, Soliman KF (2009) In vitro screening for the tumoricidal properties of international medicinal herbs. Phytother Res 23:385

373. Pan MH, Chang WL, Lin-Shiau SY, Ho CT, Lin JK (2001) Induction of apoptosis by garcinol and curcumin through cytochrome c release and activation of caspases in human leukemia HL-60 cells. J Agric Food Chem 49:1464
374. Di Micco S, Masullo M, Bandak AF, Berger JM, Riccio R, Piacente S, Bifulco G (2019) Garcinol and related polyisoprenylated benzophenones as topoisomerase II inhibitors: biochemical and molecular modeling studies. J Nat Prod 82:2768
375. Venkateswara Rao G, Karunakara AC, Santhosh Babu RR, Ranjit D, Chandrasekara Reddy G (2010) Hydroxycitric acid lactone and its salts: preparation and appetite suppression studies. Food Chem 120:235
376. Anton SD, Shuster J, Leeuwenburgh C (2011) Investigations of botanicals on food intake, satiety, weight loss and oxidative stress: study protocol of a double-blind, placebo-controlled, crossover study. Chin J Integr Med 9:1190
377. Xu SL, Choi RCY, Zhu KY, Leung K-W, Guo AJ, Bi D, Xu H, Lau DTW, Dong TTX, Tsim KWK (2012) Isorhamnetin, a flavonol aglycone from *Ginkgo biloba* L., induces neuronal differentiation of cultured PC12 cells: potentiating the effect of nerve growth factor. Evid Based Complement Altern Med 2012:278273
378. Chen T-R, Wei L-H, Guan X-Q, Huang C, Liu Z-Y, Wang F-J, Hou J, Jin Q, Liu Y-F, Wen P-H, Zhang S-J, Ge G-B, Guo W-Z (2019) Biflavones from *Ginkgo biloba* as inhibitors of human thrombin. Bioorg Chem 92:103199
379. Cho Y-L, Park J-G, Kang HJ, Kim W, Cho MJ, Jang J-H, Kwon M-G, Kim S, Lee S-H, Lee J, Kim Y-G, Park Y-J, Kim WK, Bae K-H, Kwon B-M, Chung SJ, Min J-K (2019) Ginkgetin, a biflavone from *Ginkgo biloba* leaves, prevents adipogenesis through STAT5-mediated PPARγ and C/EBPα regulation. Pharmacol Res 139:325
380. Liu P-K, Weng Z-M, Ge G-B, Li H-L, Ding L-L, Dai Z-R, Hou X-D, Leng Y-H, Yu Y, Hou J (2018) Biflavones from *Ginkgo biloba* as novel pancreatic lipase inhibitors: inhibition potentials and mechanism. Int J Biol Macromol 118:2216
381. Song Y-Q, He R-J, Pu D, Guan X-Q, Shi J-H, Li Y-G, Hou J, Jia S-N, Qin W-W, Fang S-Q, Ge G-B (2021) Discovery and characterization of the biflavones from *Ginkgo biloba* as highly specific and potent inhihibitors against human carboxylesterase 2. Front Pharmacol 12:655659
382. Bate C, Tayebi M, Williams A (2008) Ginkgolides protect against amyloid-$\beta_{1\text{-}42}$-mediated synapse damage in vitro. Mol Neurodegener 3:1
383. Wang T, Bai S, Wang W, Chen Z, Chen J, Liang Z, Qi X, Shen H, Xie P (2020) Diterpene ginkgolides exert an antidepressant effect through the NT3-TrkA and Ras-MAPK pathways. Drug Des Devel Ther 14:1279
384. Chatterjee SS, Kondratskaya EL, Krishtal OA (2003) Structure-activity studies with *Ginkgo biloba* extract constituents as receptor-gated chloride channel blockers and modulators. Pharmacopsychiatry 36:68
385. Chandrasekran K, Mehrabian Z, Spinnewyn B, Chinopoulos C, Drieu K, Fiskum G (2002) Bilobalide, a component of the *Ginkgo biloba* extract (EGb 761), protects against neuronal death in global brain ischemia and in glutamate-induced excitotoxicity. Cell Mol Biol 48:663
386. Brüggemann P, Sória MG, Brandes-Schramm J, Mazurek B (2021) The influence of depression, anxiety, and cognition on the treatment effects of *Ginkgo biloba* extract EGb 761® in patients with tinnitus and dementia: a mediation analysis. J Clin Med 10:3151
387. Lemon JA, Boreham DR, Rollo CD (2003) A dietary supplement abolishes age-related cognitive decline in transgenic mice expressing elevated free radical processes. Exp Biol Med 228:800
388. Christen Y (2004) *Ginkgo biloba* and neurogenerative disorders. Front Biosci 9:3091
389. Rai GS, Shovlin C, Wesnes KA (1991) A double-blind, placebo controlled study of *Ginkgo biloba* extract ('Tanakan') in elderly outpatients with mild to moderate memory impairment. Curr Med Res Opin 12:350
390. Laakmann G, Dienel A, Kieser M (1998) Clinical significance of hyperforin for the efficacy of *Hypericum* extracts on depressive disorders of different severities. Phytomedicine 5:435
391. Tanaka N, Takaishi Y, Shikishima Y, Nakanishi Y, Bastow K, Lee K-H, Honda G, Ito M, Takeda Y, Kodzhimatov OK, Ashurmetov O (2004) Prenylated benzophenones and xanthones from *Hypericum scabrum*. J Nat Prod 67:1870

392. Winkelmann K, Heilmann J, Zerbe O, Rali T, Sticher O (2001) Further prenylated bi- and tricyclic phloroglucinol derivatives from *Hypericum papuanum*. Helv Chim Acta 84:3380
393. Hashida W, Tanaka N, Kashiwada Y, Sekiya M, Ikeshiro Y, Takaishi Y (2008) Tomoenones A-H, cytotoxic phloroglucinol derivatives from *Hypericum ascyron*. Phytochemistry 69:2225
394. Chen X-Q, Li Y, Li K-Z, Peng L-Y, He J, Wang K, Pan Z-H, Cheng X, Li M-M, Zhao Q-S, Xu G (2011) Spirocyclic acylphloroglucinol derivatives from *Hypericum beanii*. Chem Pharm Bull 59:1250
395. Winkelmann K, San M, Kypriotakis Z, Skaltsa H, Bosilij B, Heilmann J (2003) Antibacterial and cytotoxic activity of prenylated bicyclic acylphloroglucinol derivatives from *Hypericum amblycalyx*. Z Naturforsch C J Biosci 58:527
396. Winkelmann K, Heilmann J, Zerbe O, Rali T, Sticher O (2000) New phloroglucinol derviatives from *Hypericum papuanum*. J Nat Prod 63:104
397. Rocha L, Marston A, Potterat O, Kaplan MAC, Hostettmann K (1996) More phloroglucinols from *Hypericum brasiliense*. Phytochemistry 42:185
398. Mamemura T, Tanaka N, Shibazaki A, Gonoi T, Kobayashi J (2011) Yojironins A-D meroterpenoids and prenylated acylphloroglucinols from *Hypericum yojiroanum*. Tetrahedron Lett 52:3575
399. Athanasas K, Magiatis P, Fokialakis N, Skaltsounis A-L, Pratsinis H, Kletsas D (2004) Hyperjovinols A and B: two new phloroglucinol derivatives from *Hypericum jovis* with antioxidant activity in cell cultures. J Nat Prod 67:973
400. Bernardi APM, Ferraz ABF, Albring DV, Bordignon SAL, Schripsema J, Bridi R, Dutra-Filho CS, Henriques AT, Lino von Poser G (2005) Benzophenones from *Hypericum carinatum*. J Nat Prod 68:784
401. Philipp M, Linde K, Kohnen R, Hiller K-O, Berner M (1999) *Hypericum* extract versus imipramine or placebo in patients with moderate depression: randomised multicentre study of treatment for eight weeks. Br Med J 319:1534
402. Shim YY, Gui B, Arnison PG, Wang Y, Reaney MJT (2014) Flaxseed (*Linum usitatissimum* L.) bioactive compounds and peptide nomenclature: a review. Trends Food Sci Technol 38:5
403. Wu H, Kelley CJ, Pino-Figueroa A, Vu HD, Maher TJ (2013) Macamides and their synthetic analogs: evalutation of in vitro FAAH inhibition. Bioorg Med Chem 21:5188
404. Hajdu Z, Nicolussi S, Rau M, Lorántfy L, Forgo P, Hohmann J, Csupor D, Gertsch J (2014) Identification of endocannabinoid system-modulating *N*-alkylamides from *Heliopsis helianthoides* var. *scabra* and *Lepidium meyenii*. J Nat Prod 77:1663
405. Zhou M, Ma H-Y, Liu Z-H, Yang G-Y, Du G, Ye Y-Q, Li G-P, Hu Q-F (2017) (+)-Meyeniins A-C, novel hexahydroimidazo[1,5-*c*]thiazole derivatives from the tubers of *Lepidium meyenii*: complete structural elucidation by biomimetic synthesis and racemic crystallization. J Agric Food Chem 65:1887
406. Cui B, Zheng BL, He K, Zheng QY (2003) Imidazole alkaloids from *Lepidium meyenii*. J Nat Prod 66:1101
407. Gonzales GF, Vasquez V, Rodriguez D, Maldonado C, Mormontoy J, Portella J, Pajuelo M, Villegas L, Gasco M (2007) Effect of two different extracts of red maca in male rats with testosterone-induced prostatic hyperplasia. Asian J Androl 9:245
408. Pino-Figeroa A, Nguyen D, Maher TJ (2010) Neuroprotective effects of *Lepidium meyenii* (Maca). Ann NY Acad Sci 1199:77
409. Imran M, Ahmad N, Anjum FM, Khan MK, Mushtaq Z, Nadeem M, Hussain S (2015) Potential protective properties of flax lignan secoisolariciresinol diglucoside. Nutr J 14:71
410. Rom S, Zuluaga-Ramirez V, Reichenbach NL, Erickson MA, Winfield M, Gajghate S, Christofidou-Solomidou M, Jordan-Sciutto KL, Persidsky Y (2018) Secoisolariciresinol diglucoside is a blood-brain barrier protective and anti-inflammatory agent: implications for neuroinflammation. J Neuroinflammation 15:25
411. Kaithwas G, Majumdar DK (2010) Evaluation of antiulcer and antisecretory potential of *Linum usitatissimum* fixed oil and possible mechanism of action. Inflammopharmacology 18:137

412. Kezimana P, Dmitriev AA, Kudryavtseva AV, Romanova EV, Melnikova NV (2018) Secoisolariciresinol diglucoside of flaxseed and its metabolites: biosynthesis and potential for nutraceuticals. Front Genet 9:641
413. Fabian CJ, Kimler BF, Zalles CM, Klemp JR, Petroff BK, Khan QJ, Sharma P, Setchell KDR, Zhao X, Phillips TA, Metheny T, Hughes JR, Yeh H-W, Johnson KA (2010) Reduction in Ki-67 in benign breast tissue of high-risk women with the lignan secoisolariciresinol diglycoside. Cancer Prev Res 3:1342
414. Zhang W, Wang X, Liu Y, Tian H, Flickinger B, Empie MW, Sun SZ (2008) Dietary flaxseed lignan extract lowers plasma cholesterol and glucose concentrations in hypercholesterolaemic subjects. Br J Nutr 99:1301
415. Li X-M (2007) Protective effect of *Lycium barbarum* polysaccharides on streptozotocin-induced oxidative stress in rats. Int J Biol Macromol 40:461
416. Niu A-J, Wu J-M, Yu D-H, Wang R (2008) Protective effect of *Lycium barbarum* polysaccharides on oxidative damage in skeletal muscle of exhaustive exercise rats. Int J Biol Macromol 42:447
417. Zhang XR, Zhou WX, Zhang YX, Qi CH, Yan H, Wang ZF, Wang B (2011) Macrophages, rather than T and B cells are principal immunostimulatory target cells of *Lycium barbarum* L. polysaccharide LBPF4-OL. J Ethnopharmacol 136:465
418. Zhang X-R, Qi C-H, Cheng J-P, Liu G, Huang L-J, Wang Z-F, Zhou W-X, Zhang Y-X (2014) *Lycium barbarum* polysaccharide LBPF4-OL may be a new toll-like receptor 4/MD2-MAPK signaling pathway activator and inducer. Int Immunopharmacol 19:132
419. Shen L, Du G (2012) *Lycium barbarum* polysaccharide stimulates proliferation of MCF-7 cells by the ERK pathway. Life Sci 91:353
420. Zhu C-P, Zhang S-H (2012) *Lycium barbarum* polysaccharide inhibits the proliferation of HeLa cells by inducing apoptosis. J Sci Food Agric 93:149
421. Luo Q, Li Z, Yan J, Zhu F, Xu R-J, Cai Y-Z (2009) *Lycium barbarum* polysaccharides induce apoptosis in human prostate cancer cells and inhibits prostate cancer growth in a xenograft mouse model of human prostate cancer. J Med Food 12:695
422. Chen W, Cheng X, Chen J, Yi X, Nie D, Sun X, Qin J, Tian M, Jin G, Zhang X (2014) *Lycium barbarum* polysaccharides prevent memory and neurogenesis impairments in scopolamine-treated rats. PLoS One 9:e88076
423. Paula de Oliviera A, Santin JR, Lemos M, Klein Júnior LC, García Couto A, Meyre da Silva Bittencourt C, Cechinel Filho V, Faloni de Andrade S (2011) Gastroprotective activity of methanol extract and marrubiin obtained from leaves of *Marrubium vulgare* L. (Lamiaceae). J Pharm Pharmacol 63:1230
424. Rhallab Said Chakir A, Elbadaoui K, Imolek Alaoui T (2015) Antidiabetic activities of methanolic extracts of *Marrubium vulgare* leaves in rats. Int J Pharm Phytopharmacol Res 4:258
425. Elmhdwi MF (2014) Hypoglycemic effects of *Marrubium vulgare* (Rubia) in experimentally induced autoimmune diabetes mellitus. Int Res J Biochem Bioinform 4:42
426. Singh O, Khanam Z, Misra N, Srivastava MK (2011) Chamomile (*Matricaria chamomilla* L.): an overview. Pharmacogn Rev 5:82
427. Hajjaj G, Bounihi A, Tajani M, Cherrah Y, Zellou A (2014) In vivo analgesic activity of essential oil and aqueous extract of *Matricaria chamomilla* L. (Asteraceae). World J Pharm Pharm Sci 3:1
428. Della Loggia R, Carle R, Sosa S, Tubaro A (1990) Evaluation of the anti-inflammatory activity of chamomile preparations. Planta Med 56:657
429. Rocha NFM, Rios ERV, Carvalho AMR, Cerqueira GS, Lopes ADA, Leal LKAM, Dias ML, Pergentino de Sousa D, Florenço de Sousa FC (2011) Anti-nociceptive and anti-inflammatory activities of (−)-α-bisabolol in rodents. Naunyn-Schmeideberg's Arch Pharmacol 384:525
430. Flemming M, Kraus B, Rascle A, Jürgenliemk G, Fuchs S, Fürst R, Heilmann J (2015) Revisited anti-inflammatory activity of matricine in vitro: comparison with chamazulene. Fitoterapia 106:122

431. Maschi O, Cero ED, Galli GV, Caruso D, Bosisio E, Dell'Agli M (2008) Inhibition of human cAMP-phosphodiesterase as a mechanism of the spasmolytic effect of *Matricaria recutita* L. J Agric Food Chem 56:5015
432. Amsterdam JD, Shults J, Soeller I, Mao JJ, Rockwell K, Newberg AB (2012) Chamomile (*Matricaria recutita*) may provide antidepressant activity in anxious, depressed humans: an exploratory study. Altern Ther Health Med 18:44
433. Han XH, Hong SS, Hwang JS, Lee MK, Hwang BY, Ro JS (2007) Monoamine oxidase inhibitory components from *Cayratia japonica*. Arch Pharm Res 30:13
434. Li J, Wang YQ, Jin H, Xu XF, Xiang CP (2010) Characterization of antioxidant polysaccarides in bitter gourd (*Momordica charantia* L.) cultivars. J Food Agric Environ 8:117
435. Gong J, Sun F, Li Y, Zhou X, Duan Z, Duan F, Zhao L, Chen H, Qi S, Shen J (2015) *Momordica charantia* polysaccharides could protect against cerebral ischemia/reperfusion injury through inhibiting oxidative stress mediated c-Jun N-terminal kinase 3 signaling pathway. Neuropharmacology 91:123
436. Chen HM, Li HX, Kan GS, Ren DM (2012) Correlation study between antioxidant activity and lowering blood glucose of momordica polysaccharide. Sci Technol Food Ind 18:349
437. Mishra A, Gautam S, Pal S, Mishra A, Rawat A, Maurya R, Srivastava AK (2015) Effect of *Momordica charantia* fruits on streptozotocin-induced diabetes mellitus and its associated complications. Int J Pharm Pharm Sci 7:356
438. Wu LW, Ke LJ, Huang XN, Liu ST, Chen H, Rao PF (2006) Separation and characterization of the active ingredients of *Momordica charantia* L. and their protective and repairing effect on HIT-T15 cells damaged by alloxan in vitro. J Chin Inst Food Sci Technol 6:24
439. Li ZT, Zhang J, Xie J, Sun RG (2013) The inhibition effect of two different *Momordica charantia* polysaccharides on the proliferation of human leukemia cell in vitro. J Shaanxi Normal Univ 41:76
440. Guan L (2012) Synthesis and anti-tumor activities of sulphated polysaccharide obtained from *Momordica charantia*. Nat Prod Res 26:1303
441. Deng Y-Y, Yi Y, Zhang L-F, Zhang R-F, Zhang Y, Wei Z-C, Tang X-J, Zhang M-W (2014) Immunomodulatory activity and partial characterisation of polysaccharides from *Momordica charantia*. Molecules 19:13432
442. Razavi BM, Hosseinzadeh H (2014) A review of the effects of *Nigella sativa* L. and its constituent, thymoquinone, in metabolic syndrome. J Endocrinol Invest 37:1031
443. Haq A, Abdullatif M, Lobo PI, Khabar KSA, Sheth KV, Al-Sedairy ST (1995) *Nigella sativa*: effect on human lymphocytes and polymorphonuclear leukocyte phagocytic activity. Immunopharmacology 30:147
444. Haq A, Lobo PI, Al-Tufail M, Rama NR, Al-Sedairy ST (1999) Immunomodulatory effect of *Nigella sativa* proteins fractionated by ion exchange chromatography. Int J Immunopharmacol 21:283
445. Mansour MA, Nagi MN, El-Khatib AS, Al-Bekairi AM (2002) Effects of thymoquinone on antioxidant enzyme activities, lipid peroxidation, and DT-diaphorase in different tissues of mice: a possible mechanism of action. Cell Biochem Funct 20:143
446. Awad AS, Kamel R, Sherief M-AE (2011) Effect of thymoquinone on hepatorenal dysfunction and alteration of CYP3A1 and spermidine/spermine *N*-1-acetyl-transferase gene expression induced by renal ischemia-reperfusion in rats. J Pharm Pharmacol 63:1037
447. Bamosa AO, Kaatabi H, Lebda FM, Al Elq A-M, Al-Sultan A (2010) Effect of *Nigella sativa* seeds on the glycemic control of patients with type 2 diabetes mellitus. Indian J Physiol Pharmacol 54:344
448. Abdelmeguid NE, Fakhoury R, Kamal SM, Al Wafai RJ (2010) Effects of *Nigella sativa* and thymoquinone on biochemical and subcellular changes in pancreatic β-cells of streptozotocin-induced diabetic rats. J Diabetes 2:256
449. Ahmad S, Beg ZH (2013) Hypolipidemic and antioxidant activities of thymoquinone and limonene in atherogenic suspension fed rats. Food Chem 138:1116
450. El Tahir KEH, Al-Ajmi MF, Al-Bekairi AM (2003) Some cardiovascular effects of the dethymoquinonated *Nigella sativa* volatile oil and its major components α-pinene and *p*-cymene in rats. Saudi Pharm J 11:104

451. Ghatak SB, Panchal SS (2012) Anti-diabetic activity of oryzanol and its relationship with the antioxidant property. Int J Diabetes Dev Ctries 32:185
452. Min S-W, Ryu S-N, Kim D-H (2010) Anti-inflammatory effects of black rice, cyanidin-3-*O*-*β*-D-glycoside, and its metabolites, cyanidin, and protocatechuic acid. Int Immunopharmacol 10:959
453. Wunjuntuk K, Kettawan A, Rungruang T, Charoenkiatkul S (2016) Anti-fibrotic and anti-inflammatory effects of parboiled germinated brown rice (*Oryza sativa* 'KDML 105') in rats with induced liver fibrosis. J Funct Foods 26:363
454. Accinni R, Rosina M, Bamonti F, Noce CD, Tonini A, Bernacchi F, Campolo J, Caruso R, Novembrino C, Ghersi L, Lonati S, Grossi S, Ippolito S, Lorenzano E, Ciani A, Gorini M (2006) Effects of combined dietary supplementation of oxidative and inflammatory status in dyslipidemic subjects. Nutr Metab Cardiovasc Dis 16:121
455. Ismail N, Ismail M, Imam MU, Szmi NH, Fathy SF, Foo JB, Bakar MFA (2014) Mechanistic basis for protection of differentiated SH-SY5Y cells by oryzanol-rich fraction against hydrogen peroxide-induced neurotoxicity. BMC Complement Altern Med 14:467
456. Chotimarkon C, Ushio H (2008) The effect of *trans*-ferulic acid and gamma-oryzanol on ethanol-induced liver injury in C57BL mouse. Phytomedicine 15:951
457. Stefanutti C, Mazza F, Mesce D, Morozzi C, Di Giacomo S, Vitale M, Pergolini M (2017) *Monascus purpureus* for statin and ezetimibe intolerant heterozygous familial hypercholesterolaemia patients: a clinical study. Atheroscler Suppl 30:86
458. Hong MY, Seeram NP, Zhang Y, Heber D (2008) Anticancer effects of Chinese red yeast rice versus monacolin K alone on colon cancer cells. J Nutr Biochem 19:448
459. Zheng Y, Zhang Y, Chen D, Chen H, Lin L, Zheng C, Guo Y (2016) *Monascus* pigment rubropunctatin: a potential dual agent for cancer chemotherapy and phototherapy. J Agric Food Chem 64:2541
460. Lee C-L, Wen J-Y, Hsu Y-W, Pan T-M (2018) The blood lipid regulation of *Monascus*-produced monascin and ankaflavin via the suppression of low-density lipoprotein cholesterol assembly and stimulation of apolipoprotein A1 expression in the liver. J Microbiol Immunol Infect 51:27
461. Shi Y-C, Pan T-M, Liao VH-C (2016) Monascin from *Monascus*-fermented products reduces oxidative stress and amyloid-β toxicity via DAF-16/FOXO in *Caenorhabditis elegans*. J Agric Food Chem 64:7114
462. Wang J, Wang G, Yi J, Xu Y, Duan S, Li T, Sun X-G, Dong L (2017) The effect of monascin on hematoma clearance and edema after intracerebral hemorrhage in rats. Brain Res Bull 134:24
463. Yokozawa T, Kobayashi T, Oura H, Kawashima Y (1985) Hyperlipidemia-improving effects of ginsenoside Rb2 in streptozotocin-diabetic rats. Chem Pharm Bull 33:3893
464. Lee K-T, Jung TW, Lee H-J, Kim S-G, Shin Y-S, Whang W-K (2011) The antidiabetic effect of ginsenoside Rb2 via activation of AMPK. Arch Pharm Res 34:1201
465. Li X, Chu S, Lin M, Gao Y, Liu Y, Yang S, Zhou X, Zhang Y, Hu Y, Wang H, Chen N (2020) Anticancer property of ginsenoside Rh2 from ginseng. Eur J Med Chem 203:112627
466. Liu Z, Liu T, Li W, Li J, Wang C, Zhang K (2021) Insights into the antitumor mechanism of ginsenosides Rg3. Mol Biol Rep 48:2639
467. Zhou T-T, Zu G, Wang X, Zhang X-G, Li S, Liang Z-H, Zhao J (2015) Immunomodulatory and neuroprotective effects of ginsenoside Rg1 in the MPTP (1-methyl-4-phenyl-1,2,3,6-tetrahydropyridine)-induced mouse model of Parkinson's disease. Int Immunopharmacol 29:334
468. Brunetti P, Lo Faro AF (2020) Pharmacology of herbal sexual enhancers: a review of psychiatric and neurological adverse effects. Pharmaceuticals 13:309
469. Reid K, Surridge DH, Morales A, Condra M, Harris C, Owen J, Fenemore J (1987) Double-blind trial of yohimbine in treatment of psychogenic impotence. Lancet 330:421
470. Teloken C, Rhoden EL, Sogari P, Dambros M, Souto CAV (1998) Therapeutic effects of high dose yohimbine hydrochloride on organic erectile dysfunction. J Urol 159:122

471. Charlebois D (2007) Elderberry as a medicinal plant. In: Janick J, Whipkey A (eds) Issues in new crops and new uses. ASHS Press, Alexandria, VA, p 284
472. Młynarczyk K, Walkowiak-Tomczak D, Łysiak GP (2018) Bioactive properties of *Sambucus nigra* L. as a functional ingredient for food and pharmaceutical industry. J Funct Foods 40:377
473. Abdramanov A, Massanyi P, Sarsembayeva N, Altay U, Alimov J, Tvrdá E (2017) The in vitro effect of elderberry (*Sambucus nigra*) extract on the activity and oxidative profile of bovine spermatozoa. J Microbiol Biotechnol Food Sci 6:1319
474. Santin JR, Benvenutti L, Broering MF, Nunes R, Goldoni FC, Patel YBK, de Souza JA, Kopp MAT, de Souza P, da Silva RCV, Pastor MVD, de Souza AB, Testoni LD, Couto AG, Bresolin TMB, Quintão NLM (2022) *Sambucus nigra*: a traditional medicine effective in reducing inflammation in mice. J Ethnopharmacol 283:114736
475. Castillo-Maldonado I, Moreno-Altamirano MMB, Serrano-Gallardo LB (2017) Anti-dengue serotype-2 activity effect of *Sambucus nigra* leaves- and flowers-derived compounds. Virol Res Rev 1:1
476. Roschek B, Fink RC, McMichael MD, Li D, Alberte RS (2009) Elderberry flavonoids bind to and prevent H1N1 infection in vitro. Phytochemistry 70:1255
477. Boroduske A, Jekabsons K, Riekstina U, Muceniece R, Rostoks N, Nakurte I (2021) Wild *Sambucus nigra* L. From north-east edge of the species range: a valuable germplasm with inhibitory capacity against SARS-CoV2 S-protein RBD and hACE2 binding in vitro. Ind Crops Prod 165:113438
478. Harnett J, Oakes K, Carè J, Leach M, Brown D, Cramer H, Pinder T-A, Steel A, Anheyer D (2020) The effects of *Sambucus nigra* berry on acute respiratory viral infections: a rapid review of clinical studies. Adv Integr Med 7:240
479. Silveira D, Prieto-Garcia JM, Boylan F, Estrada O, Fonseca-Bazzo YM, Jamal CM, Magalhães PO, Pereira EO, Tomczyk M, Heinrich M (2020) COVID-19: is there evidence for the use of herbal medicines as adjuvant symptomatic therapy? Front Pharmacol 11:581840
480. da Silva RFR, Barreira JCM, Heleno SA, Barros L, Calhelha RC, Ferreira ICFR (2019) Anthocyanin profile of elderberry juice: a natural-based bioactive colouring ingredient with potential food application. Molecules 24:2359
481. Anderson ML (2005) A preliminary investigation of the enzymatic inhibition of 5-alpha-reductase and growth of prostatic carcinoma cell line LNCap-FGC by natural astaxanthin and saw palmetto lipid extract in vitro. J Herb Pharmacother 5:17
482. Raynaud J-P, Cousse H, Martin P-M (2002) Inhibition of type 1 and type 2 5α-reductase activity by free fatty acids, active ingredients of Permixon®. J Steroid Biochem Mol Biol 82:233
483. Abe M, Ito Y, Oyunzul L, Oki-Fujino T, Yamada S (2009) Pharmacologically relevant receptor binding characteristics and 5α-reductase inhibitory activity of free fatty acids contained in saw palmetto extract. Biol Pharm Bull 32:646
484. Silvestri I, Cattarino S, Agilanó A, Nicolazzo C, Scarpa S, Salciccia S, Frati L, Gentile V, Sciarra A (2013) Effect of *Serenoa repens* (Permixon®) on the expression of inflammation-related genes: analysis in primary cell cultures of human prostate carcinoma. J Inflamm 10:11
485. Latil A, Pétrissans M-T, Rouquet J, Robert G, de la Tialle A (2015) Effects of hexanic extract of *Serenoa repens* (Permixon® 160 mg) on inflammation biomarkers in the treatment of lower urinary tract symptoms related to benign prostatic hyperplasia. Prostate 75:1857
486. Vela-Navarrete R, Escribano-Burgos M, Farré AL, García-Cardoso J, Manzarbeitia F, Carrasco C (2005) *Serenoa repens* treatment modifies Bax/Bcl-2 index expression and caspase-3 activity in prostatic tissue from patients with benign prostatic hyperplasia. J Urol 173:507
487. Klippel KF, Hiltl DM, Schipp B (1997) A multicentric, placebo-controlled, double-blind clinical trial of β-sitosterol (phytosterol) for the treatment of benign prostatic hyperplasia. Br J Urol 80:427
488. Berges RR, Windeler J, Trampisch HJ, Senge T (1995) Randomised, placebo-controlled, double-blind clinical trial of β-sitosterol in patients with benign prostatic hyperplasia. Lancet 345:1529

489. Carbin BE, Larsson B, Lindahl O (1990) Treatment of benign prostatic hyperplasia with phytosterols. Br J Urol 66:639
490. Federico A, Dallio M, Loguercio C (2017) Silymarin/silybin and chronic liver disease: a marriage of many years. Molecules 22:191
491. Marmouzi I, Bouyahya A, Ezzat SM, Jemli ME, Kharbach M (2021) The food plant *Silybum marianum* (L.) Gaertn.: phytochemistry, ethnopharmacology and clinical evidence. J Ethnopharmacol 265:113303
492. Ghaffari AR, Noshad H, Ostadi A, Ghojazadeh M, Asadi P (2011) The effects of milk thistle on hepatic fibrosis due to methotrexate in rat. Hepat Mon 11:464
493. Raškovic A, Stilinović N, Kolarović J, Vasović V, Vukmirović S, Mikov M (2011) The protective effects of silymarin against doxorubicin-induced cardiotoxicity and hepatotoxicity in rats. Molecules 16:8601
494. Vessal G, Akmali M, Najafi P, Moien MR, Sagheb MM (2010) Silymarin and milk thistle extract may prevent the progression of diabetic nephropathy in streptozotocin-induced diabetic rats. Ren Fail 32:733
495. Wang X, Zhang Z, Wu S-C (2020) Health benefits of *Silybum marianum*: phytochemistry, pharmacology, and applications. J Agric Food Chem 68:11644
496. Mina PR, Kumar Y, Verma AK, Khan F, Tandon S, Pal A, Darokar MP (2020) Silymarin, a polyphenolic flavanoid impede *Plasmodium falciparum* growth through interaction with heme. Nat Prod Res 34:2647
497. Camini FC, da Silva TF, da Silva Caetano CC, Almeida LT, Ferraz AC, Vitoreti VMA, de Mello SB, de Queiroz SS, de Magalhães JC, de Brito Magalhães CL (2018) Antiviral activity of silymarin against Mayaro virus and protective effect in virus-induced oxidative stress. Antiviral Res 158:8
498. Bosch-Barrera J, Queralt B, Menendez JA (2017) Targeting STAT3 with silibinin to improve cancer therapeutics. Cancer Treat Rev 58:61
499. Loguercio C, Festi D (2011) Silybin and the liver: from basic research to clinical practice. World J Gastroenterol 17:2288
500. Dastpeyman M, Motamed N, Azadmanesh K, Mostafavi E, Kia V, Jahanian-Najafabadi A, Shokrgozar MA (2012) Inhibition of silibinin on migration and adhesion capacity of human highly metastatic breast cancer cell line, MDA-MB-231, by evaluation of β1-integrin and downstream molecules, Cdc42, Raf-1 and D4GDI. Med Oncol 29:2512
501. Song X, Liu B, Cui L, Zhou B, Liu W, Xu F, Hayashi T, Hattori S, Ushiki-Kaku Y, Tashiro S-I, Ikejima T (2017) Silibinin ameliorates anxiety/depression-like behaviors in amyloid β-treated rats by upregulating BDNF/TrkB pathway and attenuating autophagy in hippocampus. Physiol Behav 179:487
502. Shenoy A, Buttar HS, Dicholkar P, Kaur G, Chintamaneni M (2022) Role of nutraceuticals, functional foods, and spices in the management of metabolic syndrome and related disorders. In: Singh RB, Watanabe S, Isaza AA (eds) Functional foods and nutraceuticals in metabolic and non-communicable diseases. Academic Press, London, p 583
503. Stefanowicz-Hajduk J, Król-Kogus B, Sparzak-Stefanowska B, Kimel K, Ochocka JR, Krauze-Baranowska M (2021) Cytotoxic activity of standardized extracts, a fraction, and individual secondary metabolites from fenugreek seeds against SKOV-3, HeLa and MOLT-4 cell lines. Pharm Biol 59:424
504. Hibasami H, Moteki H, Ishikawa K, Katsuzaki H, Imai K, Yoshioka K, Ishii Y, Komiya T (2003) Protodioscin isolated from fenugreek (*Trigonella foenum-graecum* L.) induces cell death and morphological change indicative of apoptosis in leukemic cell line H-60, but not in gastric cancer cell line KATO III. Int J Mol Med 11:23
505. Bachran C, Bachran S, Sutherland M, Bachran D, Fuchs H (2014) Preclinical studies of saponins for tumor therapy. In: Atta-ur-Rahman, Choudhary MI, Perry G (eds) Recent advances in medicinal chemistry. Elsevier, Amsterdam, p 272
506. Jesus M, Martins APJ, Gallardo E, Silvestre S (2016) Diosgenin: recent highlights on pharmacology and analytical methodology. J Anal Methods Chem 2016:4156293

507. Haeri MR, Limaki HK, White CJB, White KN (2012) Non-insulin dependent anti-diabetic activity of (2*S*,3*R*,4*S*)-4-hydroxyisoleucine of fenugreek (*Trigonella foenum graecum*) in streptozotocin-induced type I diabetic rats. Phytomedicine 19:571
508. Ahmadiani A, Javan M, Semnanian S, Barat E, Kamalinejad M (2001) Anti-inflammatory and antipyretic effects of *Trigonella foenum-graecum* leaves extract in the rat. J Ethnopharmacol 75:283
509. Kaur G, Mukundan S, Wani V, Kumar MS (2015) Nutraceuticals in the management and prevention of metabolic syndrome. Austin J Pharmacol Ther 3:1063
510. Trask LE, Chaidarun SS, Platt D, Parkin CG (2014) Treatment with novel galactomannan derivative reduces 2-hour postprandial glucose excursions in individuals with type 2 diabetes treated with oral medications and/or insulin. J Diabetes Sci Technol 8:1018
511. Reeder C, Legrand A, O'Connor-Von SK (2013) The effect of fenugreek on milk production and prolactin levels in mothers of preterm infants. Clin Lact 4:159
512. Dao CA, Patel KD, Neto CC (2012) Phytochemicals from the fruit and foliage of cranberry (*Vaccinium macrocarpon*)—potential benefits for human health. In: Patil BS, Jayaprakasha GK, Murthy KNC, Seram NP (eds), Emerging trends in dietary components for preventing and combating disease. American Chemical Society Books, Washington, DC. Symposium series, vol 1093, p 79
513. He X, Liu RH (2006) Cranberry phytochemicals: isolation, structure elucidation, and their antiproliferative and antioxidant activities. J Agric Food Chem 54:7069
514. Xie L, Xie J, Xu Y, Chen W (2020) Discovery of anthocyanins from cranberry extract as pancreatic lipase inhibitors using a combined approach of ultrafiltration, molecular simulation and spectroscopy. Food Funct 11:8527
515. Howell AB, Reed JD, Krueger CG, Winterbottom R, Cunningham DG, Leahy M (2005) A-type cranberry proanthocyanidins and uropathogenic bacterial anti-adhesion activity. Phytochemistry 66:2281
516. Lavigne JP, Bourg G, Combescure C, Botto H, Sotto A (2008) In-vitro and in-vivo evidence of dose-dependent decrease of uropathogenic *Escherichia coli* virulence after consumption of commercial *Vaccinium macrocarpon* (cranberry) capsules. Clin Microbiol Infect 14:350
517. Neto CC (2007) Cranberry and its phytochemicals: a review of in vitro anticancer studies. J Nutr 137:186S
518. Déziel B, MacPhee J, Patel K, Catalli A, Kulka M, Neto C, Gottschall-Pass K, Hurta R (2012) American cranberry (*Vaccinium macrocarpon*) extract affects human prostate cancer cell growth via cell cycle arrest by modulating expression of cell cycle regulators. Food Funct 3:556
519. Skemiene K, Pampuscenko K, Rekuviene E, Borutaite V (2020) Protective effects of anthocyanins against brain ischemic damage. J Bioenerg Biomembr 52:71
520. Liang Z, Liang H, Guo Y, Yang D (2021) Cyanidin 3-*O*-galactoside: a natural compound with multiple health benefits. Int J Mol Sci 22:2261
521. Wen H, Cui H, Tian H, Zhang X, Ma L, Ramassamy C, Li J (2020) Isolation of neuroprotective anthocyanins from black chokeberry (*Aronia melanocarpa*) against amyloid-β-induced cognitive impairment. Foods 10:63
522. Chen H-W, Wei B-J, He X-H, Liu Y, Wang J (2015) Chemical components and cardiovascular activities of *Valeriana* spp. Evid Based Complement Altern Med 2015:947619
523. Lunz K, Stappen I (2021) Back to the roots-an overview of the chemical composition and bioactivity of selected root-essential oils. Molecules 26:3155
524. Taavoni S, Ekbatani N, Kashaniyan M, Haghani H (2011) Effect of valerian on sleep quality in postmenopausal women: a randomized placebo-controlled clinical trial. Menopause 18:951
525. Ahmadi M, Khalil H, Abbasian L, Ghaeli P (2017) Effect of valerian in preventing neuropsychiatric adverse effects of efavirenz in HIV-positive patients: a pilot randomized, placebo-controlled clinical trial. Ann Pharmacother 51:457
526. Andreatini R, Sartori VA, Seabra MLV, Leite JR (2002) Effect of valepotriates (valerian extract) in generalized anxiety disorder: a randomized placebo-controlled pilot study. Phytother Res 16:650

527. Maurmann N, Reolon GK, Rech SB, Fett-Neto AG, Roesler R (2011) A valepotriate fraction of *Valeriana glechomifolia* shows sedative and anxiolytic properties and impairs recognition but not aversive memory in mice. Evid Based Complement Altern Med 2011:720853
528. Sah SP, Mathela CS, Chopra K (2011) Antidepressant effect of *Valeriana wallichii* patchouli alcohol chemotype in mice: behavioural and biochemical evidence. J Ethnopharmacol 135:197
529. Zhou X-Z, Kang L, Kang Y, Li L, Xiong S-H (2009) Effect of *Valeriana officinalis* var. *latifolia* Miq. on heart rat and arterial blood perssure of rabbit. J Liaoning Univ Tradit Chin Med 11:188
530. Fields AM, Richards TA, Felton JA, Felton SK, Bayer EZ, Ibrahim IN, Kaye AD (2003) Analysis of responses to valerian root extract in the feline pulmonary vascular bed. J Altern Complement Med 9:909
531. Yang S-H, Chen F, Ma H-M, Wang T (2012) Protection of *Valeriana officinalis* L. extract preconditioning on ischemia-reperfusion injury in rat hearts in vitro. Med J Wuhan Univ 33:639
532. Khom S, Baburin I, Timin E, Hohaus A, Trauner G, Kopp B, Hering S (2007) Valerenic acid potentiates and inhibits $GABA_A$ receptors: molecular mechanism and subunit specificity. Neuropharmacology 53:178
533. Shi S-N, Shi J-L, Liu Y, Wang Y-L, Wang C-G, Hou W-H, Guo J-Y (2014) The anxiolytic effects of valtrate in rats involves changes of corticosterone levels. Evid Based Complement Alternat Med 2014:325948
534. Fernández S, Wasowski C, Paladini AC, Marder M (2004) Sedative and sleep-enhancing properties of linarin, a flavonoid-isolated from *Valeriana officinalis*. Pharmacol Biochem Behav 77:399
535. Zhang X, Li X, Liu N, Zheng P, Ma L, Guo F, Sun T, Zhou R, Yu J (2019) The anticonvulsant effects of baldrinal on pilocarpine-induced convulsion in adult male mice. Molecules 24:1617
536. Wang Q, Wang C, Zuo Y, Wang Z, Yang B, Kuang H (2012) Compounds from the roots and rhizomes of *Valeriana amurensis* protect against neurotoxicity in PC12 cells. Molecules 17:15013
537. Begum VH, Sadique J (1987) Effect of *Withania somnifera* on glycosaminoglycan synthesis in carrageenin-induced air pouch granuloma. Biochem Med Metab Biol 38:272
538. Maitra R, Porter MA, Huang S, Gilmour BP (2009) Inhibition of NFκB by the natural product withaferin A in cellular models of cystic fibrosis inflammation. J Inflamm 6:15
539. Dhuley JN (1998) Effect of ashwagandha on lipid peroxidation in stress-induced animals. J Ethnopharmacol 60:173
540. Wube AA, Wenzig E-M, Gibbons S, Asres K, Bauer R, Bucar F (2008) Constituents of the stem bark of *Discopodium penninervium* and their LTB_4 and COX-1 and -2 inhibitory activities. Phytochemistry 69:982
541. Bhattacharya SK, Bhattacharya A, Sairam K, Ghosal S (2000) Anxiolytic-antidepressant activity of *Withania somnifera* glycowithanolides: an experimental study. Phytomedicine 7:463
542. Dhuley JN (2001) Nootropic-like effect of ashwagandha (*Withania somnifera* L.) in mice. Phytother Res 15:524
543. Ghosal S, Lal J, Srivastava R, Bhattacharya SK, Upadhyay SN, Jaiswal AK, Chattopadhyay U (1989) Immunomodulatory and CNS effects of sitoindosides IX and X, two new glycowithanolides from *Withania somnifera*. Phytother Res 3:201
544. Bhattacharya SK, Goel RK, Kaur R, Ghosal S (1987) Anti-stress activity of sitoindosides VII and VIII, new acylsterylglucosides from *Withania somnifera*. Phytother Res 1:32
545. Kumar S, Harris RJ, Seal CJ, Okello EJ (2012) An aqueous extract of *Withania somnifera* root inhibits amyloid β fibril formation in vitro. Phytother Res 26:113
546. Kuboyama T, Tohda C, Komatsu K (2006) Withanoside IV and its active metabolite, sominone, attenuate Aβ (25–35)-induced neurodegeneration. Eur J Neurosci 23:1417
547. Zhao J, Nakamura N, Hattori M, Kuboyama T, Tohda C, Komatsu K (2002) Withanolide derivatives from the roots of *Withania somnifera* and their neurite outgrowth activities. Chem Pharm Bull 50:760

548. Pandey A, Bani S, Dutt P, Satti NK, Suri KA, Qazi GN (2018) Multifunctional neuroprotective effect of withanone, a compound from *Withania somnifera* roots in alleviating cognitive dysfunction. Cytokine 102:211
549. Singh N, Singh SP, Nath R, Singh DR, Gupta ML, Kohli RP, Bhargava KP (1986) Prevention of urethane-induced lung adenomas by *Withania somnifera* (L.) Dunal in albino mice. Int J Crude Drug Res 24:90
550. Mohan R, Hammers H, Bargagna-Mohan P, Zhan X, Herbstritt C, Ruiz A, Zhang L, Hanson A, Conner B, Rougas J (2004) Withaferin A is a potent inhibitor of angiogenesis. Angiogenesis 7:115
551. Jayaprakasam B, Zhang Y, Seeram NP, Nair MG (2003) Growth inhibition of human tumor cell lines by withanolides from *Withania somnifera* leaves. Life Sci 74:125
552. Cordero CP, Morantes SJ, Páez A, Rincón J, Aristizábal FA (2009) Cytotoxicity of withanolides isolated from *Acnistus arborescens*. Fitoterapia 80:364
553. Ortega AMM, Campos MRS (2019) Medicinal plants and their bioactive metabolites in cancer prevention and treatment. In: Campos MRS (ed) Bioactive compounds. Woodhead Publishing, Cambridge, MA, p 85
554. Jin Z, Lee G, Kim S, Park C-S, Park YS, Jin Y-H (2014) Ginger and its pungent constituents non-competitively inhibit serotonin currents on visceral afferent neurons. Korean J Physiol Pharmacol 18:149
555. Zhang F, Thakur K, Hu F, Zhang J-G, Wei Z-J (2017) Cross-talk between 10-gingerol and its anti-cancerous potential: a recent update. Food Funct 8:2635
556. Akimoto M, Iizuka M, Kanematsu R, Yoshida M, Takenaga K (2015) Anticancer effect of ginger extract against pancreatic cancer cells mainly through reactive oxygen species-mediated autotic cell death. PLoS One 10:e0126605
557. Qi LW, Zhang Z, Zhang CF, Anderson S, Liu Q, Yuan CS, Wang CZ (2015) Anti-colon cancer effects of 6-shogaol through G2/M cell cycle arrest by p53/p21-cdc2/cdc25A crosstalk. Am J Chin Med 43:743
558. Sang S, Hong J, Wu H, Liu J, Yang CS, Pan MH, Badmaev V, Ho CT (2009) Increased growth inhibitory effects on human cancer cells and anti-inflammatory potency of shogaols from *Zingiber officinale* relative to gingerols. J Agric Food Chem 57:10645
559. Lee SH, Cekanova M, Baek SJ (2008) Multiple mechanisms are involved in 6-gingerol-induced cell growth arrest and apoptosis in human colorectal cancer cells. Mol Carcinog 47:197
560. El-Naggar MH, Mira A, Abdel Bar FM, Shimizu K, Amer MM, Badria FA (2017) Synthesis, docking, cytotoxicity, and LTA_4H inhibitory activity of new gingerol derivatives as potential colorectal cancer therapy. Bioorg Med Chem 25:1277
561. Zhang F, Thakur K, Hu F, Zhang J-G, Wei Z-J (2017) 10-Gingerol, a phytochemical derivative from "tongling white ginger", inhibits cervical cancer: insights into the molecular mechanism and inhibitory targets. J Agric Food Chem 65:2089
562. Sakulnarmrat K, Srzednicki G, Konczak I (2015) Antioxidant, enzyme inhibitory and antiproliferative activity of polyphenolic-rich fraction of commercial dry ginger powder. Int J Food Sci Technol 50:2229
563. Lantz RC, Chen GJ, Sarihan M, Sólyom AM, Jolad SD, Timmermann BN (2007) The effect of extracts from ginger rhizome on inflammatory mediator production. Phytomedicine 14:123
564. Young HY, Luo YL, Cheng HY, Hsieh WC, Liao JC, Peng WH (2005) Analgesic and anti-inflammatory activities of [6]-gingerol. J Ethnopharmacol 96:207
565. Nurtjahja-Tjendraputra E, Ammit AJ, Roufogalis BD, Tran VH, Duke CC (2003) Effective anti-platelet and COX-1 enzyme inhibitors from pungent constituents of ginger. Thromb Res 111:259
566. Thiese MS, Arnold ZC, Walker SD (2015) The misuse and abuse of statistics in biomedical research. Biochem Med 25:5
567. Khan IA, Smillie T (2012) Implementing a "quality by design" approach to assure the safety and integrity of botanical dietary supplements. J Nat Prod 75:1665
568. Carcache de Blanco EJ, Kinghorn AD (2020) Botanical dietary products. In: Adejare A (ed) Remington: the science and practice of pharmacy, 23rd edn. Academic, San Diego, p 45

Eric D. Salinas-Arellano obtained his Bachelor's degree in Pharmaceutical Chemistry in 2011 from the Michoacan University of Saint Nicolas of Hidalgo (Morelia, Michoacán, México). He then gained his Master's degree in Pharmacology from the Center for Research and Advanced Studies of the National Polytechnic Institute (CINVESTAV, Mexico City). In 2021, he received his Ph.D. degree in Chemistry from the National Autonomous University of Mexico (UNAM, Mexico City). Since August 2021, he has been a Postdoctoral Scholar in the group of Dr. Esperanza Carcache de Blanco at the College of Pharmacy, The Ohio State University, working on the search for bioactive compounds of natural origin with potential application in cancer, with an emphasis on drug discovery.

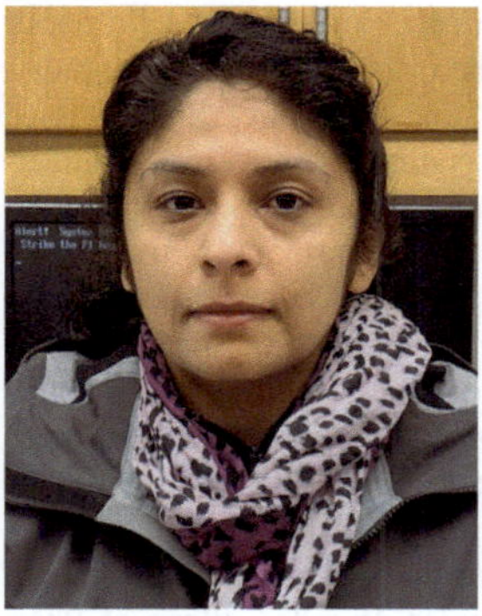

Ines Y. Castro-Dionicio received a B.S. degree in Pharmacy and Biochemistry from the National University of Trujillo (Trujillo, Peru) in 2012. She next obtained her M.S. degree in Biochemistry and Molecular Biology from the Cayetano Heredia Peruvian University (Lima, Peru) and in 2021, she was awarded her Ph.D. degree from the Paul Sabatier Toulouse III University (Toulouse, France). Since January 2022, she has been working as a Postdoctoral Scholar in Dr. Esperanza Carcache de Blanco's group at the College of Pharmacy at the Ohio State University. Her research interests are focused on the quality control of dietary supplements, LC-MS based dereplication analysis of natural products, and natural product drug discovery for potential antitumor agents.

Jonathan G. Jeyaraj was awarded a B.S. degree in Biochemistry and Molecular Biology from Michigan State University in 2005. Following the completion of his undergraduate degree, he worked in the pharmaceutical industry from 2005 to 2018 for the companies Perrigo (Allegan, Michigan), Microdose Technologies (Monmouth, New Jersey), and Celgene (Summit, NJ), taking on the roles of Senior Quality Control Analyst, Research Assistant, and Senior Global Quality Systems Specialist. In 2019, he received an M.S. degree in Physiology from North Carolina State University and currently, since 2020, has been enrolled for a Ph.D. degree in Pharmaceutical Sciences in the Medicinal Chemistry and Pharmacognosy track at the College of Pharmacy, The Ohio State University. His research interests and dissertation work are focused on natural product drug discovery for potential antitumor agents from understudied tropical plants using both phytochemical techniques and biological assays. He is the advisee of both Prof. Esperanza Carcache de Blanco and Prof. A. Douglas Kinghorn.

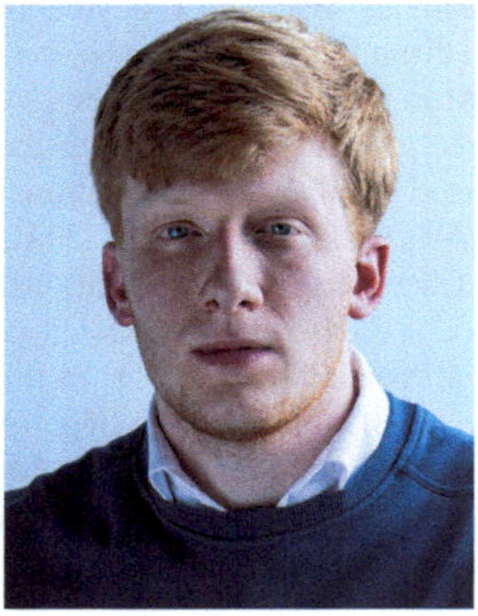

Nathan P. Mirtallo Ezzone obtained his bachelor's degree in Pharmaceutical Sciences from the Ohio State University in 2020. Since then, he has been working towards a Ph.D. degree in Pharmaceutical Sciences in the Medicinal Chemistry and Pharmacognosy track from the College of Pharmacy, The Ohio State University advised by Prof. Esperanza Carcache de Blanco and Prof. A. Douglas Kinghorn. His research interests include the isolation and biological evaluation of potential anticancer agents from natural sources, particularly for the purposes of drug discovery.

Esperanza J. Carcache de Blanco is an Associate Professor at the Division of Medicinal Chemistry and Pharmacognosy, College of Pharmacy, The Ohio State University. She obtained her Pharmacy Licensure from the National Autonomous University of Nicaragua. She was awarded her M.S. degree as a Fulbright Scholar from the College of Pharmacy, University of Illinois at Chicago (UIC), Chicago, Illinois. In 2003, she also gained her Ph.D. degree from the College of Pharmacy, University of Illinois at Chicago, Chicago, Illinois. During her graduate studies, she received three awards including a Student Research Award from the American Society of Pharmacognosy (at the 44th Annual Meeting, Chapel Hill, NC), a UIC Health Science Auxiliary Award Fellowship from the University of Illinois at Chicago, and the Van Doren Scholar Award, College of Pharmacy, UIC, Chicago, IL (as an Outstanding Graduate Student in the Department of Medicinal Chemistry and Pharmacognosy). In 2005, she became an Assistant Professor at The Ohio State University. Her scientific interests are directed towards the study of compounds from natural sources for the benefit of human health, including those for cancer and diabetes drug discovery as well as botanicals products used as dietary supplements. As a specialist in these areas, she teaches courses in Pharmacognosy, such as those in natural products drug discovery and dietary supplements/herbal products.

Quality Consistency of Herbal Products: Chemical Evaluation

Ahmed Osman, Amar G. Chittiboyina, Bharathi Avula, Zulfiqar Ali, Sebastian J. Adams, and Ikhlas A. Khan

Contents

A. Osman (✉) · A. G. Chittiboyina · B. Avula · Z. Ali · S. J. Adams · I. A. Khan
School of Pharmacy, National Center for Natural Products Research, The University of Mississippi, University, MS 38677, USA
e-mail: amgalalv@olemiss.edu

A. G. Chittiboyina
e-mail: amar@olemiss.edu

B. Avula
e-mail: bavula@olemiss.edu

Z. Ali
e-mail: zulfiqar@olemiss.edu

S. J. Adams
e-mail: jasabest@olemiss.edu

I. A. Khan
e-mail: ikhan@olemiss.edu

A. D. Kinghorn, H. Falk, S. Gibbons, Y. Asakawa, J.-K. Liu, V. M. Dirsch (eds.), *Progress in the Chemistry of Organic Natural Products 122*, Progress in the Chemistry of Organic Natural Products 122, https://doi.org/10.1007/978-3-031-26768-0_2

1 Introduction

Centuries of human use have established the benefits of herbal products for improving health [1]. In recent years, the cytoprotective effects of numerous medicinal herbs as determined in the laboratory have shown their potential to protect against the development of cancer and oxidative stress [2–4]. The traditional systems of herbal products using ground plant materials or plant extracts as remedial or health improvement agents have provided healthcare benefits to many of the world's population [5, 6]. Numerous herbal products are included in Traditional Chinese Medicine (TCM) and Ayurvedic medicine as well as Western herbal medicine.

The widespread utility of herbal products has been rising worldwide, both in developed and developing countries, from the rapid growth of global markets for medicinal herbs [7]. The World Health Organization estimates that 80% of humans worldwide still rely on botanical-based treatments [8]. Sales of botanical dietary supplements in the United States alone have increased by 17.3% in 2020 to reach a total of $11.3 billion (with these sales not including herbal teas, beverages, or natural ingredients used as components of personal care and cosmetic products) [9].

In this chapter, the term "herbal products" will encompass herbal medicines, herbal beverages, spices, and herbal personal care/cosmetic products. Herbal products may also refer to botanical medicines or botanical dietary supplements. The terms "plant raw materials" and "herbal raw materials" include the plants or plant parts harvested to supply material for the production of herbal products, while "herbs" refer to the plant materials themselves.

This substantial increase in the consumption of herbal products has increased the focus on the adverse effects experienced with certain of these products. Adverse effects may include hepatotoxicity, renal failure, and carcinogenicity [10–12]. These adverse effects primarily originate from the poor quality of plant raw materials, or of the production processing, or of the finished products. In addition to adverse effects, poor-quality products have the potential to create other safety or efficacy issues. Inadequate quality is overwhelmingly due to a lack of appropriate quality assurance and quality control practices beginning with the plant raw material and continuing through to the production of the finished herbal product.

The United States Pharmacopeia (USP) and the Herbal Medicines Compendium (HMC) stipulate the quality of botanical ingredients by three main criteria: identity, purity, and chemical composition. According to the Good Manufacturing Practices (GMP) regulations related to dietary supplements, "Quality means that the dietary supplement consistently meets the established specifications for identity, purity, strength, and composition and limits on contaminants, and has been manufactured, packaged, labeled, and held under conditions to prevent adulteration under section 402(a)(1), (a)(2), (a)(3), and (a)(4) of the Act." (21 Code of Federal Regulations, Part 111).

Data generated from analyzing these components of quality should meet the standards set by the USP and the HMC to ensure consistency in the quality of herbal products.

Purity specifications are set to minimize the levels of contamination and adulteration, as well as the presence of harmful substances up to specified limits [13]. However, the high demand that surpasses the production capabilities for herbal products, combined with a desire for maximizing profit and the lack of rigorous quality control within manufacturing facilities, have led to the reporting of herbal product quality inconsistency (see Table 1 in Sect. 2.3). The causes of quality inconsistency of herbal products may involve species misidentification, substitution, adulteration, or contamination with harmful ingredients. Analytical assessments have revealed frequent and significant compositional variations between marketed herbal products. Quality consistency of herbal products has long been a challenge and a subject of central importance for academia, industry, and regulatory agencies [14–16]. Quality and consistency are interconnected: good quality plant raw materials and adherence to GMP will lead to high quality finished products and together these will contribute to improving and maintaining the consistency of the herbal product.

A common cause of deficiencies in the quality of herbal products is the inconsistency of the plant raw material quality used to manufacture the herbal products. The quality assurance and the quality control of the plant raw materials contribute significantly to improving the quality consistency of the end products.

The inconsistency in the chemical composition of herbal products can be ascribed to a set of situations that can be classified broadly into preharvest factors and postharvest factors. These factors affect the quality of plant raw materials in different ways in addition to their impact on processing the raw material through to the finished products. The first line of maintaining consistent quality of herbal products is the quality assurance and the quality control of the plant raw materials, which largely depends on employing multiple analytical techniques to comprehensively assess the chemical composition of the plant materials from the field to the pre-processing stage. To assure consistent quality, this must be followed by the quality assessment of the plant material processing forward to the end-products. There are three general approaches for evaluating herb quality and consistency: morphological/microscopic, DNA-based/DNA barcoding, and chemical evaluation.

The current chapter is focused on the chemical evaluation of herb or herbal product quality and consistency. Different techniques, instrumentation, applications, and methods used in acquiring chemical fingerprints and chemical profiles of herbal products will be described. The advantages and limitations of the various techniques will be addressed. Before discussing the roles, techniques, and methodologies of chemical evaluation in herb quality consistency, the limitations of non-chemical approaches, namely, morphological/microscopic and DNA barcoding methods, will be presented.

1.1 Limitations of Morphological or Microscopic Analysis in the Evaluation of Herb Quality Consistency

Taxonomy-based morphological analysis necessitates that the flowering parts of the plant be attached in order to allow proper and unequivocal identification of the target species, particularly if there are several species with closely related morphological characteristics under consideration.

Microscopic examination for discriminating plant species suffers from the limitation of relying on the presence of diagnostic histological features, such as diagnostic glandular or non-glandular trichomes, the presence of characteristic sclereids (stone cells) or crystals, or a combination of these histological elements [17–19]. If these diagnostic elements are absent, microscopic methods will not be able to differentiate between different plant species or different plant organs. This limitation can be exemplified by the finding that combined morphological and microscopic inspection of *Glycyrrhiza* (licorice) species was insufficient to discriminate the different species within this genus. All licorice roots or rhizomes are similar morphologically and microscopically, leading to potential misidentification if the analysis is limited to these two considerations [20].

1.2 Limitations of DNA Barcoding Analysis in the Evaluation of Herb Quality Consistency

Deoxyribonucleic acid barcoding (see the chapter on "Deoxyribonucleic Acid Barcoding for the Identification of Botanicals", in this volume) exhibits certain limitations. The major limitation is the necessity to extract DNA immediately after collecting the plant material to avoid decomposition of the DNA molecules under storage conditions, such as exposure to light, temperature, and microorganisms. In addition, DNA barcoding techniques often produce inconsistent results when applied to highly processed plant materials such as botanical dietary supplements, since DNA may be severely degraded or eliminated in the production of herbal products. There are, however, a few reports on retrieving and analyzing the DNA of highly processed herbal products such as plant extracts (e.g., [21]). Deoxyribonucleic acid in processed herbal products may be degraded into fragments of around 100–200 bp in length [22]. The degradation of DNA in extensively processed herbal products is attributed mainly to subjecting the plant material to storage, drying, grinding, organic solvents, filtration, excessive heat, and UV light [23]. Further, herbal products often contain pharmaceutical excipients of botanical origin that are reported to hinder DNA extraction through adsorption [24]. Such filler material is often added in the processing of herbal products to increase their volume. The DNA structure of the filler material may remain intact and lead to preferential amplification of the DNA from these excipients. Consequently, false-negative results affecting the identity of the herbal species are likely to be generated [22]. Deoxyribonucleic acid barcoding and metabarcoding may also show false positives, for instance due to analyzing the DNA of fillers rather than of the botanical species or false negatives arising from degraded DNA or post-harvest processing [25, 26].

Beside degradation, failure to retrieve or amplify DNA can significantly decrease the yield of DNA available for analysis. Additionally, certain natural products such as polysaccharides, phenolic comounds, and terpenoid lactones reportedly interfere with DNA extraction. Polysaccharides and certain natural products inhibit enzymatic activity required for further amplification [27]. Plant phenols were also reported to hinder PCR amplification [27].

Conventional DNA barcoding exhibits practical limitations that restrict its application to the authentication of single-species herbal products with unprocessed plant material. Hence, using conventional DNA barcodes of approximately 500 bp represents a limitation of DNA barcoding when applied to the analysis of samples derived from highly processed plant material [24]. Unlike microscopic techniques, DNA-based approaches cannot distinguish between different parts or organs of the same plant, since DNA is not tissue specific. Deoxyribonucleic acid-based methods cannot differentiate between solvent exhausted and non-exhausted herbs, since in both cases the genetic material (DNA) will remain in the plant tissues and organic solvents will extract only the phytochemical constituents. In such cases, chemical analysis will be able to distinguish between the solvent exhausted and the non-exhausted plant materials.

In contrast to chemical marker-based approaches, DNA-based methods target the genotypes and therefore cannot discriminate the phenotypes of plants of different ages, plant parts, different geographical sources, and those harvested at distinct seasons (phenotypic species). Deoxyribonucleic acid barcoding deals with genetic composition and, as such, is unable to determine the chemical identity or the concentration of the bioactive substances in a plant species. This is important since genotypes neither fully determine the plant phenotypes nor control the plant metabolomic profile. Deoxyribonucleic acid barcoding data is valid only for qualitative assessment and does not provide information on the percentage level of a specific species in multi-species products [22, 28–30].

For instance, it was observed that two species of licorice, namely *Glycyrrhiza glabra* (Plate 1) and *Glycyrrhiza inflata* (Plate 2) share the same TG2 genotype. Additionally, morphologic or microscopic identification is insufficient to discriminate *Glycyrrhiza* species or their hybrids, since as mentioned previously, all licorice roots/rhizomes are similar in this regard. The authentication of *Glycyrrhiza* species has been performed accurately based on the metabolic profiles of each species, using separation techniques, usually chromatography combined with detection techniques including mass spectrometry or nuclear magnetic resonance spectroscopy [20].

Towards this end, DNA barcoding techniques cannot be considered as being universally appropriate for the quality control of herbal products. DNA barcoding even with next-generation DNA sequencing (NGS) should be employed in conjunction as a complementary technique with other methodologies, including instrumental chemical analysis (LC-MS, GC-MS, NMR) and botanical analysis (morphological

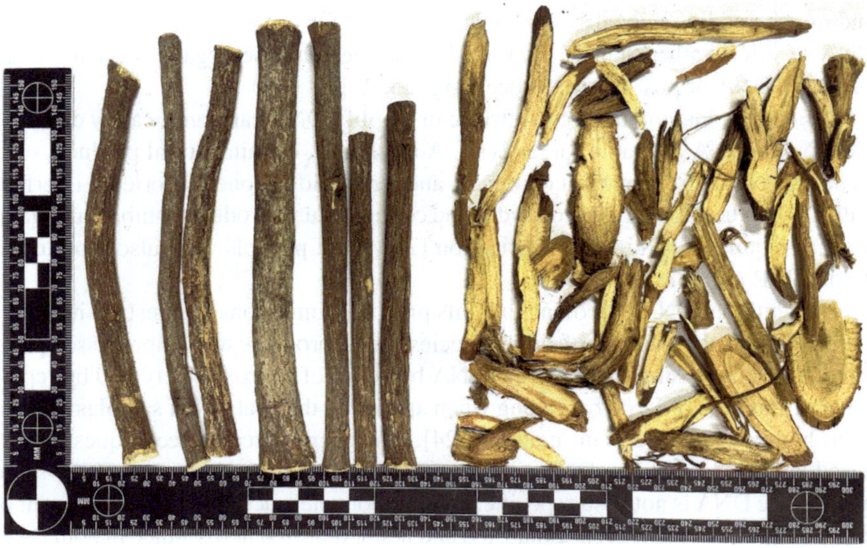

Plate 1 *Glycyrrhiza glabra* (photograph S. J. Adams)

Plate 2 *Glycyrrhiza inflata* (photograph S. J. Adams)

and microscopic analysis) [21, 31]. The application of DNA-based species discrimination for highly processed herbal products has been demonstrated to be of limited utility in identifying or authenticating of the botanical origin of the plant material used in manufacturing highly processed products. Herbal products containing plant extracts with only traces of DNA or severely damaged DNA cannot be analyzed by DNA-based methods.

2 Chemical Evaluation of Herb Quality Consistency

Environmental factors and growing conditions correlate strongly with the chemical composition of herbs [32–34]. The chemical composition at the qualitative and quantitative levels of genetically identical plants can be modified by environmental factors and growing conditions in addition to drying and processing. Remedial or therapeutic effects and the safety of herbal products are associated with their chemical profiles. Since the quality of these products is determined largely by their chemical composition, chemical evaluation of herb quality and the consistency of this quality is necessary to assure the safety and efficacy of herbs and their commercial products.

2.1 Factors Responsible for Variation of the Chemical Composition of Herbal Products

2.1.1 Preharvest Factors

Preharvest factors responsible for the variation of the chemical composition of plant raw material are divided into growing conditions and environmental factors [35].

Growing conditions include both wild-crafted (wild-collected) and cultivated plants [33, 36–38], edaphic factors (soil properties) [39–41], and harvesting activities [42–44]. Differences in soil nutrients and minerals and harvest timing can impact the resultant phytochemical content of herbs.

Major environmental factors include seasonal variations [45, 46], stress (both biotic and abiotic) [47], and altitude [32, 48], again for both wild-collected and cultivated plants.

Seasonal Variation

The influence of seasonal variation on the profile of the chemical constituents in medicinal plants can be explained based on environmental factors associated with the different seasons of the year. These factors primarily include temperature, rainfall, UV radiation, humidity, and sunshine duration [49]. A study on the composition of the essential oil of *Artemisia princeps* (Plate 3) collected at different seasons in Japan, using gas chromatography combined with multivariate analysis, indicated that the seasonal variation led to the fluctuation of the chemical composition of the essential oil from one season to another, and also affected the ratio between the monoterpenes and sesquiterpenes produced. This ratio is a detrimental factor in assessing the quality of the essential oil of *Artemisia princeps* [50].

A study on coriander seeds (*Coriandrum sativum*) (Plate 4) observed that alterations in the planting season resulted in the variation of in the chemical constituents of the seeds [51]. An investigation of the factors affecting *Buxus sempervirens* (Plate 5) found that seasonal changes affected the alkaloid content of this species, which increased in the summer when compared to the other seasons [52].

Stress Conditions

Plants have developed several mechanisms to respond to various environmental stress conditions, whether biotic (plant pathogens such as bacteria, fungi, viruses, nematodes, or herbivorous insects) or abiotic including high salinity and drought. The primary response of plants against stress is through activation of the biosynthesis of

Plate 3 *Artemisia princeps* (photograph Qwert1234,Wikimedia Commons, https://commons.wikimedia.org/wiki/File:Artemisia_princeps_2.JPG)

Plate 4 *Coriandrum sativum* seeds (photograph S. J. Adams)

Plate 5 *Buxus sempervirens* leaves (photograph S. J. Adams)

natural products, since the plant's survival, productivity, and capability for defense against stress conditions all depend on the generation of natural products [53–55].

The shikimate pathway, which is the primary pathway for the biosynthesis of aromatic amino acids, is activated under stress conditions to augment the production of the aromatic amino acids. These are the precursors for biosynthesis of the aromatic natural products such as phenylpropanoids from which many antimicrobial metabolites are derived [53].

The concentration levels of the natural products in medicinal plants grown in a moderate Atlantic climate are lower than in plants grown in semi-arid regions. The latter is attributable to a deficiency of water supply-related stress [54]. Extensive research has shown that stress inflicted by drought leads to an increasing accumulation of naturaö products in plants. Such an augmentation in secondary metabolite biosynthesis reportedly has been observed for almost all classes of natural products, such as phenolic compounds, terpenoids, alkaloids, and cyanogenic glucosides, and glucosinolates [54]. For example, *Mangifera indica* leaves responded to the damage caused by the grasshopper *Tropidacris collaris* activating the biosynthesis of stress-induced volatile compounds, primarily phenylpropanoids, such as myristicin, dillapiole, eugenol, and eugenol acetate, which inhibited the insect attack [56].

Altitude

Altitude significantly affects the composition of phytochemicals in plants through the intensity of solar radiation and temperature. An increase in altitude is accompanied by a rise in light intensity and a decrease in temperature [42, 43]. The increase in light intensity led to an increase in the content of free anthraquinones and sennosides in *Rhamnus purshiana* (Plate 6) and *Cassia angustifolia*, respectively. Moreover, environmental factors and growing conditions are correlated with the resultant pharmacological effects of herbal products [57–59]. It was shown that an ethanol extract of rhubarb (*Rhamnus tanguticum*) grown at an elevation of 4,500 m exhibited a more potent inhibitory effect on the proliferation of adenocarcinoma cells than an extract of this same plant grown at an elevation of 3,200 m [60].

2.1.2 Postharvest Factors

Postharvest factors include the drying methods of the plant raw material, storage conditions, and processing from extraction through to the finished products. In addition, there are external factors that include substitution, adulteration, and contamination, as exemplified by the accidental inclusion of one or more species containing the toxic pyrrolizidine alkaloids **1–3** [61] (Fig. 1, entries 24 and 25 of Table 1).

Adulteration, or substituting material that reduces quality, may include adding inert materials to increase the bulk of the product or the inclusion of a non-labeled pharmacologically active compound. Whatever the source, these actions compromise

Plate 6 *Rhamnus purshiana*, branch with leaves and buds (photograph J. Taylor, Wikimedia Commons, https://commons.wikimedia.org/wiki/File:Rhamnus_purshiana_--_"Cascara"_--_leaf_and_buds.png)

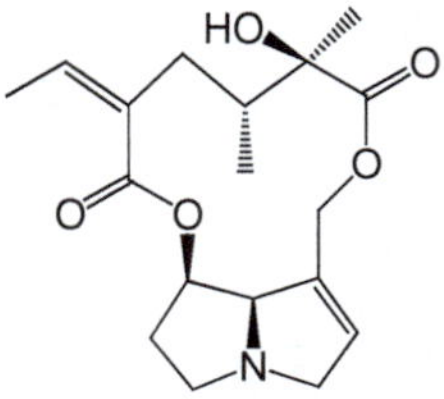

senecionine (**1**) (retronecine-type)

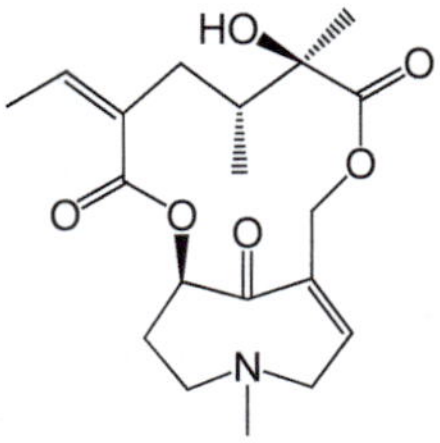

senkirkine (**2**) (otonecine-type)

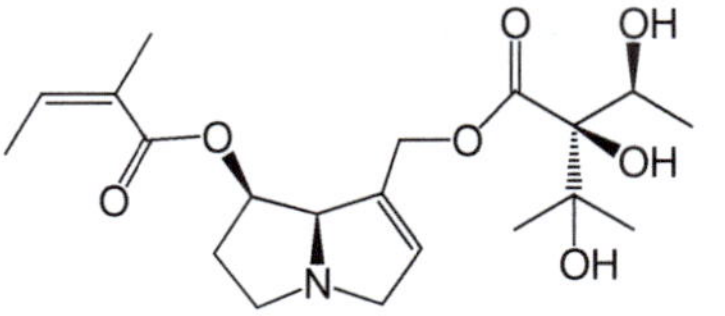

echimidine (**3**) (heliotridine-type)

Fig. 1 Representatives of three types of toxic pyrrolizidine alkaloids (**1**–**3**)

the safety and efficacy of the herbal product. Chemical analysis of the raw material and of the finished product is required to establish quality consistency of the plant material.

The most frequently used instrumental techniques, liquid chromatography-mass spectrometry, gas chromatography-mass spectrometry, proton magnetic resonance spectroscopy, and high-performance thin-layer chromatography will be described in

Table 1 Representative examples of the development and use of chemical analysis to evaluate the quality consistency of herbs and herbal products

Entry	Method of identification[a]	Role of chemical evaluation, cause of inconsistency, incorrect ingredient(s)	Listed ingredient(s)	References
1	HPLC-diode array detector/botanical analysis (morphological and microscopic analysis)	Cause: Misidentification *Mikania glomerata*	*Mikania laevigata*	[82]
2	^{1}H NMR	Cause: Adulteration *Zingiber montanum*	*Curcuma xanthorrhiza*	[83]
3	HPLC-UV/HRMS	Cause: Adulteration *Sophora japonica* fruits on an extract level	*Ginkgo biloba* leaves	[84]
4	Colorimetric sensor array	Cause: Adulteration Asian ginseng (*Panax ginseng* C. A. Meyer)	American ginseng (*Panax quinquefolium*)	[85]
5	LC-MS	Cause: Adulteration with morphologically similar species such as *Platycodon grandiflorum, Codonopsis lanceolata*, and *Pueraria lobata*	*Panax ginseng*	[72]
6	HPTLC and GC	Cause: Adulteration with synthetic dyes and unconventional flavonoid	*Hypericum perforatum* L. (St. John's wort)	[86]
7	^{1}H NMR/CCS[b]	Cause: Adulteration Synthetic curcumin	Curcumin	[87]
8	UHPLC-UV-QToF/MS	Cause: Adulteration *Camellia sinensis*	*Eleutherococcus senticosus*	[88]
9	UHPLC-DAD-QToF-MS	Cause: Adulteration, with the synthetic antimicrobial agent benzethonium chloride	Dietary supplements of grapefruit seed extract	[75]

(continued)

Table 1 (continued)

Entry	Method of identification[a]	Role of chemical evaluation, cause of inconsistency, incorrect ingredient(s)	Listed ingredient(s)	References
10	^{1}H NMR/TLC	Cause: Adulteration *Curcuma aeruginosa*	*Curcuma xanthorrhiza*	[89]
11	HPTLC/MS/^{1}H NMR	Cause: Substitution Other *Rhodiola* species	*Rhodiola rosea*	[90]
12	^{1}H NMR	Cause: Adulteration Phenylethylamines	Sports dietary supplements	[91]
13	LC-MS	Cause: Adulteration Possible adulterants *Berberis vulgaris, Mahonia aquifolium, Coptis chinensis*	*Hydrastis canadensis* (Goldenseal)	[92]
14	^{1}H NMR/UHPLC-MS	Cause: Substitution *Harpagophytum zeyheri*	*Harpagophytum procumbens*	[66]
15	UPLC-C^{18}-FT-MS/MS GC-ToF-MS	Role: Discrimination of fresh and dried samples, The influence of geographical distribution of ginger on metabolites content	*Zingiber officinale*	[93]
16	HPLC	Role: Evaluation of quality consistency	*Sambucus nigra* berries	[94]
17	DNA metabarcoding, TLC and HPLC-MS	Role: Authentication of herbal products	*Hypericum perforatum*	[21]
18	LC-MS	Role: method validation for quality control	Fenugreek (*Trigonella foenum-graecum*) seeds	[95]
19	UHPLC	Role: Method validation for quality control	Green tea (*Camellia sinensis*)	[96]
20	HPLC/DAD	Role: Method validation for quality control of CBD	Cannabidiol (*Cannabis sativa*)	[97]

(continued)

Table 1 (continued)

Entry	Method of identification[a]	Role of chemical evaluation, cause of inconsistency, incorrect ingredient(s)	Listed ingredient(s)	References
21	HPLC-CAD, HPTLC, LC-MS/MS, DNA barcoding	Role: Characterization of phytochemicals in *Echinacea purpurea* extract for safety testing	*Echinacea purpurea* root extract	[98]
22	UHPLC/ESI-MS[c], GC-MS	Role: Method validation for quality control of Avocado oils	Avocado (*Persea americana*)	[99]
23	UHPLC-HRAM-MS	Cause: Adulteration: Asian species (*Actaea foetida, A. dahurica*) and two American species (*A. pachypoda, A. podocarpa*)	*Actaea racemosa* Black cohosh	[100]
24	LC-MS/MS	Cause: Contamination with pyrrolizidine alkaloids	Marjoram, savory, oregano, cumin, parsley, thyme and ginger	[61]
25	UHPLC–QToF-MS	Cause: Contamination with pyrrolizidine alkaloids	In 44 plants/herbal products, pyrrolizidine alkaloids were identified in all	[101]
26	HPLC	Role: Authentication Geographic origin of ginger (*Zingiber officinale*)	*Zingiber officinale*	[65]
27	GC-MS	Cause: Adulteration With cabitol and diethyl phthalate (health hazard adulterants)	Essential oils: tea tree, lavender, sandalwood, rose, eucalyptus, and lemongrass	[102]
28	^{1}H NMR and UPLC-MS	Cause: Chemotypic variation, different geographic origin, *Sceletium tortuosum*	*Sceletium tortuosum*	[63]
29	NGS, LC-QToF-MS, LC-DAD, GC-NPD/MS	Cause: Adulteration *Trifolium, Musa,* Rosaceae	*Hypericum perforatum*	[79]
30	NGS, LC-QToF-MS, LC-DAD, GC-NPD/MS	Cause: Adulteration *Valeriana officinalis, Humulus lupulus, Melissa officinalis*	*Humulus lupulus,* Amaranthaceae, PACMAD clade[d]	[79]

(continued)

Table 1 (continued)

Entry	Method of identification[a]	Role of chemical evaluation, cause of inconsistency, incorrect ingredient(s)	Listed ingredient(s)	References
31	NGS, LC-QToF-MS, LC-DAD, GC-NPD/MS	Cause: Adulteration *Hypericum*, Glycine, *Descurainia sophia*, Triticeae	*Hypericum perforatum*	[79]
32	NGS, LC-QToF-MS, LC-DAD, GC-NPD/MS	Cause: Mislabeling *Valeriana officinalis*	*Valerian*	[79]
33	^{1}H NMR	Cause: Adulteration with pharmaceuticals; sildenafil, tadalafil, and vardenafil	Herbal products marketed for enhancing sexual performance	[103]

[a] UPLC and UHPLC are synonymous (UPLC is a Waters trademark for UHPLC)
[b] Countercurrent separation
[c] Electrospray ionization-mass spectrometry
[d] PACMAD clade: switchgrass (*Panicum virgatum*), Family: grasses (Poaceae or Gramineae)

this chapter. The advantages of the chemical evaluation of herb quality consistency over botanical and DNA-based methods will also be presented. The methods used and instrumentation employed for chemical analysis are described below.

The quality parameters/components described in GMP regulations, in the USP, and in the HMC, including identity, purity, strength, chemical composition, and limits on contaminants can be evaluated using these chemical tools. In this context, purity refers to the absence of contamination, adulteration, or substitution with plant species other than the target species.

While the identification or authentication of the botanical ingredients at the genotype level can be achieved using morphological or microscopic analysis and/or DNA-based barcoding, the identification or authentication at the phenotype level must be achieved by instrumental chemical analysis, such as by LC-MS, GC-MS, ^{1}H NMR, and HPTLC.

2.2 Chemical Markers

Efficient and reliable chemical assessment of herbal products relies substantially on the identification and availability of chemical markers for monitoring the quality, efficacy, and safety of these products. Chemical markers are diagnostic constituents that are characteristic of the chemical profile of specific species and can be described generally as either bioactive markers or analytical markers. Bioactive markers, which may also serve as analytical markers, are responsible for the pharmacological or toxicological effects of the herbal product. Analytical markers are involved primarily in the development of analytical methodologies aimed at identifying and authenticating the botanical sources of the raw materials, the extracts, and the botanical ingredients of herbal products. Analytical markers are employed to detect substitution with plant species other than the target species, identify adulterants or contaminants in the herbal products, and evaluate the safety and efficacy of these products [62]. It is noteworthy that conventional quality control techniques utilize only a limited number of analytical or bioactive markers [62].

2.3 Roles of Chemical Evaluation in the Management of Herb Quality Consistency

Chemical evaluation of herbs and herbal products extends to the following applications:

1. Assessing the influence of environmental factors and growing conditions on the chemical profile of the plants [32, 57, 63].

2. Determination of the optimal environmental factors and growing conditions for maximizing accumulation of the bioactive plant metabolites utilized for medicinal, cosmetic, or nutritional purposes [64].
3. Authentication of herbal raw materials and herbal products [65, 66].
4. Discrimination of phenotypes collected from different geographic regions [65, 67].
5. Discrimination of plant species [59, 68].
6. Identification of the origin of plant material at the plant part level [69, 70].
7. Identification of chemical marker compounds for different functions, including quality markers [71], adulteration markers [72], and potential age-dependent markers [73].
8. Standardization of herbal products [74].
9. Detection of synthetic preservatives such as benzethonium chloride, methylparaben, triclosan, other synthetic compounds, or pharmaceuticals in commercial herbal products [75].
10. Determination of the effect of processing on the quality of herbal products [76, 77].
11. Detection of adulteration, substitution, and contamination (Table 1).

Chemical evaluation is characterized by its ability to stand alone as a robust approach with wide-ranging technologies and the capacity to play multiple roles in assessing herbs or herbal products. However, in a few cases, it was reported to be complemented with one of the major alternative analytical approaches, namely, morphology or microscopy or DNA-based techniques (Table 1, entries 29–32), principally due to the unavailability of standard reference samples of plant material [20, 78–81]. Chemical analytical tools have been utilized particularly to assess the quality consistency of herbs or herbal products.

2.4 *Major Objectives of Chemical Evaluation*

Several advanced techniques of instrumental chemical analysis have been developed to maximize the capability of achieving the following major goals:

1. Identification and authentication of the plant raw materials.
2. Detection and identification of contaminants, adulterants, and substitution with inferior quality plant species, either related to the target species or unrelated (with the species belonging to different genera).
3. Establishing fingerprints, chemical profiles, and carrying out metabolomics analysis of raw plant materials as well as finished products.

Techniques using modern instrumental chemical analysis (such as LC-MS, GC-MS, ^{1}H NMR, and HPTLC) combined with chemometrics for performing statistical data analysis are the backbone of chemical analysis. The details of the techniques, methods, and instrumentation used are described in the following sections.

3 Methods of Instrumental Chemical Evaluation

3.1 Liquid Chromatography-Mass Spectrometry

Liquid chromatography-mass spectrometry (LC-MS) is a highly sensitive, selective, and powerful separation and structure confirmation hyphenated technique. High sensitivity is an essential feature that determines the limit of detection with mass spectrometers, allowing detection of analyzed substances down to the parts-per-trillion (ppt) range. Combined LC-MS shows high capability of separation, combined with accuracy and a broad spectrum for confirming the identity of known compounds along with elucidating the identity of unknown compounds [104].

Most mass spectrometric analyses are performed by a mass analyzer attached to a separation instrument (hyphenation technology) such as HPLC or UHPLC and are assisted with chemometrics to validate interpretation of the acquired data [105]. Liquid chromatography-mass spectrometry and LC-MS/MS are considered techniques for targeted and untargeted (or non-targeted) analyses and depend fundamentally on including selected analytical markers or reference standards to construct calibration curves. However, obtaining reference compounds is costly, and, in some circumstances, it is not possible to obtain all those required [106].

Chemometric analysis has proven to be a powerful tool for the classification of metabolomics data. Figure 2 shows the Principal Component Analysis (PCA) and Partial Least Squares Discriminant Analysis (PLS-DA) utilized for distinguishing and classifying the metabolites of *Salvia* species. The PCA scores plot is divided into five groups based on the levels and occurrence of natural products. Chemometric analysis was also used in the differentiation of *Epimedium* species [107].

Further, ultra-high performance liquid chromatography (UHPLC) has been widely coupled to mass spectrometry to enhance the chromatographic resolving power and sensitivity, and reduce the analytical run time. A recent study applied UHPLC combined with to MS for quantification of the phenolic constituents of *Fadogia*

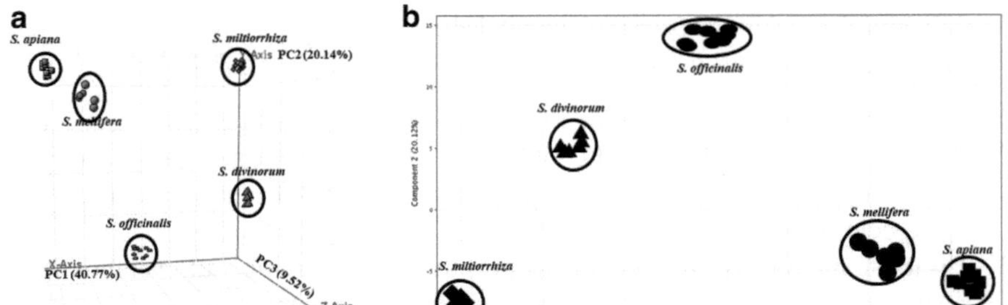

Fig. 2 **a** 3D PLS-DA, **b** 2D PCA scores plot showing the similarities and differences between five *Salvia* species samples illustrating good separation according to the positive-ionization mode data

Plate 7 *Fadogia agrestis* stems (photograph S. J. Adams)

agrestis (Plate 7) in dietary supplements claiming to contain *Fadogia*, for a study on the quality control of these herbal products [108].

3.1.1 Advantages of Liquid Chromatography-Mass Spectrometry

Liquid chromatography-mass spectrometry has a high level of sensitivity (n*M*), wide spectrum metabolite coverage, moderately expensive instrumentation, less costly maintenance compared to NMR spectroscopy, large spectral databases, and different commonly available resources of software.

Recently, ambient mass spectrometry, a rapidly growing and very successful technique for determining traceability and authenticity, has been introduced as an advanced tool to detect and quantify spice adulterants [109]. Ambient mass spectrometry has the advantage over related techniques by enabling sample analysis in the solid state. Ultra-high/high performance liquid chromatography-quadrupole-time-of-flight mass spectrometry (UH/HPLC-UV-QToFMS) has been used for the quality control, standardization, and detection of adulterants in herbal products [110–112].

A LC-DAD chemical fingerprinting method was employed for the identification of ginkgo plant material and herbal products containing this species. Various samples of ginkgo leaf or leaf extracts, *Sophora japonicum* flowers or fruits, and selected herbal products were compared (Fig. 3). Twenty-five herbal products tested showed the profile of ginkgo, but twenty contained adulterants, including *S. japonicum* and three flavonoid aglycones, with the most characteristic being genistein (**4**) (Fig. 4).

Direct Analysis in Real Time Time of Flight Mass Spectrometry (DART-ToFMS), based on direct analysis in real-time ionization in combination with time-of-flight MS

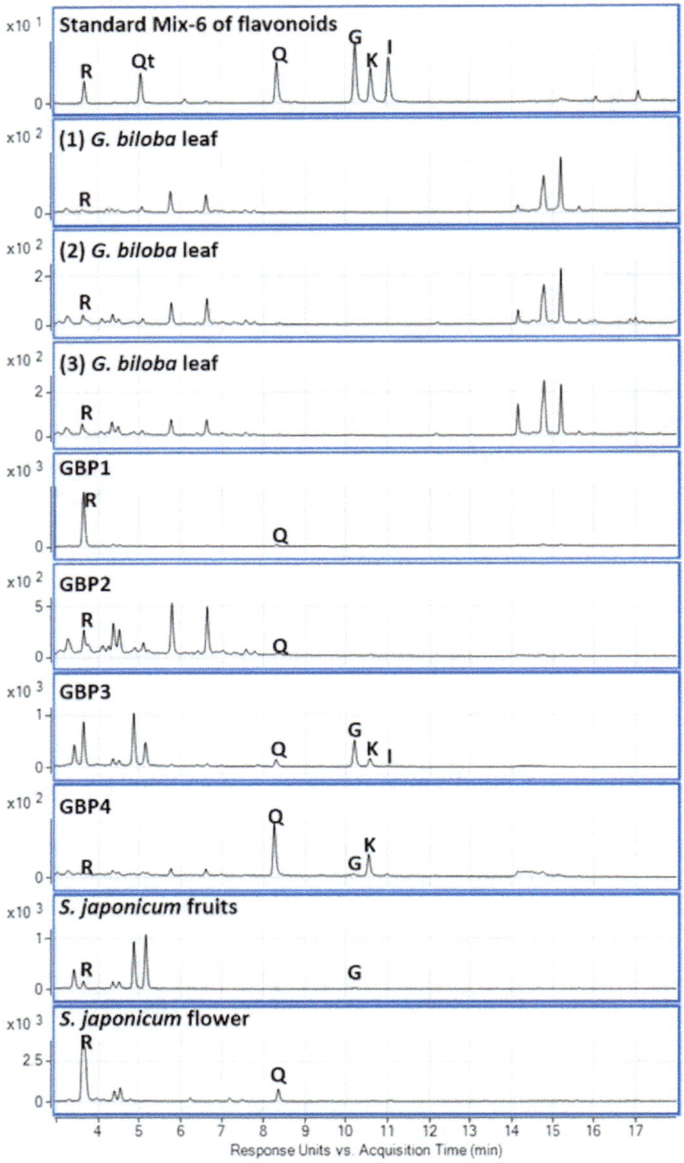

Fig. 3 LC-DAD chromatograms of a standard flavonoid mixture [rutin (R), quercitrin (Qt), quercetin (Q), genistein (G), kaempferol (K), isorhamnetin (I)], *Ginkgo biloba* leaf (#1–3), *Ginkgo* supplements (GBP1–GBP4), and flowers or fruits of *Sophora japonicum* at 280 nm

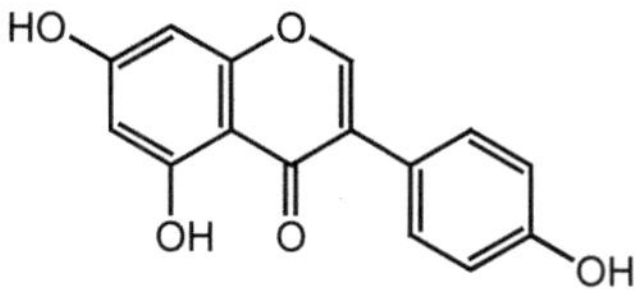

genistein (**4**)

Fig. 4 Genistein (**4**), a common adulterant in *Ginkgo biloba* herbal products

followed by principal component analysis (PCA) was employed to establish a method for discriminating different species of cinnamon with differing concentrations of coumarin (**5**) (Fig. 5), a compound with potential liver toxicity [113]. Two main varieties of cinnamon make up the cinnamon market in western countries. Ceylon cinnamon (*Cinnamomum verum* (Plate 8) or *Cinnamomum zeylanicum*) commonly is referred to as true cinnamon. Cassia cinnamon is the most widely marketed product and may refer to several species of cinnamon from different geographical regions, including *Cinnamomum cassia* (Chinese cassia) (Plate 9), *Cinnamomum loureirii* (Saigon cassia), or *Cinnamomum burmannii* (Indonesian cassia) [114]. Coumarin (**5**), a hepatotoxic compound occurs at up to a 1% concentration in cassia cinnamon but only at about 0.004% in Ceylon cinnamon [114]. The European Food Safety Authority has recommended a maximum intake of **5** of 0.1 mg/kg of body weight per day [114]. Identification and quantification of the level of **5** between Ceylon cinnamon and cassia cinnamon to minimize the risk of excessive coumarin consumption require the use of chemical analytical methods.

Direct analysis in real-time-of flight mass spectrometry was also exploited to detect the presence of Japanese star anise as an adulterant in Chinese star anise, using anisatin (**6**) (Fig. 6) as a marker. Chinese star anise, commonly used as a spice and in herbal teas, is morphologically similar to Japanese star anise, but the latter species contains **6**, a neurotoxin acting as a non-competitive GABA antagonist that may cause convulsions [115].

Another popular variant of ambient mass spectrometry is atmospheric solid analysis probe-mass spectrometry (ASAP-MS), a technique that was applied successfully to detect and quantify coumarin (**5**) in different samples of cinnamon [116].

The LC-MS technology platform has long been exploited expansively for a wide range of applications. Liquid chromatography-mass spectrometry has been utilized extensively in studying herb metabolomics as the basis for quality assurance and

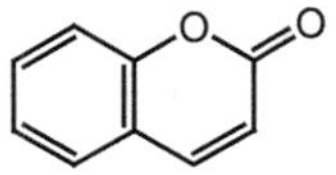

coumarin (**5**)

Fig. 5 The hepatotoxic compound coumarin (**5**)

Plate 8 *Cinnamomum verum* (photograph S. J. Adams)

quality control of herbal products and assessing their quality consistency when acquired from different sources and from different batches.

3.1.2 Limitations of Liquid Chromatography-Mass Spectrometry

Liquid chromatography-mass spectrometric instrumentation is not as robust as that used for NMR spectroscopy, and its reproducibility is moderate. The sample preparation is lengthy, with the analysis preceded by chromatographic separation, such as HPLC and UHPLC, and the spectra are not predictable unless purposely modified. The MS methods alone do not provide quantitative data and also do not provide complete structural identifications [117].

3.1.3 Metabolomics in the Application of Quality Assurance and Quality Control of Herbal Products

Metabolomics is the comprehensive qualitative and quantitative chemical profiling of a plant part or extract for a specific range of metabolites (targeted analysis) or for the complete domain of metabolites (untargeted analysis) with a molecular

Plate 9 *Cinnamomum cassia* (photograph S. J. Adams)

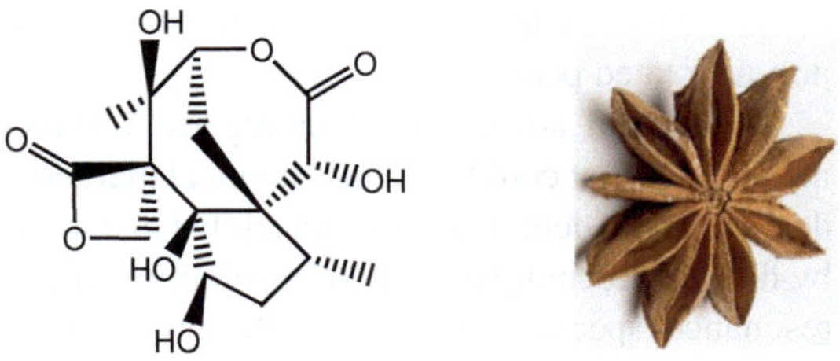

6 (anisatin)

Fig. 6 The neurotoxin anisatin (**6**), a constituent of Chinese star anise

weight range of 30–3,000 Da [114, 118, 119]. The metabolome (small-molecule metabolites) changes in response to environmental factors, growing conditions, and stress conditions and differs from plant part to plant part and among different plant tissues [120, 121]. Thus, the metabolome is associated with the plant phenotype, and with the metabolites that are the end products of cellular activity [115]. Usually, metabolomics analysis is carried out with the assistance of chemometrics allowing statistical analysis of the extensive and highly complex data [105].

The liquid chromatography-mass spectrometry-based metabolomics approach has been adopted extensively as a powerful tool for evaluation of the quality and quality consistency of herbal products. A large number of published studies show the versatility of this approach in various applications, including identification of marker metabolites [122], quality assessment of plant raw material, and quality assurance and quality control of herbal products. Metabolomics provides wide-ranging chemical profiles of the metabolites (qualitative identification combined with quantitative estimation) contained in a single herb or a multiple ingredient herbal product. In combination with chemometrics, liquid chromatography/high resolution mass spectrometry-based metabolomics (LC/HRMS-based metabolomics) can identify the complete spectrum of herbs and herbal product metabolites. The data generated from metabolomics-based analysis of a large number of metabolites allows accurate authentication of plant species and provides a basis for distinguishing between closely related species, samples of different geographic origins, and between the same species collected at different harvest times or seasons. This technique has found applications in identifying potential metabolites that can be employed as target markers for species identification and for discriminating different species [123].

3.1.4 Comprehensive Untargeted Metabolomics for High-Throughput Detection of Novel Metabolites

Herbal products may be composed of several hundred constituents or metabolites, from those that are highly polar to highly lipophilic compounds [124]. Untargeted metabolomics with LC-MS is based on simultaneously identifying such a large array of metabolites. Analysis using untargeted metabolomics depends on employing a scanning mode for comprehensively identifying all detected peaks above the noise threshold, coupled with statistical analysis, and including annotated and non-annotated peaks.

The central advantage of untargeted metabolomics is the discovery of novel metabolites that could serve as chemical markers for quality assessment. The task of detection and identification of the global metabolites has been performed extensively by different techniques of high-resolution mass spectrometry and nuclear magnetic resonance spectroscopy [125]. Recent improvements in analytical approaches, including separation techniques and mass spectrometric technology, have resulted in substantial broadening in the spectrum of metabolite coverage.

The advent of ultra-high performance liquid chromatography-mass spectrometry (UHPLC-MS) has expanded the capability of metabolite detection by approximately 20% compared to high-performance liquid chromatography-mass spectrometry (HPLC-MS). Continuous development in MS technology has led to wide-ranging metabolite identification capability [126].

Metabolome analysis is often coupled to electronic databases to provide standard spectral data for identifying the detected constituents or metabolites. Quantifying bioactive metabolites is essential for standardization and therefore for quality consistency and assurance of safety and efficacy. In addition to quantification of the bioactive metabolites, broad-spectrum chemical profiling of the metabolites extracted from a particular plant part may be carried out to generate a comprehensive assessment of botanical quality [126, 127].

Liquid chromatography-mass spectrometric techniques have gained widespread applications in the realm of fingerprinting, chemical profiling, standardization, metabolomics, and assurance of herb quality consistency. An UHPLC-DAD-QToF-MS method was developed for the quality control of grapefruit and other *Citrus* seed extracts and herbal products. The method also was designed to detect the presence of synthetic preservatives [75]. In another study, a similar method (TLC and UHPLC-DAD-QToF-MS) was used for identifying adulterants of *Ginkgo biloba* (Plate 10) in supplements [110]. A recent publication described a UHPLC-UV-QToF-MS/MS method coupled with an LC-MS informatics platform for the identification of the characteristic metabolites of *Eleutherococcus senticosus* in the leaves of the plant as well as in commercial products. Green tea (*Camellia sinensis*) was detected as an adulterant not listed on the labels of these products [88]. In conjunction with chemometrics, UHPLC-QToF-MS was employed in untargeted analysis for the generation of broad-spectrum chemical fingerprints of three herbs—*Hoodia, Chamaemelum,* and *Terminalia* species (Plates 11, 12 and 13)—in order to identify marker metabolites for discriminating German chamomile from Roman chamomile and for discriminating the different species of *Terminalia* (Plates 13, 14, 15, and 16) In addition, the authors used this method for the authentication and quality assessment of *Hoodia gordonii* (Plate 11) commercial products [71].

A recently developed UHPLC-QToF-MS method was described for assessment of the quality of paw paw dietary supplements, using targeted and non-targeted analysis leading to comprehensive chemical profiling of the alkaloid and acetogenin constituents of *Asiminia* and *Annona* species [112]. For the targeted analysis, 35 standard references were included in the research. In the different plant parts of *Asiminia triloba* (Plate 17) and *Asiminia parviflora* (Plate 18) samples, 131 metabolites were identified. These metabolites can serve as marker compounds to distinguish between the different species of *Asimina.* There are numerous examples of the chemical analysis of herbal products to demonstrate the utility of liquid chromatography-mass spectrometry hyphenated techniques with broader application than ^{1}H NMR spectroscopy [88, 99, 101, 110, 112]. Among the approaches adopted for fingerprinting and metabolic chemical profiling of herbal products, mass spectrometry connected to chromatographic instrumentation represents one of the most widely used general approaches for this purpose.

3.2 Gas Chromatography-Mass Spectrometry

Gas chromatography-mass spectrometry (GC-MS) is another major procedure that uses mass spectrometry for the identification of volatile substances such as essential oils, derivatized fatty acids and sterols, and other derivatized compounds. Gas chromatography-mass spectrometry complemented with additional techniques has been employed to authenticate aromatic plants, spices, and essential oils and for the

Plate 10 *Ginkgo biloba* leaves (photograph H. Zell, Wikimedia Commons, https://commons.wikimedia.org/wiki/File:Ginkgo_biloba_010.JPG)

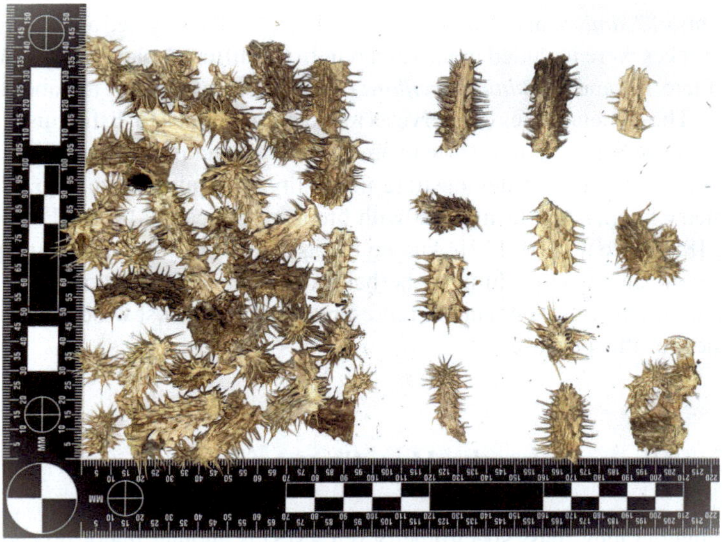

Plate 11 *Hoodia gordonii* (photograph S. J. Adams)

Plate 12 *Chamaemelum nobile* (photograph S. J. Adams)

Plate 13 *Terminalia paniculata* (photograph S. J. Adams)

Plate 14 *Terminalia alata* (photograph S. J. Adams)

Plate 15 *Terminalia arjuna* (photograph S. J. Adams)

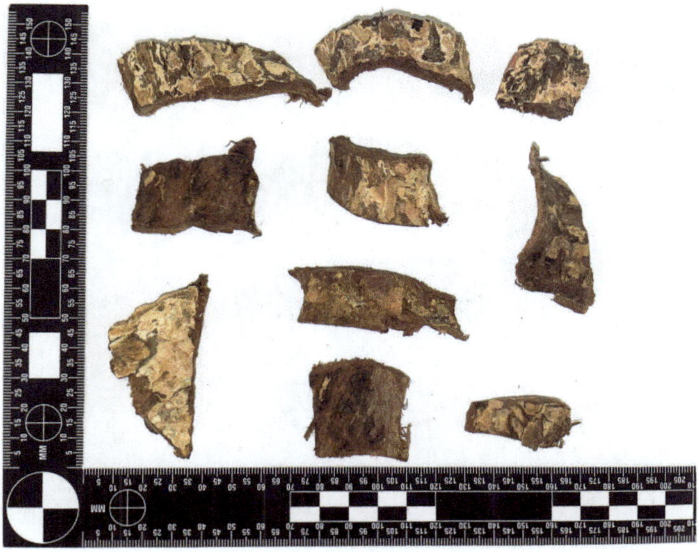

Plate 16 *Terminalia catappa* (photograph S. J. Adams)

Plate 17 *Asimina triloba* (photograph K. Ziarnek, Wikimedia Commons, https://de.m.wikipedia.org/wiki/Datei:Asimina_triloba_kz1.jpg)

Plate 18 *Asimina parviflora* (photograph Bruce Winter, Wikimedia Commons, https://commons.wikimedia.org/wiki/File:Asimina_parviflora_fruit.jpg)

detection and identification of adulterants [102]. Additionally, gas chromatography connected to combustion/pyrolysis-isotope ratio mass spectrometry (GC-C/P-IRMS) and combined with TC/EA-IRMS, GC-FID (FID = flame-ionization detector), GC-MS, and ^{14}C radiocarbon dating was able to discriminate synthetic from natural methyl salicylate, the major component of wintergreen oil. The technique was also applied to authenticate birch essential oil since methyl salicylate also represents its major component [128]. A literature search supported the utility of GC-MS in this area, since among the applications was the characterization of adulterants in fennel fruit by profiling its essential oil against the two suspected adulterants (cumin and dill), and also the addition of olive leaves to oregano and sage [129].

The utility of GC-MS with enantioselective chromatographic columns has proved a more efficient method for identifying adulterants based on the generation of chemical profiles involving the chiral characteristics of the essential oil constituents [130]. GC-MS comprised of chiral or non-chiral gas column chromatography coupled to an advanced mass spectrometry technique was employed for quality control of marketed products of bay leaves [131].

The GC-MS methodology has a limited applicability attributable to its inherent design to analyze only volatile compounds. Recently, a low-field NMR technique (60 MHz ^{1}H NMR), has been proposed as a complementary tool to GC-MS for evaluating essential oils. The analysis involved authentication, quality, and inspection for the presence of adulteration with less expensive oils or spiking with synthetic compounds. This NMR technique has an advantage over GC-MS in being simple, rapid, reproducible, and less expensive, and in having the capacity to identify non-volatile adulterants [132].

3.3 *Proton Nuclear Magnetic Resonance Spectroscopy*

Proton nuclear magnetic resonance spectroscopy (^{1}H NMR) has gained widespread popularity as a tool for the quality control, standardization, and targeted and untargeted metabolomics of herbal products [117].

Proton nuclear magnetic resonance spectroscopy is often used in association with chemometrics when applied to the quality control of herbal products [105]. It has been reported extensively for its utility in the quality assessment of traditional medicinal herbs, including fingerprinting, chemical profiling, and quantification of the bioactive phytochemicals in medicinal herbs and herbal products.

3.3.1 Advantages of Proton Nuclear Magnetic Resonance Spectroscopy

Proton nuclear magnetic resonance spectroscopy has the advantages of being highly reproducible, robust, non-destructive, and highly accurate. Other advantages include analytical time efficiency, a simple sample preparation, the lack of any need for a prior separation technique like chromatography, or for any chemical derivatization before analysis. Furthermore, ^{1}H NMR spectra can be predictable and provide accurate structure determination and quantitative analysis.

Although less sensitive than MS, ^{1}H NMR spectroscopy provides a powerful complementary technique for the identification and quantification of metabolites in plant extracts. It is utilized commonly in conjunction with mass spectrometry, with or without an associated chromatographic procedure. Adding to the aforementioned strengths, ^{1}H NMR spectroscopy can identify unknown compounds or compounds unrelated to the reported metabolites of the target herbs.

Other advantages of ^{1}H NMR methodology encompass its capability to comprehensively record signals from all chemical classes and to be able to identify unrelated substances, such as adulterants, which are not natural components of the authentic material. Proton nuclear magnetic resonance spectroscopy can also distinguish between chemical compounds with minor structural differences adding to the accuracy of quantitative analysis of multiple samples simultaneously using a single reference [133].

3.3.2 Limitations of Proton Nuclear Magnetic Resonance Spectroscopy

A main limitation of using ^{1}H NMR spectroscopy for routine analysis is the high cost of operating the instrument. Other limitations of ^{1}H NMR procedures include poor to moderate sensitivity (μM), a range of metabolite identification less than that for mass spectrometry, expensive instrumentation, costly maintenance, a limited number of available databases containing ^{1}H NMR spectra, and in having comparatively few software resources [117]. In addition, this approach may be characterized by occasional signal overlap and lower sensitivity relative to mass spectrometry and HPLC [134].

3.3.3 Applications of Proton Nuclear Magnetic Resonance Spectroscopy in the Quality Assessment of Medicinal Herbs and Herbal Products

The ^{1}H NMR technology platform has been employed for various applications in the arena of quality assurance and quality control of herbs and herbal products as well as associated aspects. These include:

- the discrimination of botanical identity based on geographic origin [70];
- the authentication of medicinal and traditional herbs [89, 135];
- the analysis of herbs using ^{1}H NMR-based metabolomics [136];
- the targeted ^{1}H NMR-based quality control of herbs [137];
- ^{1}H NMR-based quantification of targeted bioactive constituents [138, 139];
- ^{1}H NMR-based quantification of substances bearing a health risk [140];
- ^{1}H NMR-based metabolite profiling to study the effects of plant age and harvest time on the accumulation of bioactive phytochemicals [141];
- the identification and quantification of adulterants in herbal products [87, 142]; and the
- ^{1}H NMR-based metabolomics in species discrimination [59, 68].

Using DNA and NMR methods to study ginseng samples, the roots of *Panax ginseng* (Plate 19) were differentiated based on their geographic origin. Sixty ginseng samples acquired from Korea and China were inspected using DNA-based techniques and found to be indistinguishable owing to the similarity of their genetic makeup and the absence of genetic markers. In contrast, a ^{1}H NMR-based metabolomics method when combined with an OPLS-DA statistical model could distinguish the samples based on their geographic origins [67].

In December 2003, the FDA banned the sales of all dietary supplements containing *Ephedra* or any of its alkaloids (e.g., ephedrine alkaloids **7** and **8**) (Fig. 7) based on these leading to well-documented cardiovascular health risks to consumers. Prior to being banned for sale, *Ephedra* dietary supplements were used in the U.S. for weight loss and to enhance athletic performance. To detect *Ephedra* or any ephedrine alkaloid in herbal products, recently reported research has led to a simple and rapid ^{1}H NMR method for the detection and quantification of ephedrine alkaloids in suspected herbal products, in order to assure their safety and consistency [143].

In a recent study on popular ethnomedicinal curcumin-containing herbal products, ^{1}H NMR spectroscopy demonstrated an ability to identify a synthetic curcumin (**9**) adulterant (Fig. 8) with potential health risks attributable to the residual impurities from the synthesis process [87]. The results indicated the capability of quantitative proton NMR (qHNMR) in the detection and characterization of untargeted adulterants or ingredients that are omitted from the labels of the products concerned [87].

Plate 19 *Panax ginseng* (= *quinquefolium*) (photograph from "American Medica Botany", Boston, 1817 per Wikimedia Commons, https://commons.wikimedia.org/wiki/File:Panax_quinquefolium,_ginseng_%283543549600%29.jpg)

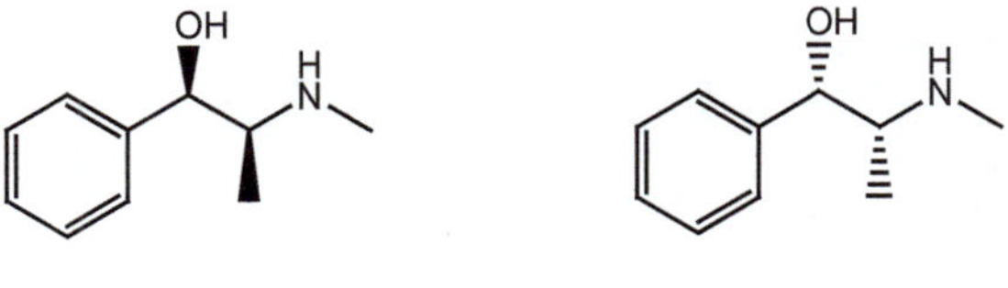

(−)-(1*R*,2*S*)-ephedrine (**7**)   (+)-(1*S*,2*R*)-ephedrine (**8**)

Fig. 7 Ephedrine alkaloids **7** and **8**

A study analyzing andrographolide (**10**), neoandrographolide (**11**), and 14-deoxyandrographolide (**12**) (Fig. 8) during the growth period of *Andrographis paniculata* (Plate 20) by ^{1}H NMR spectroscopy revealed that these bioactive compounds were most abundant in the young leaves during the pre-flowering stage of growth, indicating an optimal harvest time of 18 weeks [140].

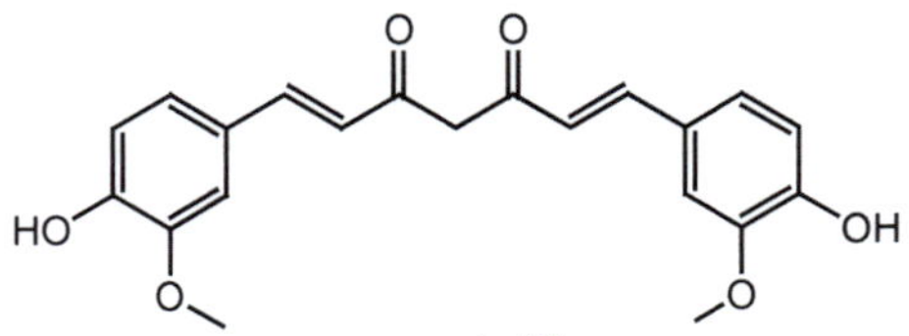

curcumin (9)

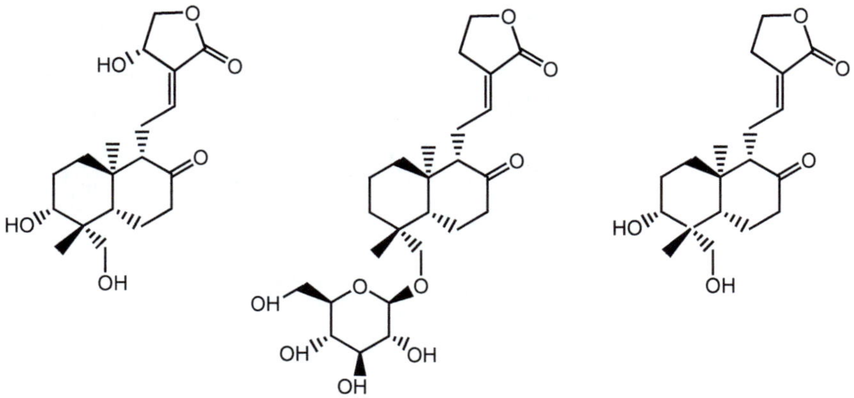

andrographolide (10) neoandrographolide (11) 14-deoxyandrographolide (12)

Fig. 8 Curcumin (**9**), andrographolide (**10**), neoandrographolide (**11**), and 14-deoxyandrographolide (**12**)

Plate 20 *Andrographis paniculata* (photograph S. J. Adams)

Proton nuclear magnetic resonance spectroscopic analysis has been combined with chemometrics for statistical analysis and validation of the generated data. Liquid chromatography-NMR hyphenated techniques and NMR spectroscopy combined with DNA barcoding methods have been used frequently to enhance the capability of NMR spectroscopic methods compared with when they are used alone [20, 135, 141].

Cranberries (*Vaccinium macrocarpon*, Plate 21) have shown multiple beneficial effects on health, with remarkable antioxidant properties [144]. A recent study was conducted using ^{1}H NMR-based metabolomics to identify the major phytochemicals present in cranberry dietary supplements. Quantitative ^{1}H NMR (qHNMR) was employed to quantify the key bioactive metabolites of cranberries, namely, anthocyanins, proanthocyanidins, and citric acid, and the constituents of cranberry peel, including ursolic acid, oleanolic acid, and hyperoside. Ursolic, oleanolic, citric, quinic, and malic acids were determined in the whole cranberry powder standard used, but their presence was minimal or they were absent in several supplements. This study, in combination with chemometrics, demonstrated that the metabolic profiles of different commercial products of cranberries exhibited significant variations when compared to a whole cranberry powder reference standard [136].

The advent of compact NMR spectrometers (benchtop NMR) with permanent magnets that provide proton NMR frequencies between 40 and 80 MHz (low field, LF NMR) has led to new applications in the area of food and herbal product quality control [145–149]. Recently, the approach of using low-field (60 MHz) NMR spectroscopy combined with chemometrics was applied to the quality control

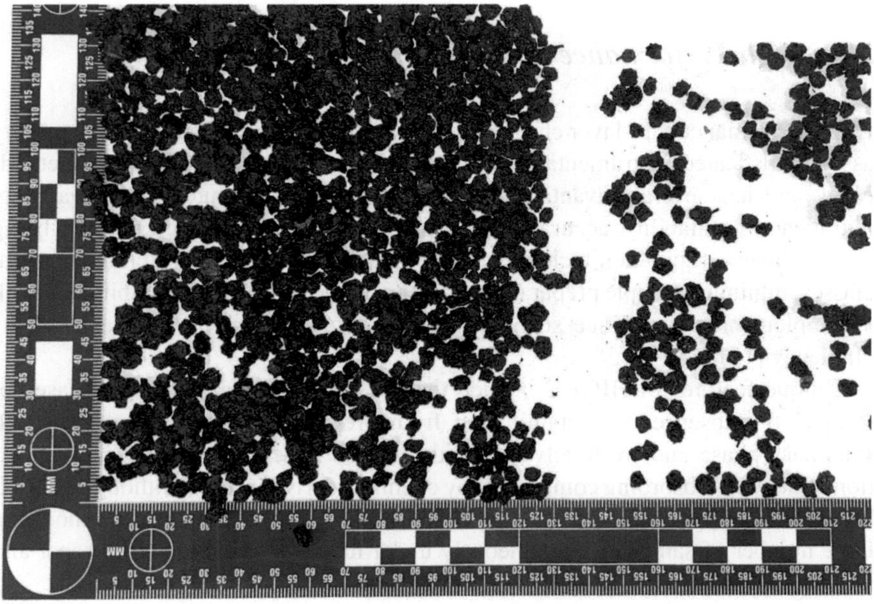

Plate 21 *Vaccinium macrocarpon* (photograph S. J. Adams)

of herbal products containing cinnamon in order to discriminate Ceylon cinnamon (*Cinnamomum verum*, Plate 8) from cassia cinnamon (*Cinnamomum cassia*, Plate 9), based on their chemical profiles. In addition, quantification of the levels of the hepatotoxic compound, coumarin (**5**) (Fig. 5), in the analyzed samples, was performed. The spectra generated from the LF NMR spectrum could not be interpreted unambiguously, due to signal overlapping. However, characteristic signals attributed to (*E*)-cinnamaldehyde and **5** could be detected when the NMR data were assessed using chemometrics, thereby considerably enhancing the resolution of these characteristic signals. The study revealed that several cassia cinnamon samples have high levels of coumarin (**5**) while Ceylon cinnamon samples are either devoid of or contain only low levels of **5**. The study showed also that the current application of LF NMR in conjunction with chemometrics should be limited to well-studied cases and cannot be considered a universal technique for quality control of plant-containing foods or herbal products [137].

Several applications for LF NMR/benchtop NMR recently were reported [146]. Low-field NMR/benchtop NMR is an emerging technique and holds promise for further utilization in the quality assurance and quality control of herbal products. Liquid chromatography-mass spectrometry (LC-MS) and liquid chromatography-proton magnetic resonance spectroscopy (LC-^{1}HNMR) are used predominantly for metabolomics analysis. Mass spectrometry and NMR spectroscopy can be used in tandem as complementary techniques for analyzing complex mixtures of plant constituents [150].

3.4 High-Performance Thin-Layer Chromatography

High-performance thin-layer chromatography (HPTLC) has gained wide popularity as a sophisticated instrumental technique operating with a standardized methodology and has several advantages that are represented by automation, scanning, optimization, reliability, accuracy, and reproducibility. It is characterized by being a cost-efficient separation technique with selective detection of different chemical classes, minimum sample preparation, and having the capacity for combination with a complementary technique such as mass spectrometry or HPLC to enhance its efficiency [151, 152].

Unique features of HPTLC include the low consumption of mobile phase per sample, the absence of contamination from previous analyses due to the fresh stationary phase and fresh solvents used for each experiment, and the visualization of non-UV absorbing compounds by chemical derivatization. Although HPTLC has a lower resolution power in comparison with GC, HPTLC can accommodate a large number of samples simultaneously under identical conditions for proper and reliable comparison and high sample throughput.

High-performance thin-layer chromatography, as a planar chromatograpic procedure, has the advantage over GC and HPLC in being able to use multiple detection methods, such as UV absorption, fluorescence, or chemical derivatization, allowing the analysis of compounds with poor UV detectability.

High-performance thin-layer chromatography includes functions that allow the generation of saved images of the chromatograms of herbal samples collected from different geographic regions, or harvested at different seasons or different stages, or from different plant parts for comparison with commercial products or for identification and authentication purposes [153, 154]. A new concept in HPTLC methodology termed "comprehensive HPTLC fingerprinting" has been introduced recently [155]. This new concept enables the conversion of HPTLC fingerprints into peak profiles (similar to HPLC chromatograms) and quantifying selected compounds based on measuring the intensities of the corresponding peaks in comparison to the intensities of standard references.

As such, HPTLC analysis can provide multiple levels of information, including identity, purity, and content of marker compounds of an herbal product, all in a single experiment.

High-performance thin-layer chromatography has been a method of choice for the identification of herbs/herbal products in several pharmacopeias. This group of techniques has been exploited extensively in assessing analytical consistency through establishing the identity and authenticity of herbs and herbal products, analyzing the raw materials and finished products, discriminating plant species based on inspecting their chemical fingerprints, and detecting adulterants. Further, this methodology has been widely utilized widely to quantify quality marker compounds or bioactive marker compounds and to generate chemical fingerprints of plant species and plant parts of individual species to be used for the identification, quality assurance and quality control of herbal products.

3.4.1 Representative Applications of High-Performance Thin-Layer Chromatography

High-performance thin-layer chromatography (Fig. 9) has been utilized for the assessment of some species used in Traditional Chinese Medicine (TCM) [107, 156, 157]. Figure 9 shows a mixture of reference standards simultaneously analyzed by HPTLC with several *Epimedium* species from different herbal products [158]. Characteristic blue fluorescent bands are distinctive for *Epimedium grandiflorum* (Plate 22) but are not seen with two other species of *Epimedium*. The method is suitable for the rapid qualitative analysis of the differences among these species and herbal products. HPTLC provides a visible or fluorescence image for parallel assessment of samples on the same plate.

The HPTLC technique was applied to discriminate 27 samples of cultivated *Angelica gigas* roots. The roots of several *Angelica* species are used as traditional medicines and are sold on the market with the same vernacular name, "Dang gui". This refers to *Angelica sinensis*, *Angelica acutiloba*, or *Angelica gigas*, despite having different chemical compositions (i.e. different chemical profiles). High-peformance

Plate 22 *Epimedium grandiflorum* (photograph KENPEI, Wikimedia Commons, https://de.m.wikipedia.org/wiki/Datei:Epimedium_grandiflorum_var_thunbergianum2.jpg)

thin-layer chromatography methods were employed to establish standard chemical fingerprints for *Angelica gigas* (Plate 23) as the authentic species while also establishing also the chemical fingerprints for the other closely related species, *Angelica sinensis* and *Angelica acutiloba,* but by considering the latter species as adulterants [157].

Further applications of HPTLC include quantification of anthocyanins in *Euterpe oleracea* (açai) [159], quantification of caffeic acid and quercetin in cranberry [160], correlation of the influence of seasonal and geographical variations on the chemical composition of *Aerva lanata* (Plate 24) [161], discrimination, when in combination with HPLC, of cultivars of Jamaican ginger (*Zingiber officinale*, Plate 25) [81, 162], and quantification of phenolic acids and withanolides in *Withania somnifera* (Plate 26) [163].

High-performance thin-layer chromatographic analysis has also been complemented with microscopic analysis to enhance its capability in verifying the identity of plant species, detecting adulterants, and evaluating quality consistency. Thus, characterization of *Dioscorea* species was accomplished by HPTLC analysis combined with microscopic analysis [164]. HPTLC chemical fingerprinting of the leaves, stems, and roots of *Fadogia agrestis* (Plate 27), for the quality assurance and quality control

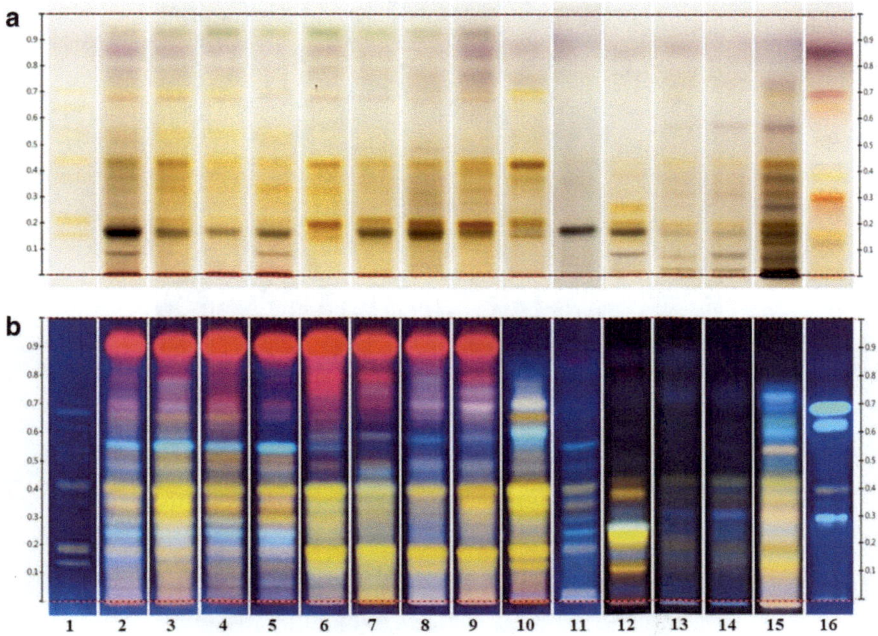

Fig. 9 High-pressure thin-layer chromatography under white light after derivatization with vanillin-sulfuric acid reagent (**a**) and UV 366 nm after derivatization with natural product/polyethylene glycol (**b**). Track 1: standard mixture; tracks 2–4: *Epimedium grandiflorum*; tracks 5–7: *E. brevicornu*; tracks 8–9: *E. sagittatum*; tracks 10–16: dietary supplements claiming to contain *Epimedium*

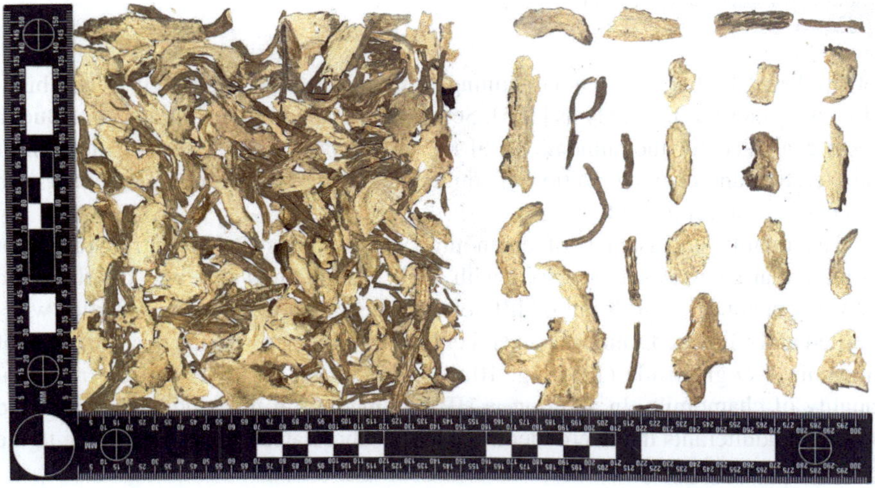

Plate 23 *Angelica gigas* (photograph S. J. Adams)

Plate 24 *Aerva lanata* (mountain knotgrass, photograph Challiyan, Creative Commons, https://commons.wikimedia.org/wiki/File:Aerva_lanata_3.jpg)

of marketed herbal products containing this species, was conducted in combination with microscopic analysis [165]. Several commercial herbal products of Buchu were evaluated for the authenticity of the target species and also for detection of adulteration and characterization of adulterants, using HPTLC complemented with microscopic analysis [166].

The quality consistency of chamomile tea products acquired from different commercial sources was assessed with a HPTLC method that was developed to generate chemical fingerprints of different samples of chamomile tea, including wild-crafted and cultivated samples (several cultivars), together with the quantification of apigenin 7-*O*-glucoside (**13**) (Fig. 10) as a biomarker for assessing the remedial quality of chamomile. In addition, a HPTLC method in this study was employed to detect adulterants that were identified as closely related species belonging to the plant family Asteraceae [167].

Based on their implication in the curative properties of *Boswellia serrata* (Plate 28), quantification and quality assessment of two boswellic acids, 11-keto-β-boswellic acid (**14**) and 3-*O*-acetyl-11-keto-β-boswellic acid (**15**) (Fig. 10), were achieved using HPTLC in an evaluation of different herbal products of *Boswellia serrata* [168].

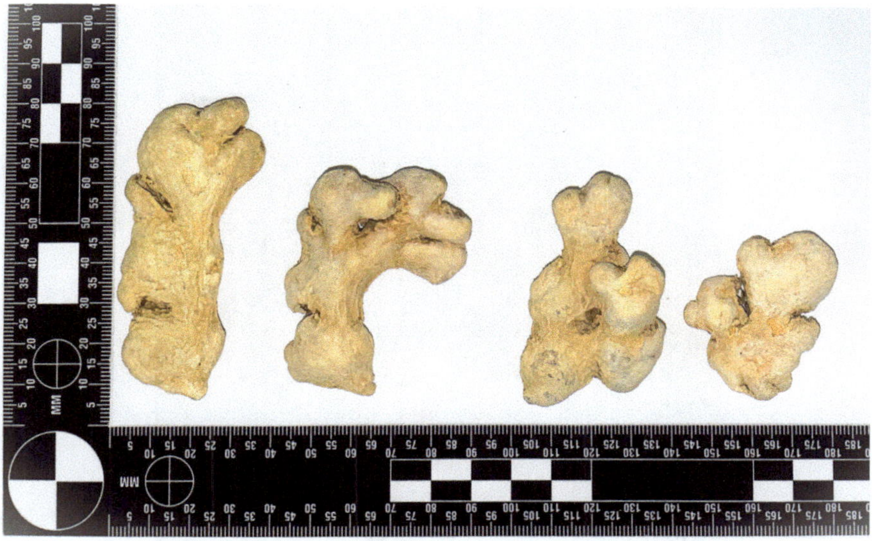

Plate 25 *Zingiber officinale* (photograph S. J. Adams)

Recently published research employed HPTLC/MS-based analysis to evaluate the product quality of ginger (*Zingiber officinale*, Plate 25) and to study the effects on its chemical composition of environmental conditions, including temperature, humidity, seasonality, soil properties, and cultivation conditions, along with postharvest factors such as grinding and extraction. In this study, two putatively bioactive constituents of ginger, namely, 6-gingerol (**16**) and 6-shogaol (**17**) (Fig. 10), were analyzed quantitatively in 17 marketed products of ginger using HPTLC-UV/Vis/FLD-bioassay (FLD = fluorescence detector) and HPTLC-ESI-HRMS methods. These techniques were utilized to assess scertain emedial claims of ginger by conducting several in vitro bioassays that led to the identification of previously undescribed bioactive components [151].

Another investigation involving HPTLC was focused on identifying the botanical origin of caffeine in commercial products and simultaneously quantifying the amount of this alkaloid included with other ingredients. The safe consumption of caffeine (**18**) (Fig. 10) is $\leq$ 400 mg per day [169]. In addition, the study described a procedure to detect caffeine when added as an adulterant [169]. In order to conduct this work, the authors developed chemical fingerprints of extracts obtained from caffeine-rich botanicals, namely, guarana seeds, tea leaves, maté leaves, coffee beans, and cola nuts, with the assistance of standard references of the constituents that occur naturally in these caffeine-rich botanicals. The HPTLC method developed allowed for the discrimination of different caffeine-containing botanicals and the selected extracts, and quantification of the caffeine contained in these samples. This method can be used in the quality control of caffeine-containing herbal products and the detection of adulteration with caffeine [169].

Plate 26 *Withania somnifera* (photograph Wowbobwow12 Wikimedia Commons, https://commons.wikimedia.org/wiki/File:WithaniaFruit.jpg)

Additional representative examples of applications of high-performance thin-layer chromatography include the development of a HPTLC method for determination of silymarin (a mixture of flavonolignan derivatives) in commercial herbal products for hepatoprotection, in order to assure the occurrence of appropriate levels of the active ingredient to provide consistent efficacy [170].

Substitution and adulteration of cinnamon, a highly popular and reputable traditional spice, provided the impetus for the development of methods to evaluate the quality of cinnamon products and the consistency of the quality of these commercial products. A HPTLC method was developed for simultaneous quantitation of the major constituents of cinnamon, cinnamaldehyde (**19**) and eugenol (**20**) (Fig. 10), as quality markers in the common species of cinnamon, *Cinnamomum zeylanicum, C. burmannii,* and *C. cassia* (Plate 9), and for their essential oils. This analytical technique, including ultrasound-assisted methanol extraction, was proposed as a method for quality assurance and quality control of herbal products containing cinnamon species [171].

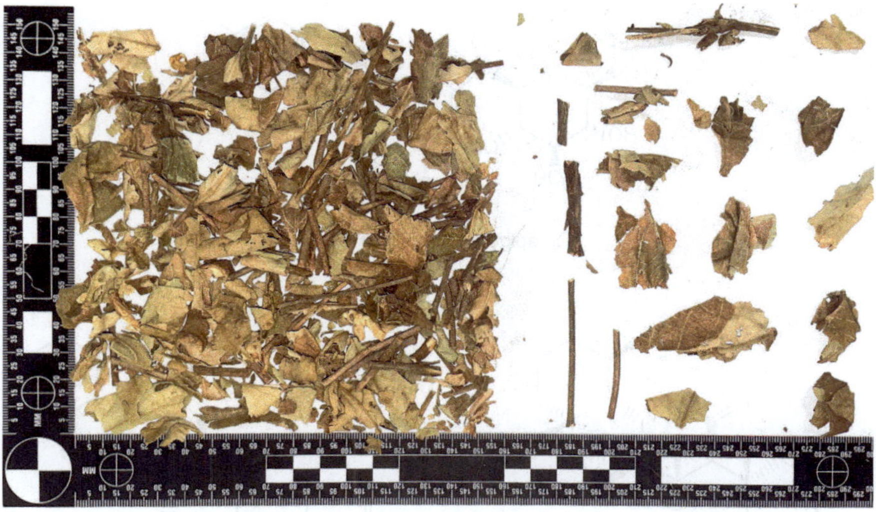

Plate 27 *Fadogia agrestis* leaves and secondary stems (photograph S. J. Adams)

Plate 28 *Boswellia serrata* (photograph S. J. Adams)

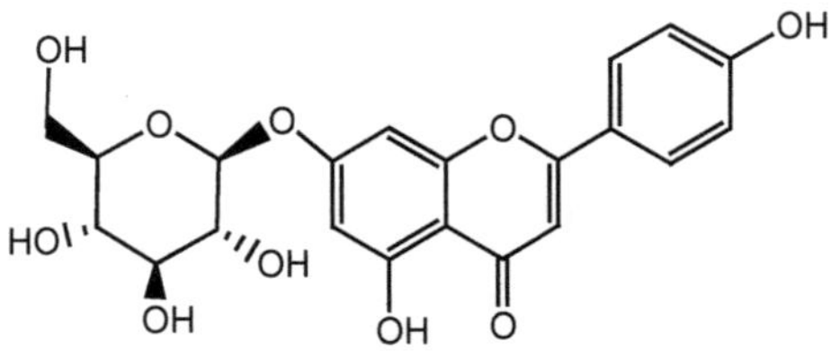

apigenin 7-*O*-glucoside (**13**)

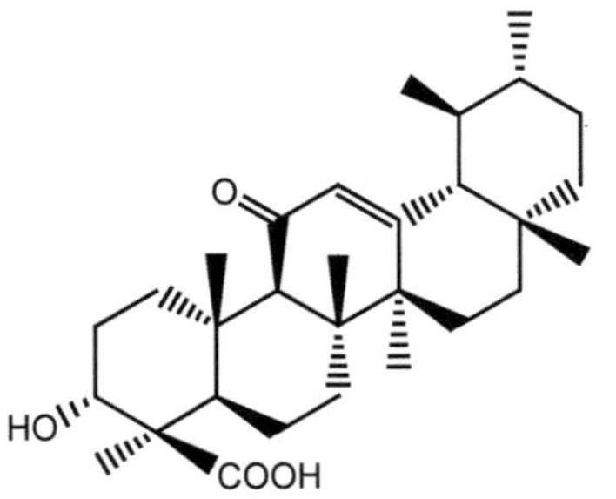

11-keto-ß-boswellic acid (**14**)

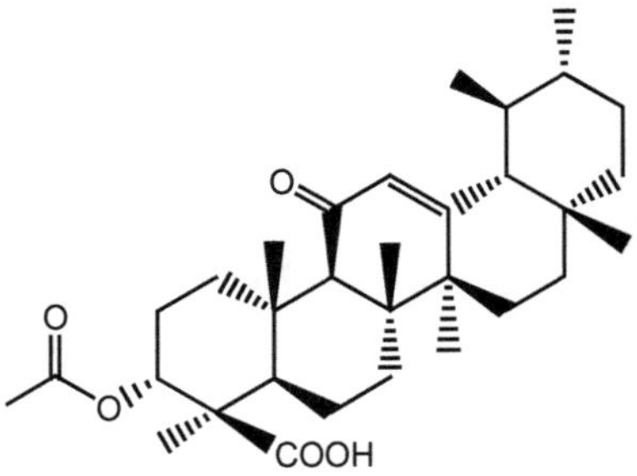

3-acetyl-11-keto-ß-boswellic acid (**15**)

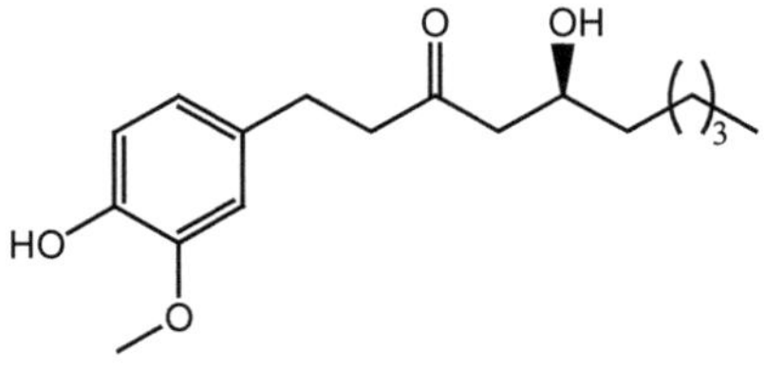

6-gingerol (**16**)

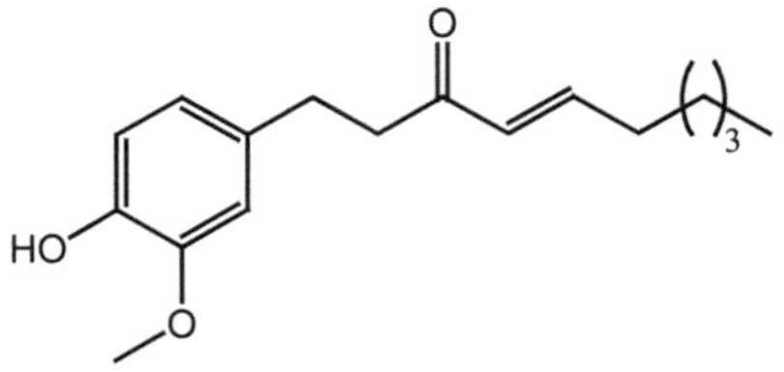

6-shogaol (**17**)

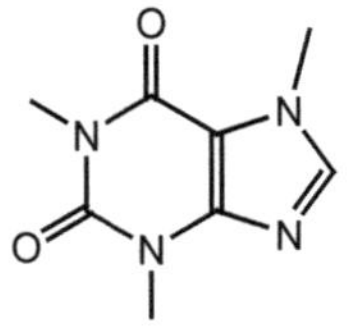

caffeine (**18**)

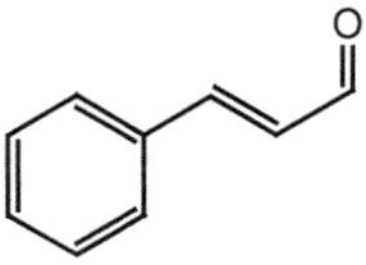

cinnamaldehyde (**19**)

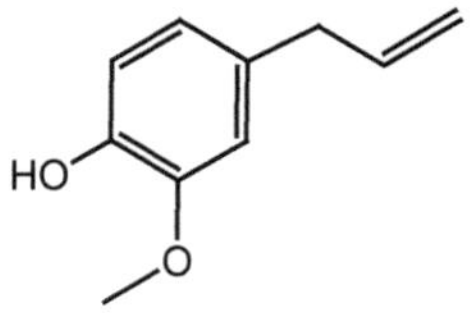

eugenol (**20**)

Fig. 10 Compounds **13–20**

4 Conclusions

Chemical composition/chemical profiles of herbal products (botanical dietary supplements), as represented by the spectrum of phytochemicals and the ratios of individual constituents in the plant, in the reported species have been linked to the remedial and therapeutic efficacy of medicinal plants. This chemical composition of a given plant species is highly responsive to environmental factors and growing conditions, leading to the development of several phenotypic species with variable chemical compositions even though they originate from one genotypic species. In addition, postharvest factors, including drying and processing methods, contribute to further inconsistencies in chemical composition, which may be reflected in the safety and efficacy of the herbal products concerned. As such, chemical analysis of herbal quality consistency has proven to be efficient in assessing all the parameters and components of quality as defined by USP, HMC, and GMP regulations. These requirements include identity/authenticity, purity (i.e., the absence of substituted species, adulterants, and contaminants), and chemical composition, at the qualitative and quantitative levels, of the herbs and herbal products being considered. Chemical methods of analysis play multiple roles in the quality assurance and quality control of herbs and herbal products, and, ultimately, in quality consistency. Chemical methods of evaluation display several advantages over morphological or microscopic and DNA-based approaches.

Unlike DNA-based methods, chemical evaluation can discriminate phenotypic species from different geographic regions, and plant parts or tissues, to determine plant age, and to investigate other variables. Further, chemical evaluation methods have the advantage over DNA-based and morphological or microscopic methods in identifying and authenticating plant species in the entire form, or as ground plant materials, or as processed plant materials, such as herbal products.

Targeted and untargeted-based metabolomics, primarily using HPLC or UHPLC coupled to high-resolution MS, have been utilized extensively as the prevailing methodologies for instrumental chemical analysis in assessing the different quality parameters. In addition, other techniques including GC-MS and HPTLC are used widely for generating chemical profiles of volatile constituents and for chemical fingerprints, respectively. However, the use of ^{1}H NMR spectroscopy has a disadvantage in the high cost of operating the required instrumentation, but has the advantages of being highly reproductive, non-destructive, highly accurate, and time efficient, and does not need prior separation techniques to be applied or prior chemical derivatization. Moreover, this procedure can be employed for quantification. In addition, it can be used to identify unknown metabolites and determine of the relative stereochemistry of compounds.

To this end, chemical methods of analysis are employed as general tools for evaluating the quality consistency of herbal products. Most importantly, these methods have been proven to be efficient for assessing the variabilities linked to the inconsistency of quality of herbal materials. Their expanded development and use will enhance consumer confidence and the knowledge of manufacturers in the quality, safety, and efficacy of medicinal, cosmetic, and other herbal products.

Acknowledgements The authors are grateful to Jeff Solomon for his constructive comments and diligent proofreading of the current book chapter.

References

1. Dias DA, Urban S, Roessner U (2012) A historical overview of natural products in drug discovery. Metabolites 2:303
2. Osman A, Chittiboyina A, Khan I (2016) Cytoprotective role of dietary phytochemicals against cancer development via induction of phase II and antioxidant enzymes. In: Fishberg JC, Heilman JM (eds) Advances in molecular toxicology, vol 10. Elsevier, Amsterdam, p 99
3. Checkouri E, Reignier F, Robert-da-Silva C, Meilhac O (2020) Evaluation of polyphenol content and antioxidant capacity of aqueous extracts from eight medicinal plants from Reunion Island: protection against oxidative stress in red blood cells and preadipocytes. Antioxidants 9:959
4. Omidifar N, Nili-Ahmadabadi A, Nakhostin-Ansari A, Lankarani KB, Moghadami M, Mousavi SM, Hashemi SA, Gholami A, Shokripour M, Ebrahimi Z (2021) The modulatory potential of herbal antioxidants against oxidative stress and heavy metal pollution: plants against environmental oxidative stress. Environ Sci Pollut Res Int 28:61908
5. Ravishankar B, Shukla VJ (2007) Indian systems of medicine: a brief profile. Afr J Trad Compl Altern Med 4:319
6. Pal SK, Shukla Y (2003) Herbal medicine: current status and the future. Asian Pac J Cancer 4:281
7. Thakkar S, Anklam E, Xu A, Uberth F, Li J, Li B, Hugas M, Sarma N, Crerar S, Swift S, Hakamatsuka T, Curtui V, Yan W, Geng X, Sikker W, Tong W (2020) Regulatory landscape of dietary supplements and herbal medicines from a global perspective. Regul Toxicol Pharmacol 114:104647
8. Farnsworth NR, Akerele O, Bingel AS, Soejarto DD, Guo Z (1985) Medicinal plants in therapy. Bull World Health Org 63:965
9. Smith T, Majid F, Eckl V, Reynolds CM (2021) Herbal supplement sales in US increase by record-breaking 17.3% in 2020. HerbalGram 131:52
10. Yang H-Y, Chen P-C, Wang J-D (2014) Chinese herbs containing aristolochic acid associated with renal failure and urothelial carcinoma: a review from epidemiologic observations to causal inference. BioMed Res Int 2014:569325
11. Ibrahim R, Nyska A, Dunnick J, Ramot Y (2021) The toxicologic pathology aspects of selected natural herbal products and related compounds. J Toxicol Pathol 34:181
12. Kilis-Pstrusinska K, Wiela-Hojenska A (2021) Nephrotoxicity of herbal products in Europe—a review of an underestimated problem. Int J Mol Sci 22:4132
13. McCoy B, Wang L, Zak M, Al-Mehmadi S, Kabir N, Alhadit K, McDonalds K, Zhang G, Sharma R, Whitney R, Sinopoli K, Snead OC III (2018) A prospective open-label trial of a CBD/THC cannabis oil in Dravet syndrome. Ann Clin Transl Neurol 5:1077
14. Avigan MI, Mozersky RP, Seeff LB (2016) Scientific and regulatory perspectives in herbal and dietary supplement associated hepatotoxicity in the United States. Int J Mol Sci 17:331
15. Geller AI, Shehab N, Weidle NJ, Lovegrove MC, Wolpert BJ, Timbo BB, Mozersky RP, Budnitz DS (2015) Emergency department visits for adverse events related to dietary supplements. N Engl J Med 373:1531
16. Zhu J, Chen M, Borlak J, Tong W (2020) The landscape of hepatobiliary adverse reactions across 53 herbal and dietary supplements reveals immune-mediated injury as a common cause of hepatitis. Arch Toxicol 94:273
17. Mishra S (2009) Understanding plant anatomy. Discovery Publishing House, New Delhi
18. Jackson BP, Snowdon DW (1990) Atlas of microscopy of medicinal plants, culinary herbs and spices. Belhaven Press, London
19. Reddy S, Chary S (2003) University botany II: (gymnosperms, plant anatomy, genetics, ecology). New Age International, New Delhi
20. Simmler C, Anderson JR, Gauthier L, Lankin DC, McAlpine JB, Chen S-N, Pauli GF (2015) Metabolite profiling and classification of DNA-authenticated licorice botanicals. J Nat Prod 78:2007

21. Raclariu AC, Paltinean R, Vlase L, Labarre A, Manzanilla V, Ichim MC, Crisan G, Brysting AK, de Boer H (2017) Comparative authentication of *Hypericum perforatum* herbal products using DNA metabarcoding, TLC and HPLC-MS. Sci Rep 7:1291
22. Parveen I, Gafner S, Techen N, Murch SJ, Khan IA (2016) DNA barcoding for the identification of botanicals in herbal medicine and dietary supplements: strengths and limitations. Planta Med 82:1225
23. Lu Z, Rubinsky M, Babajanian S, Zhang Y, Chang P, Swanson G (2018) Visualization of DNA in highly processed botanical materials. Food Chem 245:1042
24. Grazina L, Amaral JS, Mafra I (2020) Botanical origin authentication of dietary supplements by DNA-based approaches. Compr Rev Food Sci Food Saf 19:1080
25. de Boer HJ, Ichim MC, Newmaster SG (2015) DNA barcoding and pharmacovigilance of herbal medicines. Drug Saf 38:611
26. Raclariu AC, Tebrencu CE, Ichim MC, Apreutesei OT, Brysting AK, de Boer HJ (2018) What's in the box? Authentication of *Echinacea* herbal products using DNA metabarcoding and HPTLC. Phytomedicine 44:32
27. Porebski S, Bailey LG, Baum BR (1997) Modification of a CTAB DNA extraction protocol for plants containing high polysaccharide and polyphenol components. Plant Mol Biol Rep 15:8
28. Staats M, Arulandhu AJ, Gravendeel B, Holst-Jensen A, Scholtens I, Peelen T, Prins TW, Kok E (2016) Advances in DNA metabarcoding for food and wildlife forensic species identification. Anal Bioanal Chem 408:4615
29. Pawar RS, Handy SM, Cheng R, Shyong N, Grundel E (2017) Assessment of the authenticity of herbal dietary supplements: comparison of chemical and DNA barcoding methods. Planta Med 83:921
30. Heinrich M, Anagnostou S (2017) From pharmacognosia to DNA-based medicinal plant authentication–pharmacognosy through the centuries. Planta Med 83:1110
31. Raclariu AC, Heinrich M, Ichim MC, de Boer H (2018) Benefits and limitations of DNA barcoding and metabarcoding in herbal product authentication. Phytochem Anal 29:123
32. Liu W, Liu J, Yin D, Zhao X (2015) Influence of ecological factors on the production of active substances in the anti-cancer plant *Sinopodophyllum hexandrum* (Royle) TS Ying. PLoS One 10:e0122981
33. Kulyal P, Acharya S, Ankari AB, Kokkiripati PK, Tetali SD, Rhagavendra AS (2021) Variable secondary metabolite profiles across cultivars of *Curcuma longa* L. and *Curcuma aromatica* Salisb. Front Pharmacol 12:659546
34. Pant P, Pandey S, Dall'Acqua S (2021) The influence of environmental conditions on secondary metabolites in medicinal plants: a literature review. Chem Biodivers 18:e2100345
35. Zhao J, Avula B, Chan M, Clément C, Kreuzer M, Khan IA (2012) Metabolomic differentiation of maca (*Lepidium meyenii*) accessions cultivated under different conditions using NMR and chemometric analysis. Planta Med 78:90
36. Dong J, Ma X, Wei Q, Peng S, Zhang S (2011) Effects of growing location on the contents of secondary metabolites in the leaves of four selected superior clones of *Eucommia ulmoides*. Ind Crops Prod 34:1607
37. Chauhan RK, Anwar M, Chand S, Patra CD (2012) Influence of different dates of planting on growth, herb, oil yield and quality of essential oil of menthol mint (*Mentha arvensis*) in the North Indian Plain. Arch Agron Soil Sci 58:223
38. Liu J, Liu Y, Zhang Z-H, Zu Y-G, Tang Z-H, Efferth T (2016) Correlation of cultivation time of *Panax ginseng* with metabolic profiles of nine ginsenosides and mRNA expression of genes encoding major biosynthetic enzymes. Acta Physiol Plant 38:51
39. Andola HC, Gaira KS, Rawal RS, Rawat MSM, Bhatt ID (2010) Habitat-dependent variations in berberine content of *Berberis asiatica* Roxb. ex DC. in Kumaon, Western Himalaya. Chem Biodivers 7:415
40. Ghimire SK, Mckey D, Aumeeruddy-Thomas Y (2006) Himalayan medicinal plant diversity in an ecologically complex high altitude anthropogenic landscape, Dolpo, Nepal. Environ Conserv 33:128

41. Rawat P, Kumar M, Srivastava A, Kumar B, Misra A, Singh SP, Srivastava S (2021) Influence of soil variation on diosgenin content profile in *Costus speciosus* from Indo-Gangetic Plains. Chem Biodivers 18:e2000977
42. Darfour B, Agbenyegah S, Ofosu DO, Okyere AA, Asare IK (2014) Gamma irradiation of *Tetrapleura tetraptera* fruit as a post-harvest technique and its subsequent effect on some phytochemicals, free scavenging activity and physicochemical properties. Radiat Phys Chem 102:153
43. Puranik V, Chauhan DK, Mishra V, Rai GK (2012) Effect of drying techniques on the physicochemical and bioactive components of selected medicinal herbs. Ann Phytomed 1:23
44. Tanko H, Carrier DJ, Duan L, Clausen E (2005) Pre- and post-harvest processing of medicinal plants. Plant Genet Resour 3:304
45. de Lazzari AC, Xavier RM, Borghi AA, dos Santos VF, Sawaya ACHF (2017) Effect of seasonality and growth conditions on the content of coumarin, chlorogenic acid and dicaffeoylquinic acids in *Mikania laevigata* Schultz and *Mikania glomerata* Sprengel (Asteraceae) by UHPLC-MS/MS. Int J Mass Spectrom 418:162
46. Alqahtani A, Tongkao-on W, Li KM, Razmovski-Naumovski V, Chan K, Li GQ (2015) Seasonal variation of triterpenes and phenolic compounds in Australian *Centella asiatica* (L.) Urb. Phytochem Anal 26:436
47. Isah T (2019) Stress and defense responses in plant secondary metabolites production. Biol Res 52:39
48. Sun M, Li L, Wang M, van Wijk E, He M, van Wijk R, Koval S, Hankmeier T, van der Greef J, Wei S (2016) Effects of growth altitude on chemical constituents and delayed luminescence properties in medicinal rhubarb. J Photochem Photobiol B Biol 162:24
49. Sultan R, Majid N, Nissar S, Rather AM (2018) Seasonal variation of phytochemicals. Int J Pharm Biol Sci 8:987
50. Nishidono Y, Chiyomatsu T, Sanuki K Tezuka Y, Tanaka K (2019) Analysis of seasonal variations of the volatile constituents in *Artemisia princeps* (Japanese mugwort) leaves by metabolomic approach. Nat Prod Commun 14:1934578X19872600
51. Wei J-N, Liu Z-H, Zhao Y-P, Zhao L-L, Xue TK, Lan QK (2019) Phytochemical and bioactive profile of *Coriandrum sativum* L. Food Chem 286:260
52. Szabó LU, Schmidt TJ (2021) Investigation of the variability of alkaloids in *Buxus sempervirens* L. using multivariate data analysis of LC/MS profiles. Molecules 27:82
53. Jan R, Asaf S, Numan M, Lubna L, Kim K-M (2021) Plant secondary metabolite biosynthesis and transcriptional regulation in response to biotic and abiotic stress conditions. Agronomy 11:968
54. Selmar D, Kleinwächter M (2013) Stress enhances the synthesis of secondary plant products: the impact of stress-related over-reduction on the accumulation of natural products. Plant Cell Physiol 54:817
55. Kovalchuk I, Kovalchuk O (2021) Genome stability: from virus to human application, 2nd edn. Academic Press/Elsevier, Amsterdam
56. da Silva RR, da Câmara CA, Almeida AV, Ramos CS (2012) Biotic and abiotic stress-induced phenylpropanoids in leaves of the mango (*Mangifera indica* L., Anacardiaceae). J Braz Chem Soc 23:206
57. Liu Y, Chen P, Zhou M, Wang T, Fang S, Shang X, Fu X (2018) Geographic variation in the chemical composition and antioxidant properties of phenolic compounds from *Cyclocarya paliurus* (Batal) Iljinskaja leaves. Molecules 23:2440
58. Satyavarapu EM, Sinha PK, Mandal C (2020) Influence of geographical and seasonal variations on carbazole alkaloids distribution in *Murraya koenigii*: deciding factor of its in vitro and in vivo efficacies against cancer cells. BioMed Res Int 2020:1
59. Liu Y, Fan G, Zhang J, Zhang Y, Li J, Xiong C, Zhang Q, Li X, Lai X (2017) Metabolic discrimination of sea buckthorn from different *Hippophaë* species by ^{1}H NMR based metabolomics. Sci Rep 7:1589
60. Yu X, Ma L, Geri L (2013) Effects of ethanol extract of different altitude of Qinghai Province *Rheum tanguticum* on Glc-82 cells. Qinghai Yixueyuan Xuebao 34:114

61. Kaltner F, Rychlik M, Gareis M, Gottschlk C (2020) Occurrence and risk assessment of pyrrolizidine alkaloids in spices and culinary herbs from various geographical origins. Toxins 12:155
62. Osman AG, Haider S, Chittiboyina AG, Khan IA (2019) Utility of alkaloids as chemical and biomarkers for quality, efficacy, and safety assessment of botanical ingredients. Phytomedicine 54:347
63. Zhao J, Khan IA, Combrinck S, Sandasi M, Chen W, Viljoen AM (2018) ^{1}H-NMR and UPLC-MS metabolomics: functional tools for exploring chemotypic variation in *Sceletium tortuosum* from two provinces in South Africa. Phytochemistry 152:191
64. Wang Y, Liang X, Li Y, Fan J, Li Y, Cao Y, An W, Shi Z, Zhao J, Guo S (2020) Changes in metabolome and nutritional quality of *Lycium barbarum* fruits from three typical growing areas of China as revealed by widely targeted metabolomics. Metabolites 10:46
65. Jiang H, Xie Z, Koo HJ, McLaughlin SO, Timmermann BN, Gang DR (2006) Metabolic profiling and phylogenetic analysis of medicinal *Zingiber* species: tools for authentication of ginger (*Zingiber officinale* Rosc.). Phytochemistry 67:1673
66. Mncwangi NP, Viljoen AM, Zhao J, Vermaak I, Chen W, Khan IA (2014) What the devil is in your phytomedicine? Exploring species substitution in *Harpagophytum* through chemometric modeling of ^{1}H-NMR and UHPLC-MS datasets. Phytochemistry 106:104
67. Nguyen HT, Lee D-K, Choi Y-G, Min J-E, Yoon SJ, Yu Y-H, Lim J, Lee J, Kwon SW, Park JH (2016) A ^{1}H NMR-based metabolomics approach to evaluate the geographical authenticity of herbal medicine and its application in building a model effectively assessing the mixing proportion of intentional admixtures: a case study of *Panax ginseng*: metabolomics for the authenticity of herbal medicine. J Pharm Biomed 124:120
68. Viljoen AM, Zhao J, Sandasi M, Chen W, Khan IA (2015) Phytochemical distinction between *Pelargonium sidoides* ("Umckaloabo") and *P. reniforme* through ^{1}H-NMR and UHPLC-MS metabolomic profiling. Metabolomics 11:594
69. Cirak C, Radusiene J, Stanius Z, Camas N, Caliscan O, Odabas MS (2012) Secondary metabolites of *Hypericum orientale* L. growing in Turkey: variation among populations and plant parts. Acta Physiol Plant 34:1313
70. Popović Z, Milošević DK, Stefanović M, Vidaković V, Matić R, Janković J, Bojović S (2021) Variability of six secondary metabolites in plant parts and developmental stages in natural populations of rare *Gentiana pneumonanthe*. Plant Biosyst 155:816
71. Avula B, Wang Y-H, Isaac G, Yuk J, Wrona M, Yu K, Khan IA (2017) Metabolic profiling of *Hoodia*, *Chamomile*, *Terminalia* species and evaluation of commercial preparations using ultrahigh-performance liquid chromatography quadrupole-time-of-flight mass spectrometry. Planta Med 83:1297
72. Choi JY, Hong JH, Dang YM, Jamila N, Khan N, Jo CH, Kim KS (2018) Identification markers of adulteration in Korean red ginseng (*Panax ginseng*) products using high-performance liquid chromatography (HPLC) and liquid chromatography–mass spectrometry (LC-MS). Anal Lett 51:2588
73. Chang X, Zhang J, Li D, Zhou D, Zhang Y, Wang J, Hu B, Ju A, Ye Z (2017) Nontargeted metabolomics approach for the differentiation of cultivation ages of mountain cultivated ginseng leaves using UHPLC/QTOF-MS. J Pharm Biomed Anal 141:108
74. Govindarajan R, Tejas V, Pushpangadan P (2019) High-performance liquid chromatography (HPLC) as a tool for standardization of complex herbal drugs. J AOAC Int 102:986
75. Avula B, Sagi S, Wang Y-H, Wang M, Gafner S, Manthey JA, Khan IA (2016) Liquid chromatography-electrospray ionization mass spectrometry analysis of limonoids and flavonoids in seeds of grapefruits, other citrus species, and dietary supplements. Planta Med 82:1058
76. Chen Z, Zhao Y, Quan L, Zhou H, Cao D, Hu C, Li W, Ang Z (2017) Study on quality standard of processed *Curcuma longa* Radix. Evid Based Complement Altern Med 2017:2830754
77. Zhao J, Wang M, Avula B, Zhong L, Song Z, Xu Q, Li S, Khan IA (2015) Effect of processing on the traditional Chinese herbal medicine Flos Lonicerae: an NMR-based chemometric approach. Planta Med 81:754

78. Seethapathy GS, Tadesse M, Kumar S, Urumarudappa J, Gunaga SV, Vasudeva R, Malterud KE, Shaanker RU, de Boer HJ, Ravikanth G, Wangensteen H (2018) Authentication of *Garcinia* fruits and food supplements using DNA barcoding and NMR spectroscopy. Sci Rep 8:10561
79. Hoban CL, Musgrave IF, Coghlan M, Power M, Byard R, Nash C, Farrington R, Maker G, Crighton E, Trengove R, Bunce M (2018) Adulterants and contaminants in psychotropic herbal medicines detected with mass spectrometry and next-generation DNA sequencing. Pharmaceut Med 32:429
80. Galal AM, Raman V, Avula B, Wang Y-H, Rumalla CS, Weerasooriya A, Khan IA (2013) Comparative study of three *Plumbago* L. species (Plumbaginaceae) by microscopy, UPLC-UV and HPTLC. J Nat Med 67:554
81. Raman V, Avula B, Galal AM, Wang Y-H, Khan IA (2013) Microscopic and UPLC–UV–MS analyses of authentic and commercial yohimbe (*Pausinystalia johimbe*) bark samples. J Nat Med 67:42
82. de Oliveira Costa VC, Borghi AA, Mayer JLS, Sawaya ACHF (2018) Comparison of the morphology, anatomy, and chemical profile of *Mikania glomerata* and *Mikania laevigata*. Planta Med 84:191
83. Windarsih A, Nisa K, Indrianingsih A, Darish C, Handayani S, Wulanjati MP, Rosyida VT, Wijayanti T, Rohman A (2021) The use of ^{1}H-NMR spectroscopy and chemometrics of pattern recognition for authentication of *Curcuma xanthorrhiza* adulterated with *Zingiber montanum*. IOP Conf Ser Mater Sci Eng 1011:012050
84. Bampali E, Germer S, Bauer R, Kulić Ž (2021) HPLC-UV/HRMS methods for the unambiguous detection of adulterations of *Ginkgo biloba* leaves with *Sophora japonica* fruits on an extract level. Pharm Biol 59:438
85. Pan Y, Yang M, Zhang M, Jia M (2019) Rapid discrimination of commercial American ginseng and Asian ginseng according to diols composition using a colorimetric sensor array. Sens Actuators B Chem 294:48
86. Frommenwiler DA, Reich E, Sudberg S (2016) St. John's wort versus counterfeit St. John's wort: an HPTLC study. J AOAC Int 99:1204
87. Kim SB, Bisson J, Friesen JB, Bucchini L, Gafner S, Lankin DC, Chen S-N, Pauli GF, McAlpine JB (2021) The untargeted capability of NMR helps recognizing nefarious adulteration in natural products. J Nat Prod 84:846
88. Wang Y-H, Meng Y, Zhai C, Wang M, Avula B, Yuk J, Smith KM, Isaac G, Khan IA (2019) The chemical characterization of *Eleutherococcus senticosus* and Ci-wu-jia tea using UHPLC-UV-QTOF/MS. Int J Mol Sci 20:475
89. Rohman A, Wijayanti T, Windarsih A, Riyanto S (2020) The authentication of Java Turmeric (*Curcuma xanthorrhiza*) using thin layer chromatography and ^{1}H-NMR based-metabolite fingerprinting coupled with multivariate analysis. Molecules 25:3928
90. Booker A, Jalil B, Frommenwiler D, Reich E, Zhai L, Kulic Z, Heinrich M (2016) The authenticity and quality of *Rhodiola rosea* products. Phytomedicine 23:754
91. Zhao J, Wang M, Avula B, Khan IA (2018) Detection and quantification of phenethylamines in sports dietary supplements by NMR approach. J Pharm Biomed Anal 151:347
92. Wallace ED, Oberlies NH, Cech NB, Kellogg JJ (2018) Detection of adulteration in *Hydrastis canadensis* (goldenseal) dietary supplements via untargeted mass spectrometry-based metabolomics. Food Chem Toxicol 120:439
93. Salem MA, Zayed A, Alseekh S, Fernie AR, Giavalisco P (2021) The integration of MS-based metabolomics and multivariate data analysis allows for improved quality assessment of *Zingiber officinale* Roscoe. Phytochemistry 190:112843
94. Viapiana A, Wesolowski M (2016) HPLC fingerprint combined with quantitation of phenolic compounds and chemometrics as an efficient strategy for quality consistency evaluation of *Sambucus nigra* berries. Nat Prod Commun 11:1449
95. Singh P, Bajpai V, Gond V, Kumar A, Tadigoppula N, Kumar B (2020) Determination of bioactive compounds of fenugreek (*Trigonella foenum-graecum*) seeds using LC-MS techniques. Meth Mol Biol 2107:377

96. Naldi M, Fiori J, Gotti R, Periat A, Veuthey J-L, Guillarme D, Andrisano V (2014) UHPLC determination of catechins for the quality control of green tea. J Pharm Biomed 88:307
97. Andonova L, Ivanov E, Pencheva I, Konstantinov S (2021) Elaboration and validation of HPLC/DAD method for quality control of products containing cannabidiol. Curr Pharm Anal 17:1218
98. Waidyanatha S, Pierfelice J, Cristy T, Mutlu E, Burback B, Rider CV, Ryan K (2020) A strategy for test article selection and phytochemical characterization of *Echinacea purpurea* extract for safety testing. Food Chem Toxicol 137:111125
99. Wang M, Yu P, Chittiboyina AG, Chen D, Zhao J, Avula B, Wang Y-H, Khan IA (2020) Characterization, quantification and quality assessment of avocado (*Persea americana* Mill.) oils. Molecules 25:1453
100. Geng P, Harnly JM, Sun J, Zhang M, Chen P (2017) Feruloyl dopamine-*O*-hexosides are efficient marker compounds as orthogonal validation for authentication of black cohosh (*Actaea racemosa*)—an UHPLC-HRAM-MS chemometrics study. Anal Bioanal Chem 409:2591
101. Avula B, Sagi S, Wang Y-H, Zweigenbaum J, Wang M, Khan IA (2015) Characterization and screening of pyrrolizidine alkaloids and *N*-oxides from botanicals and dietary supplements using UHPLC-high resolution mass spectrometry. Food Chem 178:136
102. Dubnicka M, Cromwell B, Levine M (2020) Investigation of the adulteration of essential oils by GC-MS. Curr Anal Chem 16:965
103. Gilard V, Balayssac S, Tinaugus A, Martins N, Martino R, Malet-Martino M (2015) Detection, identification and quantification by ^{1}H NMR of adulterants in 150 herbal dietary supplements marketed for improving sexual performance. J Pharm Biomed Anal 102:476
104. Martin MM (2019) Aspects of gradient elution in LC-MS analysis. Gradient HPLC for practitioners. In: Kromidas S (ed) RP, LC-MS, ion analytics, biochromatography, SFC, HILIC. Wiley-VCH, Hoboken, NJ, p 189
105. Ikhsan AN, Syifa F, Mustafidah M, Abdul R (2021) Implementation of chemometrics as a solution to detecting and preventing falsification of herbal medicines in Southeast Asia: a review. J Appl Pharm Sci 11:139
106. Simmler C, Graham JG, Chen S-N, Pauli GF (2018) Integrated analytical assets aid botanical authenticity and adulteration management. Fitoterapia 129:401
107. Morlock G, Schwack W (2010) Hyphenations in planar chromatography. J Chromatogr A 1217:6600
108. Avula B, Bae J-Y, Raman V, Fantoukh O, Wang Y-H, Osman A, Wang M, Ali Z, Khan IA (2019) Quantification of phenolic compounds from *Fadogia agrestis* and dietary supplements using UHPLC-PDA-MS. Planta Med 85:145
109. Black C, Chevallier OP, Elliott CT (2016) The current and potential applications of ambient mass spectrometry in detecting food fraud. Trends Anal Chem 82:268
110. Avula B, Sagi S, Gafner S, Upton R, Wang Y-H, Wang M, Khan IA (2015) Identification of *Ginkgo biloba* supplements adulteration using high performance thin layer chromatography and ultra high performance liquid chromatography-diode array detector-quadrupole time of flight-mass spectrometry. Anal Bioanal Chem 407:7733
111. Avula B, Bae J-Y, Chittiboyina AG, Wang Y-H, Wang M, Khan IA (2019) Liquid chromatography-quadrupole time of flight mass spectrometric method for targeted analysis of 111 nitrogen-based compounds in weight loss and ergogenic supplements. J Pharm Biomed 174:305
112. Avula B, Bae J-Y, Majrashi T, Wu T-Y, Wang Y-H, Wang M, Ali Z, Wu Y-C, Khan IA (2018) Targeted and non-targeted analysis of annonaceous alkaloids and acetogenins from *Asimina* and *Annona* species using UHPLC-QToF-MS. J Pharm Biomed Anal 159:548
113. Avula B, Smillie TJ, Wang Y-H, Zweigenbaum J, Khan IA (2015) Authentication of true cinnamon (*Cinnamomum verum*) utilising direct analysis in real time (DART)-QToF-MS. Food Addit Contam Part A Chem Anal Expo Risk Assess 32:1
114. Hall RD (2006) Plant metabolomics: from holistic hope, to hype, to hot topic. New Phytol 169:453

115. Shen Y, Van Beek TA, Claassen FW, Zuilhof H, Chem B, Nielen MWF (2012) Rapid control of Chinese star anise fruits and teas for neurotoxic anisatin by direct analysis in real time high resolution mass spectrometry. J Chromatogr A 1259:179
116. Wang Y-H, Avula B, Nanayakkara ND, Zhao J, Khan IA (2013) Cassia cinnamon as a source of coumarin in cinnamon-flavored food and food supplements in the United States. J Agric Food Chem 61:4470
117. Wishart DS (2019) NMR metabolomics: a look ahead. J Magn Reson 306:155
118. Augustijn D, de Groot HJ, Alia A (2021) HR-MAS NMR applications in plant metabolomics. Molecules 26:931
119. Shinbo Y, Nakamura Y, Altaf-Ul-Amin M, Asahi H, Kurokawa K, Arita M, Saito K, Ohta D, Shibata D, Kanaya S (2006) KNApSAcK: a comprehensive species-metabolite relationship database. In: Saito K, Dixon RA, Willmitzer L (eds) Biotechnology in agriculture and forestry, vol 57. Springer, Berlin, p 165
120. Flores IS, Martinelli BC, Pinto VS, Queiroz LHK Jr, Lião LM (2019) Important issues in plant tissues analyses by HR-MAS NMR. Phytochem Anal 30:5
121. Kitazaki K, Fukushima A, Nakabayashi R, Okazaki Y, Kobayashi M, Mori T, Nishizawa T, Reyes-Chin-Wo S, Michelmore RW, Saito K, Shoji K, Kusano M (2018) Metabolic reprogramming in leaf lettuce grown under different light quality and intensity conditions using narrow-band LEDs. Sci Rep 8:7914
122. Guijas C, Montenegro-Burke JR, Warth B, Spiker ME, Siudzak G (2018) Metabolomics activity screening for identifying metabolites that modulate phenotype. Nat Biotechnol 36:316
123. Avula B, Wang Y, Isaac G, Yuk J, Wrona M, Yu K, Khan IA (2016) Metabolomics based UHPLC-QToF-MS approach for the authentication of various botanicals and dietary supplements. Planta Med 82:OA13
124. Li X, Zhang X, Ye L, Khang Z, Jia D, Yang L, Zhang B (2019) LC-MS-based metabolomic approach revealed the significantly different metabolic profiles of five commercial truffle species. Front Microbiol 10:2227
125. Bingol K, Brüschweiler R (2015) NMR/MS translator for the enhanced simultaneous analysis of metabolomics mixtures by NMR spectroscopy and mass spectrometry: application to human urine. J Proteome Res 14:2642
126. Wang XJ, Ren JL, Zhang AH, Sun H, Yan G-L, Han Y, Liu L (2019) Novel applications of mass spectrometry-based metabolomics in herbal medicines and its active ingredients: current evidence. Mass Spectrom Rev 38:380
127. Zhao J, Ma S-C, Li S-P (2018) Advanced strategies for quality control of Chinese medicines. J Pharm Biomed 147:473
128. Murphy BJ, Carlson RE, Howa JD, Wilson TM, Buch M (2021) Determining the authenticity of methyl salicylate in *Gaultheria procumbens* L. and *Betula lent*a L. essential oils using isotope ratio mass spectrometry. J Essent Oil Res 33:442
129. Osman AG, Raman V, Haider S, Ali Z, Chittiboyina AG, Khan IA (2019) Overview of analytical tools for the identification of adulterants in commonly traded herbs and spices. J AOAC Int 102:376
130. Paraschos S, Magiatis P, Gikas E, Smyrnioudis I, Skaltsounis A-L (2016) Quality profile determination of Chios mastic gum essential oil and detection of adulteration in mastic oil products with the application of chiral and non-chiral GC-MS analysis. Fitoterapia 114:12
131. Wang M, Raman V, Zhao J, Avula B, Wang Y-H, Wylie PL, Khan IA (2018) Application of GC/Q-TOF combined with advanced data mining and chemometric tools in the characterization and quality control of bay leaves. Planta Med 84:1045
132. Krause A, Wu Y, Tian R, van Beek TA (2018) Is low-field NMR a complementary tool to GC-MS in quality control of essential oils? A case study: patchouli essential oil. Planta Med 84:953
133. Hu Y, Wang S, Wang S, Lu X (2017) Application of nuclear magnetic resonance spectroscopy in food adulteration determination: the example of Sudan dye I in paprika powder. Sci Rep 7:2637

134. Holzgrabe U (2010) Quantitative NMR spectroscopy in pharmaceutical applications. Prog Nucl Magn Reson Spectrosc 57:229
135. Seethapathy GS, Tadesse M, Urumarudappa SKJ, Gunaga SV, Malterud KE, Shaanker RU, de Boer HJ, Ravikanth G, Wangensteen H (2018) Authentication of *Garcinia* fruits and food supplements using DNA barcoding and NMR spectroscopy. Sci Rep 8:10561
136. Turbitt JR, Colson KL, Killday KB, Milstead A, Neto CC (2020) Application of ^{1}H-NMR-based metabolomics to the analysis of cranberry (*Vaccinium macrocarpon*) supplements. Phytochem Anal 31:68
137. Wu N, Balayssac S, Assemat G, Danoun S, Déjean S, Malet-Martino M, Gilard V (2021) Evaluation of low-field versus high-field proton NMR spectroscopy for quality control of cinnamon samples. J Food Comp Anal 96:103706
138. Duangdee N, Chamboonchu N, Kongkiatpaiboon S, Prateeptongkum S (2019) Quantitative ^{1}H NMR spectroscopy for the determination of oxyresveratrol in *Artocarpus lacucha* heartwood. Phytochem Anal 30:617
139. Cicek SS, Girreser U, Zidorn C (2018) Quantification of the total amount of black cohosh cycloartanoids by integration of one specific ^{1}H NMR signal. J Pharm Biomed Anal 155:109
140. Tajidin N, Shaari K, Maulidiani M, Salleh NS, Ketaren BR, Mohamed M (2019) Metabolite profiling of *Andrographis paniculata* (Burm. f.) Nees. young and mature leaves at different harvest ages using ^{1}H NMR-based metabolomics approach. Sci Rep 9:16766
141. Ioanna V, Ioannis P (2020) Quality assessment and quantification of genistein in dietary supplements by high-performance liquid chromatography, quantitative nuclear magnetic resonance, and two-dimensional diffusion ordered spectroscopy. Asian J Pharm Clin Res 13:132
142. Hachem R, Assemat G, Martins N, Balayssac S, Gilard V, Martino R, Malet-Martino M (2016) Proton NMR for detection, identification and quantification of adulterants in 160 herbal food supplements marketed for weight loss. J Pharm Biomed Anal 124:34
143. Hung H-Y, Lin S-M, Li C-Y, Lam S-H, Chan Y-Y, Liou M-J, Shieh P-C, Chen F-A, Kuo P-C, Wu T-S (2021) A rapid and feasible ^{1}H-NMR quantification method of ephedrine alkaloids in *Ephedra* herbal preparations. Molecules 26:1599
144. Zhao S, Liu H, Gu L (2020) American cranberries and health benefits–an evolving story of 25 years. J Sci Food Agric 100:5111
145. Blümich B (2016) Introduction to compact NMR: a review of methods. Trends Anal Chem 83:2
146. Blümich B (2019) Low-field and benchtop NMR. J Magn Reson 306:27
147. van Beek TA (2021) Low-field benchtop NMR spectroscopy: status and prospects in natural product analysis. Phytochem Anal 32:24
148. Kim JH, Lee HJ, Kwon K, Chun SC, Ahn S, Kim BH (2018) A 43 MHz low-field benchtop ^{1}H nuclear magnetic resonance method to discriminate perilla oil authenticity. J Oleo Sci 67:507
149. Assemat G, Dubocq F, Balayssac S, Lamoureux C, Malet-Martino M, Gilard V (2017) Screening of "spice" herbal mixtures: from high-field to low-field proton NMR. Forensic Sci Int 279:88
150. Bingol K, Bruschweiler-Li L, Cao Y, Somogy A, Zhang F, Brüschweiler R (2015) Metabolomics beyond spectroscopic databases: a combined MS/NMR strategy for the rapid identification of new metabolites in complex mixtures. Anal Chem 87:3864
151. Krueger S, Bergin A, Morlock GE (2018) Effect-directed analysis of ginger (*Zingiber officinale*) and its food products, and quantification of bioactive compounds via high-performance thin-layer chromatography and mass spectrometry. Food Chem 243:258
152. Guzelmeric E, Cevik D, Yesilada E (2020) Quality assessment of marketed food supplements and herbal tea products of olive leaf in Turkey. J Res Pharm 24:159
153. Galal AM, Avula B, Khan IA (2015) Utility of thin-layer chromatography in the assessment of the quality of botanicals. In: Gala AM (ed) Instrumental thin-layer chromatography. Elsevier, Amsterdam, p 479

154. Lalhriatpuii T (2020) HPTLC fingerprint in herbal drug formulations. In: Sen S, Chakraborty R (eds) Herbal medicine in India. Springer Nature, Singapore, p 337
155. Frommenwiler DA, Kim J, Yook C-S, Tran TTT, Cañigueral S, Reich E (2018) Comprehensive HPTLC fingerprinting for quality control of an herbal drug—the case of *Angelica gigas* root. Planta Med 84:465
156. Fengyan H, He Y, Zheng X, Wang R, Lu J, Dai Z, Ma S (2018) Screening of chemical dyes in traditional Chinese medicine by HPTLC-MS. J AOAC Int 101:686
157. Mishra J, Bhardwaj A, Pal M, Rajput R, Misra K (2018) High performance thin layer chromatography hyphenated with electrospray mass spectrometry for evaluation of nucleobases in two traditional Chinese medicinal mushrooms: a metabolomic approach. J Liq Chromatogr Relat Technol 41:910
158. Bae J-Y, Avula B, Zhao J, Raman V, Wang YH, Wang M, Zulfiqar F, Feng W, Park JH, Abe N, Ali Z, Khan IA (2020) Analysis of prenylflavonoids from aerial parts of *Epimedium grandiflorum* and dietary supplements using HPTLC, UHPLC-PDA and UHPLC-QToF along with chemometric tools to differentiate *Epimedium* species. J Pharm Biomed Anal 177:112843
159. Rumalla C, Avula B, Wang Y-H, Smillie T, Khan IA (2012) Densitometric-HPTLC method development and analysis of anthocyanins from açai (*Euterpe oleracea* Mart.) berries and commercial products. J Planar Chromatogr Mod TLC 25:409
160. Boudesocque L, Dorat J, Pothier J, Gueiffier A, Enguehard-Gueiffier C (2013) High performance thin layer chromatography–densitometry: a step further for quality control of cranberry extracts. Food Chem 139:866
161. Mammen D, Daniel M, Sane R (2011) Rapid and parallel analysis using HPTLC to detect seasonal and geographical variation in *Aerva lanata* Juss. ex Schultes. J Planar Chromat 24:388
162. Salmon CN, Bailey-Shaw YA, Hibbert S, Green C, Smith AM, Williams LAD (2012) Characterisation of cultivars of Jamaican ginger (*Zingiber officinale* Roscoe) by HPTLC and HPLC. Food Chem 131:1517
163. Tomar V, Beuerle T, Sircar D (2019) A validated HPTLC method for the simultaneous quantifications of three phenolic acids and three withanolides from *Withania somnifera* plants and its herbal products. J Chromatogr B Anal Technol Biomed Life Sci 1124:154
164. Raman V, Galal AM, Avula B, Sagi S, Smillie TJ, Khan IA (2014) Application of anatomy and HPTLC in characterizing species of *Dioscorea* (Dioscoreaceae). J Nat Med 68:686
165. Raman V, Budel JM, Zhao J, Bae J-Y, Avula B, Osman AG, Ali Z, Khan IA (2018) Microscopic characterization and HPTLC of the leaves, stems and roots of *Fadogia agrestis*—an African folk medicinal plant. Rev Bras Farmacogn 28:631
166. Raman V, Sagi S, Galal AM, Avula B, Viljoen A, Khan IA (2015) Adulteration in commercial buchu dietary supplements: analyses of commercial and authentic buchu samples and comparative studies of *Agathosma betulina* and *Agathosma crenulata* by microscopy and HPTLC. S Afr J Bot 100:122
167. Guzelmeric E, Ristivojević P, Vovk I, Milojković-Opsenica D, Yesilada E (2017) Quality assessment of marketed chamomile tea products by a validated HPTLC method combined with multivariate analysis. J Pharm Biomed Anal 132:35
168. Mehta M, Garg M, Dua K, Satija S (2019) Simultaneous HPTLC densitometric estimation of KBA and AKBA from *Boswellia serrata*. Curr Anal Chem 15:84
169. Do T, Santi I, Reich E (2019) A harmonized HPTLC method for identification of various caffeine containing herbal drugs, extracts, and products, and quantitative estimation of their caffeine content. J Liq Chromatogr Relat Technol 42:274
170. Deore SL, Tidke HP, Khadabadi SS, Baviskar BA, Kide AA, Shende BP (2021) Validated HPTLC method to estimate silymarin in marketed herbal formulation. Int J Pharm Invest 11:52
171. Foudah AI, Shakeel F, Alqarni MH, Ross SA, Salkini MAA, Alam P (2021) Simultaneous estimation of cinnamaldehyde and eugenol in essential oils and traditional and ultrasound-assisted extracts of different species of cinnamon using a sustainable/green HPTLC technique. Molecules 26:2054

Ahmed Osman is a senior research scientist at the National Center for Natural Products Research (NCNPR), School of Pharmacy, University of Mississippi, USA. He obtained his Ph.D. degree from the Department of Pharmaceutical Sciences, School of Pharmacy, University of Pittsburgh, USA. After graduation, he worked at Washington State University, USA, as a postdoctoral researcher the group of Prof. Rodney Croteau at the Institute of Biological Chemistry. In 2007, he moved to NCNPR to work with Prof. Ikhlas Khan to conduct various phytochemical investigations on botanicals. His research areas include the isolation and structural elucidation of metabolites from plants used in traditional plants, adopting targeted and untargeted isolation techniques for quality control, and pursuing compound biological activities. In addition, he has performed research on the identification and authentication of selected plant samples and the determination of their botanical adulterants using microscopic techniques. He also has worked on the synthesis of derivatives of bioactive natural products.

Amar G. Chittiboyina is an Assistant Director at the NCNPR, University of Mississippi. He has a broad background in organic chemistry, with specific training and expertise in medicinal chemistry, natural products, and analytical chemistry. He received his doctoral (Ph.D.) degree from the National Chemical Laboratory, Pune, India, by working on the asymmetric total synthesis of (+)-biotin, under the supervision of Dr. Subhash P. Chavan. He completed his postdoctoral training at the Department of Medicinal Chemistry, University of Mississippi, under the mentorship of Prof. Mitchell A. Avery, by investigating the structure-activity relationships of various natural products potentially useful for maintaining human health. Dr. Chittiboyina holds several national and international patents, and has contributed to book chapters on natural products and medicinal chemistry, including the quality and safety of phytochemicals in various matrices. He is an active member of several professional organizations and recently was inducted as a Fellow of the Royal Society of Chemistry (London, U.K.). He has authored over 180 peer-reviewed publications, and his group collaborates with biomedical researchers and other interdisciplinary researchers.

Bharathi Avula received her Master's and Ph.D. degrees in Pharmacy at the Birla Institute of Technology and Science (BITS), Pilani, India. She joined the NCNPR at the University of Mississippi, MS, in 2002 as a Postdoctoral Research Associate and is currently employed as a Principal Research Scientist. She has contributed directly to more than 200 peer-reviewed research articles in various high-impact scientific journals. She has been instrumental in developing hyphenated analytical methods to ensure the overall quality of various botanical products. She has multifaceted expertise in several techniques representing a broad range of analytical instrumentation [HPLC/UHPLC with PDA/ELS/FLR/SQD/QDa/QToF-MS/QqQ-MS detectors, ICP-MS, GC/FID/MS, SFC/PDA/MS, capillary electrophoresis (CE)/DAD, UV-visible spectrophotometry, and TLC densitometry (HPTLC)], covering many applications pertinent to natural products, biological samples and pharmaceuticals. This expertise involves the qualitative and quantitative analysis, chemical profiling, characterization, and standardization of chemical constituents from various botanicals, dietary supplements, and biological samples.

Zulfiqar Ali is a natural products scientist with over 20 years of research expertise in natural products research. He obtained a Ph.D. degree in Organic Chemistry in 2000 from HEJ Research Institute of Chemistry, University of Karachi, Pakistan. He was awarded a scholarship from the Science and Technology Agency, Japan, to work as STA fellow at Gifu Pharmaceutical University, Japan in 2000. He joined NCNPR at The University of Mississippi in 2004 as a postdoctoral research associate, where he became a Research Scientist in 2011 and was promoted to Senior Research Scientist in 2017. His research focuses on the isolation and structure elucidation of constituents from herbal supplements that may be useful as therapeutics or chemical markers. Dr. Ali has authored or co-authored over 260 original research and review articles.

Sebastian J. Adams is a plant biologist currently serving as a postdoctoral research associate at the National Center for Natural Products Research (NCNPR), University of Mississippi, where he is involved in the identification and authentication of botanicals and the quality control of finished natural products. He received his Ph.D. degree in Botany from the Department of Botany, Bharathidasan University, Tiruchirappalli, Tamil Nadu, India. He conducted his earlier higher education studies at St. Joseph's College, Tiruchirappalli, India and Madras Christian College, Chennai, Tamil Nadu, India. Dr. Adams worked previously for four years as a research fellow at the Foundation of Revitalization of Local Health and Traditions (FRLHT), Bangalore, India, and subsequently as a research associate at Himalaya Wellness Company, Bangalore, for a year and as a scientist in Phyto-Pharmacognosy, Research and Development, Sami-Sabinsa group, Bangalore, for four years. From this pharmaceutical industrial experience, Dr. Adams was named as a co-inventor on two international patent applications. He specializes in botanical authentication using taxonomy, macro- and micro-morphology, and HPTLC analysis. His research findings have appeared in over 20 book chapters in "Plant Biology and Biotechnology" (Springer, India), and in "Ethnobotany of India", Volumes 1–5 (CRC: Apple Academic Press, USA). Also, he has co-authored 25 research articles in peer-reviewed journals on the topics of plant anatomy and histochemistry, as well as plant endophytic fungi and their use in the bioconversion of lead compounds. In 2020, he co-edited a book titled "Asymmetry in Plants: Biology of Handedness", published by Routledge/CRC Press, Boca Raton, FL, USA.

Ikhlas A. Khan is Research Professor and Director of the both the National Center for Natural Products Research (NCNPR) and the FDA Center of Excellence, as well as Distinguished Professor of Pharmacognosy in the Department of BioMolecular Sciences, all at the University of Mississippi. He also serves as Coordinator for Natural Products Research in the Center for Water and Wetland Diseases. Prof. Khan received B.S. (Chemistry, 1980) and M.S. (Organic Chemistry, 1982) degrees from the Aligarh Muslim University, Aligarh, India and a Ph.D. degree in Pharmacy from the Institute für Pharmaceutische Biology in Munich, West Germany in 1987. He then performed postdoctoral studies at the Swiss Federal Institute of Technology (ETH) in Zurich, and was appointed initially as a Research Scientist at NCNPR in 1992. Prof. Khan's primary research interests include analytical fingerprinting for the standardization of herbal products, bioanalytical approaches to the improvement of product quality and safety, microbial biotransformation, and the isolation of natural products of potential commercial value. Prof. Khan has authored or co-authored over 840 original research and review articles. He has presented invited lectures in many countries, and has served on numerous committees, scientific boards, and task forces.

Prof. Khan has received many awards and honors throughout his career, including fellowships of the American Institute of Chemists and of the Royal Society of Chemistry (London), and is also an Honorary Member of the American Society of Pharmacognosy (ASP). He was awarded the Varro E. Tyler Award of ASP (2011) and the U.S. FDA Center for Food Safety and Applied Nutrition (CFSAN) Directors' Special Citation Award (2012) as well as the Wiley Award of the Association of Official Agricultural Chemists International (2018). He holds honorary or adjunct professor titles at several universities, including the Chinese University of Hong Kong, Heilongjiang University of Chinese Medicine, Hunan University of Chinese Medicine, and Soochow University in the People's Republic of China, and King Saud University in Saudi Arabia.

Nomenclature: Herbal Taxonomy in the Global Commerce of Botanicals

Roy Upton

Contents

1 Introduction

All aspects of botanical medicine commence with knowing which plants are the most appropriate for their intended medicinal purpose, and which quality standards are to be met. Historically, the botanicals used in the practice of medicine were codified in early materia medica texts of various cultures. In the west, this was most well documented in "De Materia Medica" by the Greek physician and herbalist Pedanius

R. Upton (✉)
American Herbal Pharmacopoeia, P.O. Box 66809, Scotts Valley, CA 95067, USA
e-mail: roy@herbal-ahp.org

A. D. Kinghorn, H. Falk, S. Gibbons, Y. Asakawa, J.-K. Liu, V. M. Dirsch (eds.), *Progress in the Chemistry of Organic Natural Products 122*, Progress in the Chemistry of Organic Natural Products 122, https://doi.org/10.1007/978-3-031-26768-0_3

Dioscorides in the first century CE (Plate 1). In China, the earliest codification of medicinal plant use occurs in the "Shennong Bencao Jing", originally considered to be written in the third century and later annotated by the Taoist physician and herbalist Tao Hong Jing in the fifth century. In the Ayurvedic system of healing of India, medical philosophy, as well as the earliest formalized records of medicinal plant use, emerged in the "Caraka Samhita", believed to have originated sometime between the second century BCE and the second century CE. Other works, such as the "Canon of Medicine" of the Persian physician Ibn Sina (ca. 1025), continued to greatly influence medical and medicinal plant knowledge for centuries. These are among the most authoritative historical medical texts that established the traditional knowledge base of medicinal plant use that predominates today. While these texts provided records of the medicinal use of plants, they are rarely discussed medicinal plant identity and quality. Later works introduced illustrations to help in the identification of medicinal plants, but the earliest of these renderings were crude or fanciful and did not allow an accurate determination according to modern taxonomic standards. These eventually evolved into more exacting and faithful botanical representations as botany developed as a discipline independent from medicine into the photographic documentation of today (Plate 2).

The naming of plants may sound simplistic but it is the foundation of all aspects of the global trade of medicinal plants. There may be also tremendous beauty and meaning in plant names that often are overshadowed by the requirements of modern regulatory compliance. The misnaming of plants conversely presents significant challenges in all aspects of medicinal plant commerce and research that affect consumer recognition, herbal product labeling, efficacy, and public safety. Although common

Plate 1 Rendering of Greek herbalist and physician Pedanius Dioscorides (first century CE) from John Gerard's "The Herball" (1636) and cover page from a sixteenth century edition of "De Medica Materia" from Italy. Source: American Herbal Pharmacopoeia, Scotts Valley, CA, USA

Plate 2 Historical evolution of the botanical and morphological identification St. John's wort (*Hypericum perforatum*). Picture Sources: **a** Bock H. "Kreutterbuch", Straßburg 1665; **b** Woodville W. "Medical Botany", 1790–1794; **c** American Herbal Pharmacopoeia, Scotts Valley, CA, USA

names are familiar to consumers, botanical taxonomy is the gold standard for naming plants with a relative degree of assurance [1, 2]. However, while botanical sciences were once inherently linked with medicine, this is no longer the case. Botanical nomenclature as applied historically, reflected a relationship between humans and the culture and science of naming plants. In contrast, botanical taxonomy, as applied today, has been developed by professional botanists to specifically address issues of taxonomic exactness and phylogenetic relationships, with minimal consideration of human-plant relationships. This may be claimed to result in a disconnect in the relationship that humans have maintained with plants for millennia.

From a more practical perspective, taxonomy focuses on the identification of wild plants, preferentially emphasizing floral parts and familial and species relationships. It was not designed for identifying plant parts, which requires specialized training beyond classical botany. Therefore, other nomenclatural systems must be considered when assigning names to plant parts used for medicinal and dietary supplement purposes, although classical botanical taxonomy will remain the primary nomenclatural system for plants.

Accordingly, when using plants medicinally, the name of the plant itself is only one aspect of establishing its identity, since one must also know the plant part to be used and the quality standards to which it must conform. Neither of these points is evident in a Latin botanical name. Rather, the nomenclature of an acceptable medicinal plant is provided and codified in pharmacopeias internationally, and typically includes all botanically relevant fields of information (see Sect. 9, Table 4). Every pharmacopeial monograph provides a suite of nomenclatural and other parameters that represent the "identity" and acceptable quality of the medicinal ingredient. No other nomenclatural system does this, and generally, in contrast to typical botanical identification, this is the gold standard for identifying and naming medicinal plant parts, as well as assuring their quality [3, 4], although numerous authoritative sources give deference to standard botanical taxonomy [5].

2 Nomenclatural Systems

There are four primary nomenclatural (naming) systems applied to medicinal plants and their parts: common names, Latin binomials, Galenic (pharmaceutical) names, and pharmacopeial definitions. When and how each should be used may differ between experts and is often not discussed with due academic candor between disciplines. There are advantages and disadvantages for each of these systems, but all are equally relevant and appropriate in the disclosure and commerce of botanical ingredients depending on the purpose for which the names are applied, and the form in which the plant material is assessed. Macroscopic and organoleptic assessment is valid for those plants for which unique identifying characters are intact to a sufficient degree that allows for the identification and observation of the apparent absence of plant adulterants. Once herbal materials are powdered or extracted, other methodologies, usually involving several assessments, are required.

2.1 Common Names

Common names are terms that are most familiar to persons and cultures using a specific commodity, in this case, plants. Communication regarding plants likely predates written language, and by far predates the use of the word taxonomy, which did not became used until the nineteenth century [6, 7]. Historically, common names were assigned according to their physical form (milk thistle), cultural relationships (passion flower), function (emetic weed, pleurisy root), organoleptic characteristics (amalaki, lemon balm), or even the combination of a common and Latin name (aloe vera). However, historically there have been no formal rules for assigning common plant names and so these appellations varied greatly from region to region and culture to culture [1]. This was partially rectified with the publication of "Herbs of Commerce" by the American Herbal Products Association [8]. This text attempts to assign a single common name to a single species, and when appropriate, assigns a single common name to multiple species. The first volume of "Herbs of Commerce" was entered into the Code of Federal Regulations as the primary source of common names legally required for the labeling of herbal supplements. While the second volume did not attain the same level of legal recognition, it is widely consulted by regulatory authorities and has been integrated into industrial practice.

Many of the plant names in use today have their origins from ancient Rome and Greece, or from Asia and have been maintained for centuries [9]. Throughout most of recorded history, a primary interest in plants was based on what they could provide to humans and this evolved into a sub-discipline of medicine [7] as plants have provided the majority of drugs until relatively recently.

Similarly, in Native American traditions, plants that were closely related but botanically distinct according to modern taxonomy, and appeared, tasted or smelled alike, or elicited analogous medicinal effects, were often grouped together and referred to by the same common name [10]. In some cases, closely related plant names may be somewhat differentiated, such as in the case of true cinnamon, Chinese cinnamon, Ceylon cinnamon, and Saigon cinnamon, among other variants [8].

Formal inquiry into botany as a discipline separate from medicine began in earnest with the Greek philosopher Theophrastus (327–287 BCE), a student of the Greek philosopher Aristotle (384–322 BCE), who in turn studied with the Greek scholar Plato (~ 428–348 BCE). It was Theophrastus who first described plants in a purely phylogenetic manner in his "Enquiry Into Plants", in which he classified plants as trees, shrubs, undershrubs, and herbs, and distinguished between flowering and non-flowering plants, and between deciduous and evergreen trees, and also noted different habitats in which different plants grew preferentially. However, this early attempt at botanical classification languished mostly disregarded for centuries thereafter. Instead, useful plants were classified primarily according to their medicinal uses. The integration of plants as a part of medicine lasted until the fifteenth and sixteenth centuries, until the emergence of botanical gardens in renaissance Italy along with global exploration expeditions. The discovery of new continents led, in turn, to the discovery of new plants that needed to be described, heralding in a new era of purely botanical enquiry [6, 7].

2.1.1 Advantages of Common Names

In the example given of cinnamon, each culture with access to *Cinnamomum* species uses them as food or medicine in a similar manner. Similarly, each culture with familiarity with cinnamon recognizes the species concerned by their color, flavor, aroma, and warming pungency. One newly discovered use of cinnamon is an ability to increase glucose intake into insulin receptor sites, and therefore its use in supporting healthy blood sugar levels and as a potential treatment for diabetes. Several species of *Cinnamomum* have demonstrated this activity [11, 12] so from a therapeutic perspective, many of the species concerned appear to be equivalent and may need not be differentiated. In contrast, cassia-type cinnamon species (e.g., *Cinnamomum cassia*) have higher concentrations of constituents (e.g., coumarins) that has raised safety concerns among some authorities (e.g., the European Commission) and so may need to be differentiated based on such parameters.

Today, the various species of *Cinnamomum* can be recognized as "cinnamon" by most consumers, irrespective of having had any training in medicinal plants or taxonomy, and regardless of the subtle or significant differences attributed to the species of concern by taxonomists and phytochemists. Conversely, the scientific names assigned to the various species, *Cinnamomum verum*, *C. aromaticum*, *C. loureiroi*, and *C. burmannii*, as well as their synonyms, are completely foreign terms to a non-botanist consumer. Other common names such as blueberry and garlic are similarly familiar to members of English-speaking cultures, while their corresponding Latin binomials, *Vaccinium corymbosum* and *Allium sativum*, respectively, are not. Nearly every consumer somewhat familiar with these botanicals can accurately identify blueberries, cinnamon, and garlic with confidence.

Due to their cultural familiarity, common names are most appropriate in the labeling and marketing of consumer herbal products. Following this precept, regulatory authorities typically require that common names be used in the consumer labeling of herbal dietary supplements [13, 14]. In the U.S., the common names used in product labeling are expected to conform with the Standard Common Names assigned to each botanical species as outlined in "Herbs of Commerce" [8], unless there is a scientifically valid reason not to do so. Each entry provides the Latin binomial along with the botanical authority and plant family name, the standardized common name to be used in labeling, and, when appropriate, synonyms, Ayurvedic, and Chinese names. While static texts such as these do not allow for real-time botanical revisions, "Herbs of Commerce" provides a foundation for the assignment of names across both the botanical and common nomenclatural systems [8]. With the inclusion of Ayurvedic and Pinyin names, this text is also of immense value cross-culturally (Plate 3).

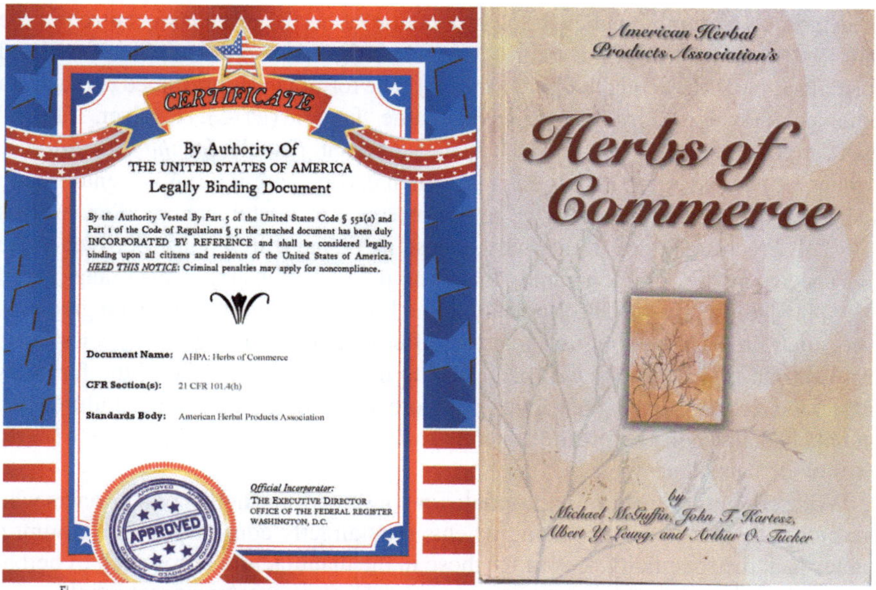

Botanical nomenclature, botanical authority, botanical family	***Cinnamomum loureirii*** **Nees** [*Lauraceae*] SCN: **Saigon cinnamon**
SCN = Standardized Common Name	***Cinnamomum tamala*** (Buch.-Ham.) T. Nees & Eber. [*Lauraceae*] SCN: **Indian cassia** AN: *tamala; tamalpatra* PN: *chai gui* (bark) OCN: Indian bark
AN: = Ayurvedic name	
PN = Chinese pinyin transliterations	
Syn: Currently accepted synonyms	
OCN = Other common names	***Cinnamomum verum*** J. Presl. [*Lauraceae*] SCN: **cinnamon** Syn: *Cinnamomum zeylanicum* Nees AN: *tvak* OCN: Ceylon cinnamon, true cinnamon
	Cinnamomum zeylanicum Nees = *Cinnamomum verum*

Plate 3 "Herbs of Commerce" (2000) of the American Herbal Products Association, the formally accepted source for common names of botanicals to be used in herbal product labeling in the United States. Publisher: CRC Press, Boca Raton, FL, USA [8] (left: vol. 1, 1998, right: vol. 2, 2000, bottom: abbreviations)

2.1.2 Disadvantages of Common Names

In contrast, the use of common names in the trade of plants or plant parts in the manufacture of finished products presents serious challenges and potential negative consequences. For example, the common name "arnica" can be applied to 32 different species in six plant families [1]. An extreme example of the serious consequences of using only common names in the production of herbal products occurred in the late 1990s and early 2000s. Bariatric clinics in Belgium integrated a Chinese herbal formula as part of a weight loss program [15]. The formula was presumed to include a botanical known in China by the common name "fang ji" (防己). However, "fang ji" was variably applied to two different species of plant, namely, *Stephania tetrandra*, which is safe, and *Aristolochia fangchi*, which carries with it a risk of renal failure and long-term carcinogenicity due to the presence of aristolochic acid derivatives [16–18]. Practitioners of traditional Chinese medicine differentiate between these two species by the common names "han fang ji" (漢防己), and "guang fang ji" (廣防己) [18], respectively. Unfortunately, some European buyers of herbal ingredients were unaware of the subtle but marked difference between the two species. This resulted in a toxic *Aristolochia* species being used in at least some of the formulas consumed, conservatively contributing to more than 200 cases of kidney failure [19], and likely many more, before the use of *Aristolochia* species was subjected to a nearly worldwide ban (Plate 4).

While common names are appropriate for consumers and herbal practitioners who are culturally familiar with the ingredients they directly consume or use, common names lack the specificity needed for those in the manufacture or research of herbal products destined for widespread consumption. Thus, to avoid nomenclatural confusion, a greater level of nomenclatural specificity is required in the trade of ingredients

Plate 4 *Stephania tetrandra* (bottom) and *Aristolochia fangji* (top), the former safe and the latter responsible for several hundred cases of kidney injury due to the presence of aristolochic acid derivatives. Source: American Herbal Pharmacopoeia, Scotts Valley, CA, USA

and manufacture of herbal products and dietary supplements, even if the finished products only present the common name for consumer recognition.

2.2 *Latinized Binomials*

Latinized binomials, also referred to as Latin, botanical, or scientific names, consist of three primary parts, a generic (genus) name (e.g., *Echinacea*) and a specific epithet (e.g., *purpurea*), along with the taxonomic authority abbreviation (e.g., L., referring to the Swedish botanist Carl Linnaeus). Binomials are typically either of Greek or Latin origin, and have been included in the plant literature since at least the early 1600s. However, it is Linnaeus (1707–1778) who is the most well-renowned for his work on codification in his "Species Plantarum" of 1753 [20] (Plate 5).

In some cases, Latin binomials are accompanied by subspecies or varietal names. The differences between species, subspecies and varieties can be so minor as to make no practical difference regarding which material is used, but, in other cases, the differences can be marked, requiring differentiation. This may be exemplified

CAROLI LINNÆI
S:Æ R:GIÆ M:TIS SVECIÆ ARCHIATRI; MEDIC. & BOTAN.
PROFESS. UPSAL; EQUITIS AUR. DE STELLA POLARI;
nec non ACAD. IMPER. MONSPEL. BEROL. TOLOS.
UPSAL. STOCKH. SOC. & PARIS. CORESP.
SPECIES
PLANTARUM,
EXHIBENTES
PLANTAS RITE COGNITAS,
AD
GENERA RELATAS,
CUM
DIFFERENTIIS SPECIFICIS,
NOMINIBUS TRIVIALIBUS,
SYNONYMIS SELECTIS,
LOCIS NATALIBUS,
SECUNDUM
SYSTEMA SEXUALE
DIGESTAS.
TOMUS I.
Cum Privilegio S. R. M:tis Sueciæ & S. R. M:tis Poloniæ ac Electoris Saxon.
HOLMIÆ,
IMPENSIS LAURENTII SALVII.
1753.

Plate 5 Noted Swedish botanist Carl Linnaeus (left) who codified the Latin binomial system for naming plants with the publication (right) of his "Species Plantarum" in 1753. Source: Rendering of Linnaeus in "A History of the Vegetable Kingdom of William Rhind" (1857). American Herbal Pharmacopoeia, Scotts Valley, CA, USA

for agricultural crops. For example, there are at least 90 different varieties of apples (*Malus domestica*). When making apple juice, apple pie, or other apple products, varietals may be of little or no importance and perhaps more significantly, would be based solely on the personal preference of the consumer. Conversely, the Latin binominal *Brassica oleracea* has been assigned collectively to broccoli, Brussel sprouts, cabbage, cauliflower, collard greens, and kale, which are markedly different, and must be further differentiated by their respective varietal names, viz., *italica, gemmifera, capitata, botrytis, viridis,* and *sabellica* (Plate 6). Similarly, although not as extensive, recognition of varietal forms exists with herbal ingredients (e.g., elder, sage, and rosemary). Sometimes "varietals" are based on specific taxonomic classifications

Plate 6 Example of the importance of discerning varieties in some species, e.g., *Brassica oleracea*; brussels sprouts, cauliflower, kale, collard greens, cabbage, broccoli. Source: American Herbal Pharmacopoeia, Scotts Valley, CA, USA

that require differentiation of the plants. In other cases, varietals are assigned uncritically and do not result in any meaningful taxonomic difference between plants, and may be used uncritically in commerce.

2.2.1 Advantages of Common Names

In contrast to common names, Latin binomials, with exceptions, are assigned to a single plant species and are recognized internationally. Latin binomials are proposed by botanists with a particular expertise in a group of species, and then adjudicated by the International Code of Nomenclature for Algae, Fungi and Plants [1]. Once accepted, an assigned specific Latin binomial, in general, is the most authoritative botanical name ascribed to the plant and supersedes other botanical names internationally. However, not all authoritative botanical sources will be completely up to date in terms of taxonomic revisions or in agreement with a particular taxonomic assignment.

As with common names, Latin names were originally based on physical form (e.g., *Dipsacus*), cultural relationships (e.g., *Achillea*), physiological function (e.g., *somniferum*), organoleptic characteristics (e.g., *Piper*), or a combination of these (e.g., *Hypericum perforatum*). *Dipsacus* refers to the manner in which the sessile leaves of the teasel plant form a cup and capture water, with *dip* referring to the Latin word root for water. *Achillea* was named after the Greek warrior Achilles, for whom it is claimed in legends that he used the common herb yarrow to stop bleeding, a use that persists today. The specific epithet *somniferum*, meaning "sleep bringing" was assigned to the opium poppy (*Papaver somniferum*) from which the alkaloid morphine was isolated, and so was named aptly after Morpheus, the Greek god of sleep and dreams. The genus name *Piper* was coined after the Sanskrit word for long pepper, "pippali", which denotes the spicy pungency of the herb. *Hypericum perforatum*, originated from the Greek ύπέρ, hyper (over) and εἰκών, eikon (image) referring to the historical hanging of St. John's wort over pictures (icons) (Plate 7). The term *perforatum* refers to the translucent glands that are visible on the plant leaf and appear as perforations when held up to the light. Thus, both common and Latin names reflect a history of morphological, cultural, and functional meaning when their origin is understood.

Except for labeling of consumer products, the use of Latin binomials in commerce is preferred whenever possible, to ensure that the correct species of plant is procured. While common names tend to be used by consumers, participants involved in the trade and manufacture of herbal products should establish specifications based on both common names and Latin binomials whenever possible. Unfortunately, not all those persons included in the supply of medicinal plants, from wild harvesters to farmers, are equally familiar with formal taxonomy.

The use of formally accepted botanical names is similarly critical in all aspects of the academic research of medicinal plants, as this is essential for establishing the evidence base for botanical medicine. In research, Latin binomials, botanical authorities, and sourcing information clearly should be provided. Moreover, technical

Plate 7 St. John's wort *Hypericum perforatum* L. (left) from "hyper" and "eikon" meaning "over an icon referring" to the historical hanging of St. John's wort over pictures; *perforatum* referring to the translucent glands on the leaves (middle), and St. John's due to the red coloring of hypericin contained in the glands and oil (right), and is said to represent the "blood of St. John". *Hypericum perforatum* has remained the single species for medicinal St. John's wort for centuries. Source: American Herbal Pharmacopoeia, Scotts Valley, CA, USA

information helpful in the characterization of a given species, such as phytochemical profiles or details on specific varieties or chemotypes, when applicable, also should be included [5].

2.2.2 Disadvantages of Common Names

There are some disadvantages in relying only on modern botanical names when referring to medicinal species used historically. First and foremost, in many cases, the exact species recorded in historical texts may or may not be in accordance with the most recent taxonomic revisions. From historical materia medica texts to current taxonomic practice, botanical specificity may be carried through. For example, there is consensus among pharmacopeial authorities that the medicinal plant species St. John's wort is comprised of the single botanical species *Hypericum perforatum* [21–23]. In other cases, the medicinal article of trade is comprised of multiple species such as willow bark (*Salix*) (Table 1) or hawthorn (*Crataegus*) (Plate 8), each of which has more than one species used interchangeably due to historical precedent and cultural practices, and, to some extent, modern phytochemistry, pharmacology, and/or clinical medicine observations [24–28] (see Tables 1, 2, 3 and 4). Contemporary taxonomy is based on phylogenetic relationships with no consideration of cultural, functional, or medicinal relationships, which represented the historical basis for naming plants.

Table 1 Varying species of *Salix* historically used as the medicinal willow bark

Black willow	*Salix nigra*	= Salicis cortex = Willow bark All accepted in most pharmacopeias internationally
Brittle willow	*Salix fragilis*	
Goat willow	*Salix caprea*	
Laurel willow	*Salix pentandra*	
Purple willow	*Salix purpurea*	
Violet willow	*Salix daphnoides*	
Weeping willow	*Salix babylonica*	
White willow	*Salix alba*	

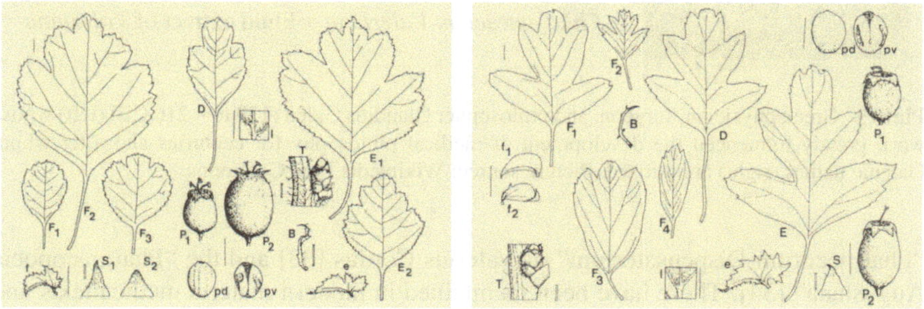

Plate 8 Hawthorn, *Crataegus laevigata* (left) and *C. monogyna* (right); an example of more than one species used as a single medicinal agent. Reproduced from Christiansen [29] with permission of the American Society of Plant Taxonomists

There are strengths and limitations regarding the real-world assignment of phylogenetic relationships [30], as reflected in contemporary research on traditional herbal drugs in which the very identity of a botanical in question may not be assured [1]. According to one review, more than 30% of the scientific plant names cited in phytochemical journals were reported as erroneous [31], while in another, only 7% followed best practices for plant nomenclature [32], suggesting that precision in botanical nomenclature, at least as it is applied to certain scientific publications, is a desired goal but not yet fully realized.

2.3 Galenic Names

Galenic names, also referred to as pharmaceutical names, so named after the Greek physician, surgeon, and philosopher Galen of Pergamon (Plate 9), are comprised of the genus (typically) and plant part that is to be used medicinally (Salicis cortex = willow bark) or finished product (Oleum Hyperici; oil of St. John's wort). Galenic names have been used since the earliest of standards-setting texts as represented in the

Galenical Nomenclature

Medicinal Plant Parts

Hyperici herba = *Hypericum* (St. John's wort) herb
Valerianae radix = *Valeriana* root
Croci stigma = *Crocus* stigma

Medicinal Preparations

Tinctura Hyperici = Tincture of *Hypericum*
Oleum Hyperici = Oil of *Hypericum*
Fluid extractum Valeriana = Fluid extract of *Valeriana*

Plate 9 Greek physician, surgeon, and philosopher Claudius Galen (129 to ~ 216 CE) (left) whose work greatly influenced the development of medical philosophy for centuries and after whom Galenic names (right) are utilized. Picture source: Wikimedia Free Commons

"Pharmacorum Dispensatorium" of Valerius Cordus [33] and the "Pharmacopoeia Augustana" [34]. These have been maintained in modern materia medica texts and other authoritative works worldwide, such as the "American Herbal Pharmacopoeia", "European Pharmacopoeial Commission", "Pharmacopoeia of the People's Republic of China", and the World Health Organization, while being omitted from others.

2.3.1 Advantages of Galenic Names

Galenic or pharmaceutical nomenclature is the only formal naming system that obviates the need for the complexities, inconsistencies, and challenges that exist with common names and Latin binomials by identifying both the acceptable taxon or taxa used and the plant part. Moreover, botanical articles to which Galenic names are applied, predominantly in pharmacopeias and materia medica texts, are further delimited to specific quality, purity, and potency criteria, all of which constitute the identity of the herbal drug.

With Galenic names, a specific epithet name is only included if it is relevant for referring to a specific singular species rather than multiple species. Galenic names are predominantly referred to primarily in pharmacopeias and volumes on materia medica, two of the primary sources of medicinal plant standards referenced worldwide. When integrated into monographs or materia medica texts, a Galenic name inherently refers to acceptance criteria that include the identity of the botanical, namely, macroscopic, microscopic, and organoleptic characteristics, and specific potency and purity considerations.

Both historically and as presently used in many countries, Galenic names are the preferred nomenclatural system for medicinal plant parts.

2.3.2 Disadvantages of Galenic Names

The uncritical application of Galenic names potentially results in the use of an inappropriate species. For example, the Galenic name for *Arctostaphylos uva-ursi* leaf, which is used as a urinary tract antiseptic, is Uvae ursi folium. The antiseptic properties of *A. uva-ursi* are due predominantly to the presence of the hydroquinone glycoside arbutin. One subspecies, *Arctostaphylos uva-ursi* subsp. *stipitata*, does not contain arbutin [35]. While arbutin is likely not the only compound contributing to the biological activity of this plant [36], species lacking arbutin may not possess the same activity, and would not be delimited by use of the Galenic name only.

In other cases, botanical revisions can cause challenges in the application of Galenic names, such as for Sambuci fructus, which refers mainly to the fruit of the European or black elder, *Sambucus nigra*. Most research supporting the clinical efficacy of *Sambucus* is with *S. nigra*. Based on taxonomic revisions of recent decades, the North American elder *Sambucus canadensis* and the California blue elder *Sambucus cerulea*, have been subsumed as subspecies of *S. nigra* [37]. While *S. nigra* subsp. *canadensis* and *S. nigra* subsp. *nigra* appear to be very much alike morphologically, chemically, and pharmacologically, *S. nigra* subsp. *cerulea* is markedly different from the others [38] (Plate 10). If only referred to by its Galenic name and not within the context of a pharmacopeial monograph or formal standard, appropriate differentiation between these species may not be made. This nomenclatural challenge is compounded by the fact that not all taxonomic authorities agree with all aspects of the *Sambucus* revision [39]. Regardless, the uncritical application of a Galenic name could suggest to those unfamiliar with their formal acceptance criteria that any fruit of any *Sambucus* species could be used, including the fruits of the potentially poisonous red elder (*Sambucus racemosus*).

The application of Galenic names might be frowned upon by some trained botanists due to this lack of taxonomic specificity, as they perhaps mistakenly

Plate 10 Elderberry, *S. nigra* subsp. *nigra* (left), and *S. nigra* subsp. *cerulea* (right)—once considered separate species but subsumed by Bolli (1994) as subspecies of *S. nigra* [37]. Source: American Herbal Pharmacopoeia, Scotts Valley, CA, USA

consider Galenic names to have no formal parameters of delimitation [2], although such parameters are established in authoritative texts.

3 Challenges with Using Botanical Type Specimens, Vouchers, and Genetics for Medicinal Plants

In the formal naming of a plant species, a "type specimen" is utilized as the basis of the name assigned. A review of the historical records of a type specimen allows botanists to trace the genesis of the botanical name ascribed so that its veracity may be either ascertained or challenged. Type specimens are archived as permanent records in formal herbarium collections as botanical vouchers. This is a dried, pressed, and preserved specimen of the plant upon which is affixed information including the nomenclature, collection date, location and possibly the habitat, the name of the collector(s), and a unique identifying number so that all aspects of that plant specimen can be traced. Type specimens and vouchers are the foundation of botanical nomenclature and remain the gold standard of plant taxonomic identification.

Botanical academicians and other authorities recommend that medicinal plant parts in commercial trade should be traceable to a botanical specimen or voucher deposited in a formal herbarium collection [1]. Others have adopted the same recommendations, including national and international pharmacopeias and other national and international standards-setting bodies [5]. The lack of such documentation may call into question the very identity of any plant material used for any purpose. Therefore, tracing a specimen back to a properly identified voucher represents a best practice that should be encouraged and applied with some flexibility, as academic perfection is often extremely difficult to achieve.

In the trading practices of medicinal plants, there are many challenges that can prevent the precise tracking of bulk material to a properly identified voucher. These are more than logistical difficulties in procuring a voucher but challenges in the discipline of taxonomy itself. In some cases, the original type specimen may have been erroneously applied to other species, as in the case of *Ganoderma* in which the name *Ganoderma lucidum*, originally proposed in England, was misapplied to an Asian species. This resulted in nearly all authorities for several decades improperly assigning the Latin binomial *G. lucidum* to what is today generally regarded as *G. lingzhi*, based on phylogenetic relationships. According to one investigation, in the absence of a detailed investigation of type specimens of the approximately 290 taxonomic names referring to an estimated 80 species of *Ganoderma*, "it will not be possible to apply a taxonomically correct name with certainty" [40].

Perhaps more relevant for the example of *Ganoderma* is that many experts recognize the medicinal species of "ling zhi" used in Asia as being comprised of a complex representing numerous, but closely related, *Ganoderma* species (Plate 11; Table 2) [41, 42]. More challengingly, the nomenclatural assignment of Asian *G. lingzhi* as distinguished from European *G. lucidum* was established by Cao et al. [43] based

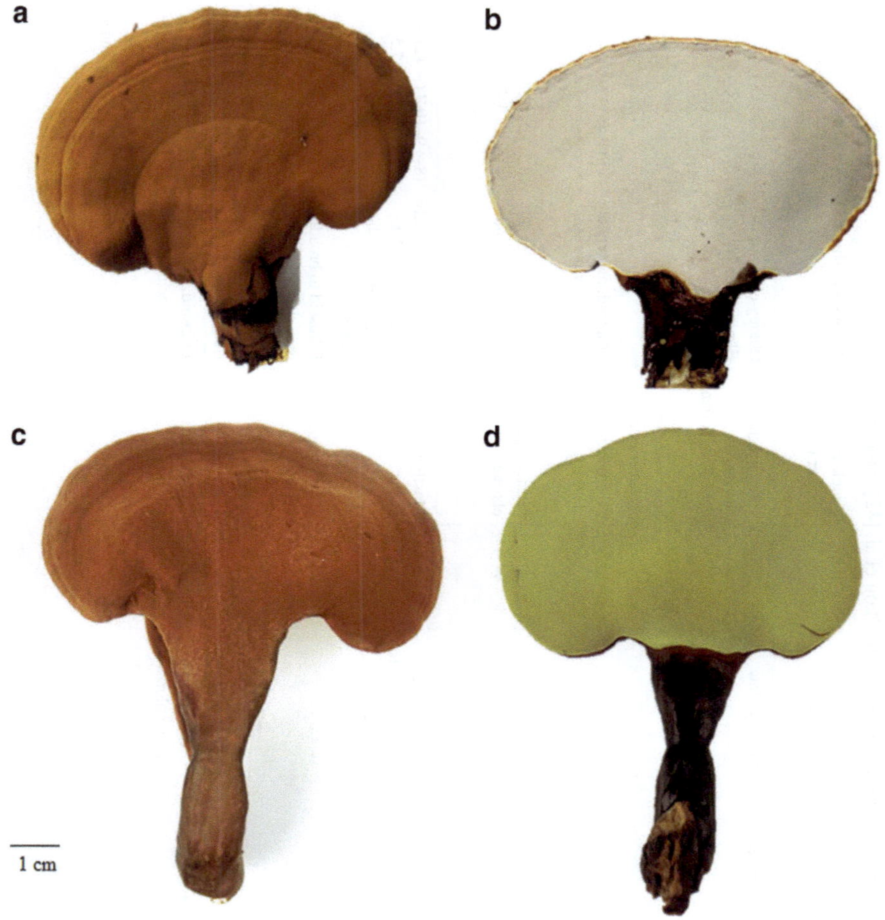

Plate 11 Examples of two different species within the *Ganoderma lingzhi* complex: *G. lingzhi* (**a** and **b**), *G. resinaceum* (**c** and **d**). Source: Chen et al. [51]

on genetic work that was challenged subsequently by other work proposing *G. sichuanense* as "the appropriate name for the Asian species due to publication precedent" [44]. In the process of discerning the 'correct' name for the Chinese fungal drug "ling zhi", a survey of more than 600 *Ganoderma* sequences available in the genetic sequence database GenBank, revealed that 65% of sequences for *Ganoderma* species were either misidentified or ambiguously labeled [44]. Additional botanical authorities have reported other similar insufficiencies of GenBank [45]. This can be expected in any endeavor to integrate large, disparate, and heterogeneous datasets [46] and where anything less than a 99% genetic match can only identify down to the genus level [30]. However, others report on the relative accuracy of GenBank [47]. Additionally, genetic differences within a single species of *Ganoderma* growing in different regions that are sometimes greater than those existing between different

Table 2 Differentiation of species within the *Ganoderma lucidum* complex according to Zhou et al. [42] (modified)

Species	Pileus surface color	Concentric growth zones	Melanoid bands in mature fruiting body	Pore color (at maturity)	Size (mm)	Cuticle cell shape of outer pileus layer	Basidospore shape	Size (μm)
G. boninense	Nearly homogeneous, dark clay-buff	Absent	Usually present	Straw-yellow	4–5	Often irregular, clavate or cylindrical, often with blunt outgrowths or protuberances	Oblong-ellipsoid to ellipsoid, finely echinulate	(8–) 8.7–12.8 (– 13.5) × (4.2–) 4.7–6 (– 6.3)
G. curtisii	Not completely homogeneous, light buff to clay-buff	Absent	Present	Yellowish	4–6	Mostly regular, clavate	Ellipsoid, moderately coarsely echinulate	(9–) 9.2–11 (– 11.2) × (5.8–) 6–6.8 (– 7)
G. flexipes	Not completely homogeneous, yellow brown to dark brown	Absent	Present	White to pale sulfur yellow	4–6	Mostly regular, clavate	Ellipsoid, moderately coarsely echinulate	(8.5–) 9–10.3 (– 11) × (5–) 5.3–7
G. lingzhi	Not completely homogeneous, light buff, buff, clay-buff to snuff-buff	Absent	Present	Pale yellow, sulfur-yellow to straw-colored	5–6	Mostly regular, clavate	Ellipsoid, moderately coarsely echinulate, sometimes with short ridges	(8–) 9–10.7 (– 12) × (5.2–) 5.8–7 (– 7.5)
G. lucidum	Not completely homogeneous, cream or pinkish buff to clay-buff	Occasionally present	Absent	White	4–5	Mostly regular, clavate	Ellipsoid to broadly pear-shaped, coarsely echinulate, but more commonly with sinuous ridges	(8.8–) 9.7–12.2 (– 13.2) × (6–) 6.3–8 (–8.5)
G. multipileum	Not completely homogeneous, clay-buff to fulvous	Present	Present	Cream to straw-colored	4–6	Mostly regular, clavate	Ellipsoid, finely and thickly echinulate	(8–) 8.8–10.5 (– 11.3) × (5–) 5.5–7 (– 7.2)

(continued)

Table 2 (continued)

Species	Pileus surface color	Concentric growth zones	Melanoid bands in mature fruiting body	Pore color (at maturity)	Size (mm)	Cuticle cell shape of outer pileus layer	Basidospore shape	Size (μm)
G. oregonense	Not completely homogeneous, cream to pinkish-buff	Absent	Absent	White	3–4	Mostly regular, clavate	Ellipsoid, coarsely echinulate	11–12.5 (– 12.8) × (6.8–) 7–8
G. resinaceum	Not completely homogeneous, straw-yellow or buff to clay-buff	Present	Absent	White	3–4	Mostly regular, clavate	Ellipsoid to ovoid, finely to moderately coarsely echinulate	(8.8–) 9–11.7 (– 12) × (6–) 6–7.5 (– 8)
G. sessile	Not completely homogeneous, pinkish-buff to clay-buff	Mostly present	Sometimes present	Straw-yellow	4–6	Mostly regular, clavate	Ellipsoid, slightly to moderate coarsely echinulate	(9–) 9.2–12 (– 12.5) × (5.3–) 6–7.8 (– 8.2)
G. sichuanense	Not completely homogeneous, buff to pale clay-buff	Absent	Present	Buff-yellow	5–6	Mostly regular, clavate	Ovoid to ellipsoid, finely to moderate-coarsely echinulate	(7–) 7.5–9.2 (– 9.3) (4.5–) 5–6.5 (– 6.8)
G. tropicum	Homogeneous, fulvous	Present	Present	Cream to pale straw-colored	4–6	Mostly irregular, clavate, often with blunt outgrowths or protuberances	Ellipsoid to broadly ellipsoid, coarsely echinulate	(8.3–) 8.8–10.7 (– 11.2) × (5–) 5.5–6.3 (– 6.8)
G. tsugae	Not completely homogeneous, white to pinkish-buff or clay-buff	Absent	Absent	White	4–5	Mostly regular, clavate	Ellipsoid, moderately coarsely echinulate	(8.8–) 9–10.8 (– 11) × (5.3–) 5.8–6.8 (– 7)
G. zonatum	Nearly completely homogeneous, umber	Present	Absent	Buff-yellow	4–5	Mostly irregular, clavate or cylindrical, with protuberances or slight branches	Oblong-ellipsoid, finely echinulate	(9.8–) 10–12 (– 12.8) × (5–) 5.3–6.3 (– 6.7)

species [48]. At the same time, the morphological characters between different specimens displaying genetic differences may be so morphologically similar (with genetic plasticity and morphological stasis) as to make morphological discrimination challenging [49]. Thus, even when the most sophisticated attempts are made to distinguish between plant species, those involved in such work acknowledge the challenges in establishing identities with certainty [40–42]. Traditionally, identification of the Chinese herbal drug "ling zhi" was based on morphological characters that distinguished species based on different colorations [50]. While one group has expressed the lack of botanical specificity of *Ganoderma* species as "chaotic" [49], many Chinese herbalists could identify "ling zhi" readily from its discernible morphological characteristics (Plate 11).

Ganoderma does not represent an anomalous nomenclatural challenge; a lack of nomenclatural consensus exists for numerous medicinal plant and fungal species including *Achillea, Cordyceps* (*Ophiocordyceps*), *Crataegus, Euphrasia, Opuntia, Sambucus, Salix*, and *Taraxacum*, among many others. Conversely, the ability to determine the authenticity of any of these genera of plants as an appropriate herbal medicine or dietary supplement is readily achievable by anyone with only a rudimentary training in plant identification and materia medica. Hence, the morphological and organoleptic characteristics for identifying these plants would be evident. This is analogous to a grandmother knowing exactly what apples to use for a pie, but without knowing their Latin name, the botanical authority who named them, their taxonomic history, or the apple's genetic sequence.

Type specimens may also not encompass the full range of characteristics within a given species. The morphological expression of the common herb yarrow (*Achillea millefolium*), considered to occur within an aggregate consisting of diploid, hexaploid, octaploid, and tetraploid types and encompassing many species, varies greatly depending on environmental conditions [52–54], but are nevertheless used similarly worldwide [55]. Thus, the splitting or grouping of a species into an aggregate or finite number of species, subspecies, or varieties, may or may not have relevance to those using the plant for a specific health benefit. In determining the identity of a plant, the botanist will focus on morphological, reproductive, and genetic relationships. In contrast, a herbal medicine practitioner determines the identity based on morphological, organoleptic, and medicinal relationships. Therefore, the acceptability criteria for using a plant species differ between disciplines. For medicinal use, ultimately, the acceptability of species must be determined either through the traditional use of practitioners or by clinical investigation. Phytochemical analytical work most certainly will be a good indication of potential benefit or acceptability, but if this is lacking any clinical investigation component, it can be argued as representing only a surrogate for what might be acceptable.

Cultivated plants present a significant challenge. Normally, type specimens are established on specimens growing in their native habitat. Cultivated plants often express morphological characteristics that are far removed from those of the type specimen used to establish the identity of the material. This is evident in terms of the cultivation of *Cannabis* spp. (Plate 12) either for hemp (fiber type) or high concentrations of Δ^9-*trans*-tetrahydrocannabinol (THC) (drug type). Historically, these plants were differentiated between *Cannabis sativa* for fiber and seed and *C. indica* for the drug type and were easily distinguished morphologically. Extensive cross-breeding of the two primary types has resulted in a nearly complete morphological melding of identifying characters that do not allow for adequate botanical differentiation between the types [56]. This is additionally challenging as marijuana and hemp have been defined in a regulatory statute based on THC content, superseding classical botany, at least in terms of regulation.

Nomenclatural challenges such as these underscore the difficulty that exists when there is a lack of consensus within a discipline itself. These are mainly academic obstacles that are inherently divorced from the role of a traditional practitioner, whose primary goal is to provide a safe and effective medicine to a patient, and may have little interest in attempting to determine which taxonomic treatment is most appropriate, currently proposed, or accepted.

Plate 12 Typical European hemp (left) and drug type cannabis (right). Source: courtesy of CV Sciences, San Diego, CA, USA (left) and American Herbal Pharmacopoeia, Scotts Valley, CA, USA (right)

4 Academic Specificity Versus Global Reality: Saffron—A Case History

To the greatest extent possible, medicinal plants used in trade should be traceable to a properly identified and archived botanical voucher. In addition to the lack of botanical consensus that exists within the discipline of taxonomy itself, there are both practical and logistical hindrances to obtaining plant vouchers internationally.

Saffron consists of the stigmas of the flower of *Crocus sativus* (Plate 13). Fortunately, saffron is a monotypic species making its plant identification relatively facile. The centuries-long cultural practice of cultivating saffron in Iran and its instantly recognizable morphological countenance adds an increased level of simplicity that allows for the ready identification of the species. Generally, each flower has three stigmas that must be plucked by hand. Saffron is the most expensive and, because of its high cost, one of the most adulterated botanicals in the world. Approximately 90% of saffron produced in the world originates in Iran, which, in 2021, reportedly produced 350 metric tons. It has been estimated that one kilo of dry saffron requires hand-picking 167,000 flowers from which approximately 500,000 stigmas are plucked [57].

There are many different regions of Iran that produce saffron. Saffron from these different regions is harvested and consolidated (mixed) to provide the world's market with this precious spice and medicinal agent, leaving open the question as to what percentage of the world's market for saffron can be traced to a botanical voucher. Obviously, it is not practical to have a botanical voucher for 167,000 flowers and so, a representative sample must be chosen. Thus, which sample from which region should be used as the representative sample? For the 350 metric tons of saffron produced annually in Iran, how many vouchers would be needed to ensure the identity of the complete range of plants making up the world's supply? Iranian saffron cultivation consists of literally thousands of small family operations or collectives. The ideal would be to have an expert botanist prepare representative vouchers at every cultivation site, every month of every year of cultivation and harvest. However, the question must also be asked, what purpose might it serve? When maintained in its whole form, saffron is so recognizable that the need for botanical vouchering for the identification is superfluous. Adulteration of saffron occurs after plant harvesting by the mixing or substitution with other materials and additives. Documentation of the identity of the plant by virtue of a voucher specimen will not reduce the propensity for subsequent saffron adulteration. Moreover, a botanical voucher is only a dried and pressed representation of a single plant. When maintained in its whole or semi-whole form, saffron represents the entirety of the batch, and can be readily assessed.

As with most botanicals, the macroscopic and organoleptic characteristics of saffron are correlated directly with the constituents responsible for its medicinal activity, specifically, red and yellow coloring (apocarotenoids), aroma (picrocrocin), and flavor (safranal) [58, 59]. While representative vouchers can be obtained for those vertically integrated with their supply chain, a single botanical voucher is not intended to capture the level of specificity required for assessing the identity of the

Plate 13 Approximately 167,000 flowers and 500,000 stigmas are required to produce 1 kg of dry saffron making it difficult or impossible to trace entire batches to a botanical voucher

medicinal plant part. Additionally, no single or multiple vouchers can be expected to represent adequately the tens of millions of plants that contribute to the saffron market. It is neither logistically possible nor necessary, as other tests, most notably macroscopic and organoleptic, are the most specific for identifying authoritatively the plant material (stigmas), as well as the relative purity, and relative quality.

As with *Ganoderma*, saffron is not an anomalous example of the challenges in having traceability of medicinal species back to a properly identified and archived voucher that represents each of the batches that make up the entirety of species in world trade. A large percentage of botanical ingredients in the world market comprises mixed lots from different harvesters from different regions, collected by wildcrafters or farmers with varying levels of experience and expertise in identifying plants. For those manufacturing products from a single or relatively few botanical ingredients, such as *Theobroma* for chocolate, *Vitis vinifera* for wine, *Camellia sinensis* for green tea, or *Vaccinium macrocarpon* for cranberries, traceability to a proper identification, including vouchered material, is possible. For the rest of the several hundred medicinal plants in commerce that are traded in metric tons from varying sources and sometimes from varying species, botanical vouchering represents an ideal towards which to strive, but this requires an acknowledgement of the appropriateness and superiority of other testing methodologies to confirm the authenticity of an herbal drug ingredient. Block chain controls have been proposed to create more robust documentation and transparency in the trade of medicinal plant parts. However, severe challenges exist in making this a reality [60].

5 Authenticity Versus Botanical Specificity

The concept of authenticity versus botanical specificity is occasionally considered in academic investigations. Simmler et al. [5] defined the authenticity of botanicals as follows: "an authentic botanical has a confirmed identity, a validated chemical composition, and is devoid of any adulteration ("pure, unadulterated") in accordance with certified guidelines and by comparison with authentic reference materials whenever possible ("conforms to an original, certifiable")". Clearly, in modern times with extensive analytical capabilities, this represents the gold standard of medicinal plant authentication in the broad sense that encompasses identity, relative purity, relative quality, traceability, and the documentation of each parameter. As reflected in the previous discussion of different species being used as a single article of trade, "confirmed identity" from a botanical perspective can be ambiguous. In contrast from a medicinal or economic perspective, authenticity can be readily assured. As indicated in the example of saffron, authenticity, relative quality, relative purity, and lack of adulteration can be attained without any formal botanical "certification" (i.e., implying a third-party oversight) or chemical or genetic assay by those with the requisite training. In historical and current pharmacognosy reference texts, an abundance of information exists for guiding analysts in the assessment of the identity and quality of herbal ingredients macroscopically and organoleptically (e.g., Refs. [61–67]) and it is recommended that the application of classical pharmacognosy methods should be utilized in routine herbal ingredient quality control.

Most gold standard assessment techniques that those in academia and regulators may wish to apply are more theoretically oriented, rather than being designed for real-world applications. Thus, an analyst without any training in botany, horticulture, culinary arts, or traditional herbal medicine may be considered as lacking the primary orientation required for directly assessing herbal ingredient quality. Accordingly, other phytochemical and genetic analytical tools are used, which may be regarded as only surrogates for assessing the true identity and quality of herbal ingredients.

6 Multiple Species-Single Medicinal Agent

As noted, a more practical limitation of the adherence to specific botanical names in commerce, is that, frequently, multiple plants may be used to provide a single article of trade. For example, the United States Department of Agriculture recognizes several species of *Vaccinium*, which can provide the agricultural and medicinal ingredient blueberry, inclusive of *Vaccinium angustifolium* (lowbush blueberry), *V. boreale* (Northern blueberry), *V. corymbosum* (Northern high-bush blueberry), among others. Similarly, there are other species recognized internationally as blueberries such as *V. myrtilloides* (Canadian blueberry) and *V. koreanum* (Korean blueberry) [68, 69]. To a general consumer looking for blueberries, or even an herbal medicine consumer expecting a rich source of anthocyanins common to blueberry, the Latin binomial, species, botanical authorities, and varieties hold little relevance as several species are used interchangeably. The same interchangeable use of multiple species exists for less common medicinal plants.

The "Pharmacopoeia of the People's Republic of China" [70] accepts the interchangeable use of four different species of *Epimedium* as the medicinal herb "yin yang huo" (淫羊藿) (*Epimedium brevicornu, E. sagittatum, E. pubescens*, and *E. koreanum*) [70], while a Japanese and Korean species, *E. grandiflorum*, is listed as the primary species used for "yin yang huo" (*Epimedium*) in the 2nd edition of an authoritative English-language materia medica compendium of Chinese herbs, "Chinese Herbal Medicine" [71] (Plate 14). In the 3rd edition of this work, *E. grandiflorum* was removed and an additional species, *E. wushanense*, was included, with acknowledgement of three other species used in local regions as being "yin yang huo" [72]. Thus, a variety of authentic species, although of varying botanical specificity, are represented in even the most authoritative of sources. There are numerous other cases of the use of multiple species in the "Pharmacopoeia of the People's Republic of China" and in other pharmacopeias and texts of materia medica, and these underscore the use of different species as a single medicinal plant of trade.

The "Pharmacopoeia of the People's Republic of China" [70], as well as other literature sources (e.g., [67]), provides details for differentiating between *Epimedium* species based on leaf characters. However, more is needed to identify the species with botanical certainty, especially if species are mixed or cut in any way, which is commonly done. Even when intact the identifying characters almost completely overlap and if cut, any differentiation is almost impossible. For those in academia desiring specificity, this would be problematic. In contrast, any capable practitioner of Chinese herbal medicine should be able to readily identify authentic "yin yang huo" visually, whether in the whole or semi-whole forms (Table 3). In the case of

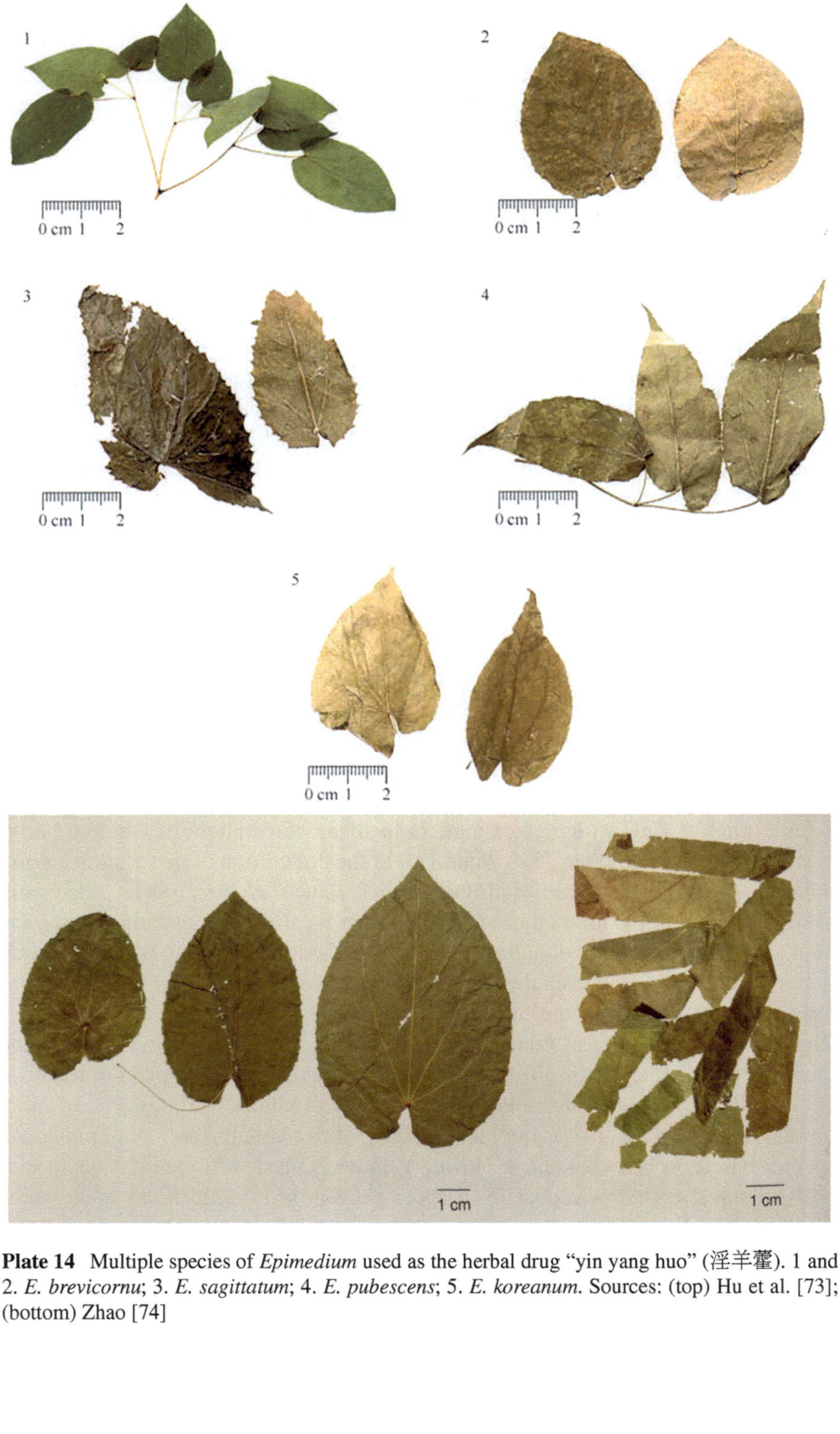

Plate 14 Multiple species of *Epimedium* used as the herbal drug "yin yang huo" (淫羊藿). 1 and 2. *E. brevicornu*; 3. *E. sagittatum*; 4. *E. pubescens*; 5. *E. koreanum*. Sources: (top) Hu et al. [73]; (bottom) Zhao [74]

Table 3 Macromorphological differentiation between five primary species of *Epimedium* used as the Chinese herbal drug "yin yang huo"

Item	*E. brevicornu*	*E. sagittatum*	*E. wushanense*	*E. koreanum*	*E. pubescens*
Shape	Biternate compound leaf, leaflets oval shaped	Ternate compound leaf, leaflets have a lanceolate to narrow lanceolate shape	Ternate compound leaf, leaflets have a lanceolate to narrow lanceolate shape	Biternate compound leaf, leaflets are relatively wide and large	Ternate compound leaf, lower surface of the leaf and petiole are densely covered in velvety soft hairs
Base of lateral leaflets	Flat heart shaped, outer side relatively large and earshaped	Obviously deviated outer side is arrow-shaped	Deviated, the inner side is small and round and the outer side is large and triangular		
Texture	Almost leathery	Leathery		Relatively thin	
Leaflet size	3–8 cm long 2–6 cm wide	4–12 cm long 2.5–5 cm wide	9–23 cm long 1.8–4.5 cm wide	4–10 cm long 3.5–7 cm wide	

Source: Zhao et al. [67]

the different species of blueberries (or any other number of plants), no distinction between the species can be discerned morphologically. However, anyone familiar with these fruits can identify them correctly as blueberries. Thus, the orientation of academicians studying plant medicines may be different than that of "medical practitioners" and encompasses naturopaths, herbalists, MDs, pharmacists, or any other practitioner using herbs.

7 The Role of Modern Pharmacognosy in Medicinal Plant Authentication

Pharmacognosy evolved as an independent subdiscipline of pharmacy due to the need for a specialized field of knowledge that was separate from the practice of medicine, the delivery of drugs (pharmaceutics), materia medica (pharmacology), and the identification of live plants (pharmaceutical botany). At the same time, pharmacognosy integrated aspects of each of these disciplines to a greater or lesser degree. Most notably, this included the optimal sourcing and processing of natural products used in medicine and pharmacy that is still relevant today.

Before the advent of modern analytical chemistry, pharmacognosy focused on assessing natural products morphologically, microscopically and organoleptically [75], with considerable emphasis placed on optimal sourcing and the detection of adulterants. As medicinal agents evolved from crude plants and other organisms and their extracts to pure chemical entities, pharmacognosy followed the same trajectory. Hence there became a new focus on pure compounds from natural products, especially after the discovery and marketing of the anticancer drugs vinblastine and vincristine in the early 1960s [76], as well as other natural product drugs targeting other diseases. In this natural evolution, the focus was primarily on phytochemistry and biological activity. However, the taxonomic authentication of the plants of origin of bioactive lead compounds has been still deemed as a very important research component [76].

Simmler et al. [5], clearly articulated the role modern pharmacognostic practice can play in medicinal plant authentication and quality assessment, as follows:

> Authentic botanicals are by definition non-adulterated, a mutually exclusive relationship that is confirmed through the application of a multilayered set of analytical methods designed to validate the (chemo) taxonomic identity of a botanical and certify that it is devoid of any adulteration. In practice, the ever-increasing sophistication in the process of intentional adulteration, as well as the growing number of botanicals entering the market, altogether necessitate a constant adaptation and reinforcement of authentication methods with new approaches, especially new technologies. This article summarizes the set of analytical methods — classical and contemporary — that can be employed in the authentication of botanicals. Particular emphasis is placed on the application of untargeted metabolomics and chemometrics. An NMR-based untargeted metabolomic model is proposed as a rapid, systematic, and complementary screening for the discrimination of authentic vs. potentially adulterated botanicals. [5]

The recent approval of the cannabis-based drug Epidiolex® (cannabidiol) for rare seizure disorders in children offers an example of one type of a modern pharmacognostic approach, which led to the subsequent formal approval of a drug of botanical origin in the USA and also the European Union and United Kingdom (where it is known as Epidylolex®).

There are frequently reported accounts of ancient Sumerian tablets suggesting the use of cannabis for "nocturnal convulsions" [77]. These were followed by similar accounts in the eleventh and nineteenth centuries [78] and again in the nineteenth century by Irish physician O'Shaughnessy [79], and beginning in the early 1970s and onwards through various cannabis use surveys [78]. These reports of empirical use eventually resulted in interest in developing a cannabis-based drug that would be approved according to modern drug approval standards.

The company GW Pharmaceuticals (UK), which eventually developed the FDA-approved Epidiolex®, was founded in 1998 specifically to advance research into the medical use of cannabis and cannabinoids. Their process, which has now led to widespread approval of this drug, involved the identification and propagation of specific cannabis cultivars, developing a comprehensive chemical profiling of the material, the establishing of complete agricultural and processing practices for overall control and characterization of the raw cannabis material, the subsequent characterization and standardization of a finished product, and the requisite pre-clinical, clinical, and safety data needed to meet with regulatory drug approval. Approval of Epidiolex® was granted in the U.S. in 2018, initially to treat Lennox-Gastaut syndrome and Dravet syndrome, and then in 2020 for tuberous sclerosis complex, all for patients over one year of age. The above steps exemplify the needs of the modern drug development process, which perhaps some investigators believe should be applied to all medicinal plants, and, in an ideal world, would be appropriate. The difficulty of this system is that for this single botanical-based drug, this process took 20 years at an estimated cost of $1.2 billion [80]. Based on relatively recent reports, the cost of drug approval across 98 countries is estimated at $350 million to $5 billion [81]. Such is the cost of the approval of a traditional medicine with an 1,800-year history of relative safety and efficacy (Plate 15).

Other investigators have presented a comprehensive model for characterizing medicinal plants with a focus on *Glycyrrhiza glabra, G. uralensis*, and *G. inflata*, three species in the genus formally recognized as authentic medicinal licorice in traditional cultures, and this epitomizes the scientific pursuit of botanical characterization [82] (Plate 16).

> The developed chemometric models enable the identification and classification of *Glycyrrhiza* species according to their chemical composition ... Further key outcomes demonstrated that DNA authentication combined with chemometric analyses enabled the characterization of mixtures, hybrids, and species outliers. This study provides a new foundation for the botanical and chemical authentication, classification, and metabolomic characterization of crude licorice botanicals and derived materials. Collectively, the proposed methods offer a comprehensive approach for the quality control of licorice as one of the most widely used botanical dietary supplements. ... The quality control of pharmacopeial licorice relies mainly on the detection and quantitation of glycyrrhizin, a constituent that

Plate 15 *Cannabis* budding tops (left) and dried bud (right) have been used for seizure disorders for more than 1,800 years before formal approval was granted in the U.S. in 2018 at an estimated cost of $1.2 billion to bring this botanically derived drug to market. Source: American Herbal Pharmacopoeia, Scotts Valley, CA, USA

Plate 16 *Glycyrrhiza inflata* (left) and roots of *G. glabra* (right) two of several species of *Glycyrrhiza* that formally comprise the herbal drug licorice root. Source: American Herbal Pharmacopoeia, Scotts Valley, CA, USA

> is present in all three *Glycyrrhiza* species and, thus, is the least a species-specific marker. … Considering that "metabolite profiling" is predominantly associated with identifying key metabolites that lead to the differentiation of samples, its usage is preferred in this study.

In this investigation by Simmler and colleagues referred to above, clear distinctions were made between the three *Glycyrrhiza* species based on genetic and chemical differences that showed gross similarities and subtle differences. These observations suggested that the species should be differentiated because they can be distinguished, despite acknowledging that each is considered an acceptable herbal drug [82]. In a previous investigation by the same research group, oriented to investigating plants to promote women's health, it was reported that different *Glycyrrhiza* species contain varying amounts of liquiritigenin and therefore "have various levels of estrogenic activities, suggesting the importance of precise labeling of botanical supplements". This may suggest the need for these species to be separated [83]. However, absent from this model is the recognition that all cultures who use licorice root do so because of its sweet flavor, which corresponds to the presence of the triterpene glycoside, glycyrrhizin, the primary qualitative compound of pharmacopeial standards [82]. Detailed pharmacological investigations of the various licorice species have demonstrated anti-inflammatory activity associated with numerous compounds, including for glycyrrhizin, which provides mechanistic support for many of the traditional uses of licorice, including digestive and respiratory support and for soothing irritated mucus membranes [84]. The estrogenic activity of compounds of licorice and other members of the plant family Fabaceae, reflects a modern pharmacological finding that is of interest to researchers on the bioactive principles of medicinal plants. This is because the observations made could have relevance to alleviating menopausal symptoms [83]. However, this same study seems to have little relevance to the quality control of licorice root within the context of its traditional uses.

8 Traditional Herbal Medicines—Emphasis on Classical Botanical Pharmacognosy

The development and approval of Epidiolex® represent a streamlined approach to producing a new drug of botanical origin. However, it may be argued that this is not a process that should be required to approve or regulate traditional herbal medicines with a long history of relatively safe and effective use. Large numbers of herbal products are already being utilized by literally tens of millions of consumers in virtually every country, representing innumerable daily exposures. Rather than approaching the authentication and quality control of herbal medicines and dietary supplements in the same way as one would be required for contemporary drug development, it may be more consistent with historical precedent to once again embrace the use of classical botanical pharmacognosy skills. These include an intimate knowledge of the sourcing and assessment criteria for crude botanical ingredients and not only a reliance on phytochemistry.

The integration of scientists with training in classical botanical pharmacognosy techniques could well provide a useful service to the herbal products industry, as well as to regulators, academic scientists, and ultimately consumers. Modern laboratory investigations of plants are based on highly sophisticated principles and practices. However, traditional plant assessment should not be considered inferior to modern techniques. The livelihood of early plant identifiers depended on their ability to successfully treat patients. The need for an emphasis to be given to both botany and phytochemistry, as well as an acknowledgment of the skills of traditional herbalists, was emphasized by noted English pharmacognosists Trease and Evans in their "A Textbook of Pharmacognosy" [85]:

> All information, both botanical and chemical, should be taken into account, but as botanists themselves point out, when primitive tribes name the plants of their environment they are almost as successful in defining limits as the trained taxonomists.

9 Pharmacopeial Definitions—The Gold Standard for Botanical Ingredient Authenticity and Quality

A pharmacopeial definition is the only nomenclatural system that establishes standards for identity, purity, quality, and specific analytical methods for medicinal plants.

Pharmacopeial definitions clearly articulate the genus, species, and part of the plant that must be used, and the quality parameters that should be met to conform to the monograph. In some cases, single species are designated as the only one appropriate for medicinal use. In other cases, multiple species including their hybrids are allowed, even for species that are among the most difficult to identify taxonomically (e.g., *Crataegus* and *Salix* species; see Table 4). Pharmacopeial standards are more consistent with the traditional use of herbal medicines historically, and in more contemporary times. As presented in pharmacopeias, the orthogonal approach to identification overcomes many of the challenges of only using classical botanical taxonomy in medicinal plant identification.

Table 4 Examples of pharmacopeial definitions used internationally

Medicinal plant	Definition	References
St. John's wort	Whole or fragmented, dried flowering tops of *Hypericum perforatum* L., harvested during flowering time. Content: minimum 0.08% of total hypericins, expressed as hypericin ($C_{30}H_{16}O_8$; M_r 504.4) (dried drug)	EP [22]
St. John's wort	St. John's Wort Flowering Tops consists of the dried flowering tops of *Hypericum perforatum* L. (family Hypericaceae) gathered shortly before or during flowering. They contain not less than 0.6% of hyperforin ($C_{13}H_{52}O_4$) and not less than 0.04% of the combined total of hypericin ($C_{30}H_{16}O_8$) and pseudohypericin ($C_{30}H_{16}O_9$), on the dried basis	USP [23]
Hawthorn berry	Dried false fruits of *Crataegus monogyna* Jacq. (Lindm.) or *C. laevigata* Poir.) DC., (synonym: *C. oxyacantha* L.), or their hybrids or a mixture of these false fruits	EP [26]
Hawthorn leaf and flower	Whole or cut, dried flower-bearing branches of *Crataegus monogyna* Jacq. (Lindm.), *C. laevigata* (Poir) DC. (synonyms: *C. oxyacanthoides* Thuill.; *C. oxyacantha* Auct.) or their hybrids or, more rarely, other European *Crataegus* species including *C. pentagyna* Waldst. et Kit. ex Willd., *C. nigra* Waldst. et Kit. and *C. azarolus* L., containing a minimum of 1.5% of total flavonoids, expressed as hyperoside (dried drug)	EP [27]
Hawthorn leaf and flower	Hawthorn leaf with flower consists of a mixture of the fresh or dried flowering tops of *Crataegus laevigata* (Poir.) DC. (syn. *Crataegus oxyacantha* L.), *Crataegus monogyna* Jacq. (Lindm.), or their hybrids, or other species of *Crataegus* containing not less than 1.5% of flavonoids calculated as hyperoside ($C_{21}H_{20}O_{12}$: M_r 464.4) as determined on a dry weight basis	AHP [24]
Willow bark	Salix Species Bark is prepared from the whole or fragmented dried bark of the young branches, or whole dried pieces of the current-year twigs, obtained from *Salix* species (family Salicaceae). Common in pharmacopeial use are *S. alba* L., *S. babylonica* L., *S. daphnoides* Vill., *S. fragilis* L., *S. chilensis* Molina, *S. pentandra* L., *S. purpurea* L., and a number of other complying willow species and their hybrids "…" Contains not less than 1.50% of total salicylate derivatives, calculated as salicin ($C_{13}H_{18}O_7$) on a dried basis	USP [28]

(continued)

Table 4 (continued)

Medicinal plant	Definition	References
Epimedium	Epimedium Leaf is the dried leaf of *Epimedium brevicornu* Maxim., *Epimedium sagittatum* Maxim., *Epimedium pubescens* Maxim., and *Epimedium koreanum* Nakai. It contains not less 5% total flavones calculated as icariin and not less than 0.40% icariin by HPLC	PPRC [70]
Epimedium	Epimedium Leaf is the dried leaf of *Epimedium sagittatum* Maxim., *Epimedium koreanum* Nakai, or *Epimedium brevicornu* Maxim. and similar species (family Berberidaceae). It contains not less than 15.0% of dilute ethanol-soluble extractives, not less than 6.0% of water extractives and not less than 0.4% of icariin	THP [86]

10 Botanical Reference Materials

Central to assuring the reliability and accuracy of any analysis is the use of botanical reference materials. According to international standards, a reference material is a "Material, sufficiently homogeneous and stable concerning one or more specified properties, which has been established to be fit for its intended use in a measurement process..." [87]. Reference materials used in chemistry are developed primarily for quantitative analysis for general or specific purposes. For botanicals, a reference material may be developed to assure the identity of a material or its constituent content. Other reference materials are developed specifically to contain fixed concentrations of selected compounds to ensure the precision of a specific analytical method (e.g., as developed by the U.S. National Institute of Standards and Technology). More commonly, botanical reference materials (see Plate 17 for the example of *Punica granatum*) are plant standards used to help verify the identity, purity, or composition of botanical products [88]. Reference materials may be developed to contain a fixed constituent profile or for discerning qualitative properties such as form, color, texture, taste, aroma, mouth feel, and skin feel, each of which is a part of the macroscopic and organoleptic analytical profile of herbal ingredients. Any reference material must be chosen based on the specific analytical goal for which it will be used (i.e., made fit for purpose).

There are several hundred botanicals in common trade in North America, of which many are of Asian, Indian, and South American origin. Obtaining commercial material of such species is relatively easy as there are few international prohibitions regarding their trade. However, there are strict restrictions regarding transferring plants from one country to the another, even as dry specimens, such as vouchers. Formal herbaria, the ideal source of most properly identified botanical specimens, similarly have strict restrictions regarding trading materials between countries. These restrictions include intellectual property agreements, adherence to the Convention on the Trade of Endangered Species (where relevant), and national agricultural restrictions from both importing and exporting countries.

Plate 17 American Herbal Pharmacopoeia-verified Botanical Reference Material (BRM) for pomegranate seed (*Punica granatum*) that begins with properly identified (upper left), documented (upper right), harvested (middle left), fresh (middle right) plant material, and verified BRM package (bottom)

As previously noted, most materials traded internationally are from varied sources, multiple batches, and sometimes various countries, which are mixed before entering commercial trade or may be co-mingled at various points in the supply chain. Attempting to trace trade materials to a particular voucher specimen is challenging logistically, if not impossible for most plants.

11 Summary

Accurate nomenclature is critical for ensuring that the most relevant article of commerce is used in the trade of plants for health or medical purposes. A variety of nomenclatural systems exist, including common names, Latin binomials, Galenic (pharmaceutical) names, and pharmacopeial definitions. Each system has advantages and disadvantages and may be appropriate or inappropriate at varying stages of botanical trade. Common names are most appropriate for products marketed to consumers familiar with these, and unfamiliar with Latin binomials. Conversely, Latin binomials should be used in all aspects of botanical trade and academic research to identify unambiguously specific plants are being traded and investigated. Latin binomials, along with botanical families and authorities, are specific for botanists and academicians wishing to trace the origin of plant nomenclature but are of no importance to consumers and of little relevance to those in trade, unless there is a specific need to research the nomenclatural history of a botanical. Classic botanical taxonomy is the gold standard for identifying live plants in the wild. However, it may be subjective with different authorities expressing varying opinions regarding the veracity of botanical classifications even when formally adjudicated. Classical botanical identification is primarily relevant for plant parts for which a botanical voucher or other documentation of botanically characteristic features (high-resolution photographs) of the unique identifying features of the plant are shown. As botanical identification is based primarily on identifying above-ground plant parts, especially the flowers, some floras and taxonomic keys underemphasize or give little guidance to identifying other plant parts of a given species (e.g., roots, rhizomes, and bark).

Cultivated plants present unique challenges as they can express characteristics far outside the parameters of type specimens. Some botanical medicine ingredients must be of a certain age before being harvested, or aged or processed in a certain way, or they may be composed of multiple species. Single or even multiple botanical vouchers typically do not represent these requirements. Plant parts used in global trade are comprised of literally millions of plants that eventually make up a single lot of material. In some cases, the plant part is distinctive enough to identify at the species level. For some botanicals, a single plant is used for a single article of trade, while, for others, multiple species are utilized. In some cases, species may be so closely related as to not be easily distinguishable botanically, genetically, chemically, or pharmacologically but can be readily identifiable as an authentic article of trade by those with the requisite experience or training. Thus, the complexity of establishing the most appropriate identity of medicinal plant parts requires more than a confirmation of a single or even multiple botanical vouchers.

Worldwide, definitions as included in formal pharmacopeias provide the optimum nomenclatural system inclusive of the true identity of the article in trade, which encompasses the identity of accepted species, the plant part used, and the specific quality parameters regarding acceptability. No other nomenclatural system is as specific for establishing the identity and quality of a medicinal plant of commerce.

References

1. Bennett BC, Balick MJ (2014) Does the name really matter? The importance of nomenclature and plant taxonomy in biomedical research. J Ethnopharmacol 152:387
2. Dauncey EA, Irving J, Allkin R, Robinson N (2016) Common mistakes when using plant names and how to avoid them. Eur J Integr Med 8:597
3. Food and Drug Administration (2016) Botanical drug development guidance for industry. U.S. Department of Health and Human Services Food and Drug Administration Center for Drug Evaluation and Research; Pharmaceutical Quality/CMC Revision 1. https://www.fda.gov/files/drugs/published/Botanical-Drug-Development--Guidance-for-Industry.pdf
4. Sarma N, Upton R, Rose U, Guo D-A, Marles R, Khan I, Giancaspro G (2023) Pharmacopeial standards for the quality control of botanical dietary supplements in the United States. J Diet Suppl. 20:485. https://doi.org/10.1080/19390211.2021.1990171
5. Simmler C, Graham JG, Chen SN, Pauli GF (2018) Integrated analytical assets aid botanical authenticity and adulteration management. Fitoterapia 129:401
6. Raven P (2004) Taxonomy: where are we now? Phil Trans R Soc Lond B 359:729
7. Rouhan G, Gaudeul M (2021) Plant taxonomy: a historical perspective, current challenges, and perspectives. In Besse P (ed) Molecular plant taxonomy: methods and protocols. Methods in molecular biology, 2nd edn, vol 2222. Humana, New York, NY, p 1
8. McGuffin M, Tucker A, Leung AY, Kartesz JT (2000) Herbs of commerce, 2nd edn. American Herbal Products Association, Silver Spring, MD
9. Petrovska BB (2012) Historical review of medicinal plants' usage. Pharmacogn Rev 6:1
10. Linares E, Bye RA (1987) A study of four medicinal plant complexes of Mexico and adjacent United States. J Ethnopharmacol 19:153
11. Bandara T, Uluwaduge I, Jansz ER (2011) Bioactivity of cinnamon with special emphasis on diabetes mellitus: a review. Int J Food Sci Nutr 63:380
12. Bernardo MA, Silva ML, Santos E, Moncada MM, Brito J, Proenca L, Singh J de Mesquita MF (2015) Effect of cinnamon tea on postprandial glucose concentration. J Diabet Res 913651
13. Code of Federal Regulations (2022) Title 21 subchapter B part 102. https://www.fda.gov/food/food-labeling-nutrition/standards-identity-food#:~:text=Standards%20of%20Identity%20for%20Food%20The%20FDA%20began,breads%2C%20peanut%20butter%2C%20and%20ketchup%20have%20a%20SOI
14. Tea and Herbal Infusions Europe (2018) Compendium of guidelines for tea (*Camellia sinensis*). Tea Herb Inf Europe 5:49
15. Shaneberg B, Khan I (2004) Analysis of products suspected of containing *Aristolochia* or *Asarum* species. J Ethnopharmacol 94:245
16. Koh HL, Wang H, Zhou S, Chan E, Woo SO (2006) Detection of aristolochic acid I, tetrandrine and fangchinoline in medicinal plants by high performance liquid chromatography and liquid chromatography/mass spectrometry. J Pharm Biomed Anal 40:654

17. Okhale SE, Egharevba HO, Okpara OJ, Ugabe GE, Ibrahim JA, Fatokum OT, Sulyman AO, Igoli JO (2019) Aristolochic acids in herbal medicine: public health concerns for consumption and poor regulation of botanical products in Nigeria and West Africa. J Med Plant Res 13:55
18. Wu KM, Farrelly JG, Upton R, Chen J (2007) Complexities of the herbal nomenclature system in traditional Chinese medicine (TCM): lessons learned from the misuse of *Aristolochia*-related species and the importance of the pharmaceutical name during botanical drug product development. Phytomedicine 14:273
19. Tankeu S, Vermaak I, Chen W, Sandasi M, Viljoen A (2016) Differentiation between two "fang ji" herbal medicines, *Stephania tetrandra* and the nephrotoxic *Aristolochia fangchi*, using hyperspectral imaging. Phytochemistry 122:213
20. Nicolson DH (1991) A history of botanical nomenclature. Ann Missouri Bot Gard 78:33
21. American Herbal Pharmacopoeia (1998) St. John's wort. In: Analytical, quality control, and therapeutic monograph, vol 32. Santa Cruz, CA
22. European Pharmacopoeia (2008) St. John's wort; Hyperici herba, 2958. Strasbourg, France
23. United States Pharmacopeia (2011) St. John's wort:1226, Rockville, MD
24. American Herbal Pharmacopoeia (1999) Hawthorn leaf and flower. In: Analytical, quality control, and therapeutic monograph, vol 29, Santa Cruz, CA
25. American Herbal Pharmacopoeia (1999) Willow bark monograph and therapeutic compendium, vol 17. Santa Cruz, CA
26. European Pharmacopoeia (2013) Hawthorn berry; Crataegus fructus, vol 1271. Strasbourg, France
27. European Pharmacopoeia (2013) Willow bark; Salicis cortex, vol 1422. Strasbourg, France
28. United States Pharmacopeia (2011) Salix species. In: Dietary supplements, vol 4850. Rockville, MD
29. Christiansen K (1992) *Crataegus* (Rosaceae). In: Systematic botany monographs, vol 35. American Society of Plant Taxonomy, Ann Arbor, MI
30. Parveen I, Gafner S, Techen N, Murch SJ, Khan IA (2016) DNA barcoding for the identification of botanicals in herbal medicine and dietary supplements: strengths and limitations. Planta Med 82:1225
31. Rivera D, Allkin R, Obón C, Alcaraz F, Verpoorte R, Heinrich M (2014) What is in a name? The need for accurate scientific nomenclature for plants. J Ethnopharmacol 152:393
32. Nesbitt M, McBurney RPH, Broin M, Beentje HJ (2010) Linking biodiversity, food and nutrition: the importance of plant identification and nomenclature. J Food Comp Anal 23:486
33. Cordus V (1556, reprinted 1934) Das Dispensatorium des Valerius Cordus. Arthur Nemayer, Mittenwald, Bavaria, Germany
34. Pharmacopoeia Augustana (1613, reprinted 1944) State Historical Society of Wisconsin, Madison, WI
35. Packer JG, Denford KE (1974) A contribution to the taxonomy of *Arctostaphylos uva-ursi*. Can J Bot 52:743
36. Kosalec I, Zovko M, Kalogjera Z, Pepeljnjak S (2007) Comparison of antimicrobial activity of traditionally used uroantiseptics against clinical isolates of uro-genital pathogens. Planta Med 73:888
37. Bolli R (1994) Revision of the genus *Sambucus*. In: Dissertationes botanicae, vol 223. J. Cramer, Berlin, Stuttgart, Germany
38. Mikulic-Petkovsek M, Ivancic A, Schmitzer V, Veberic R, Stampar F (2016) Comparison of major taste compounds and antioxidative properties of fruits and flowers of different *Sambucus* species and interspecific hybrids. Food Chem 200:134
39. Applequist W (2015) A brief review of recent controversies in the taxonomy and nomenclature of *Sambucus nigra* sensu lato. Acta Hortic 1061:25
40. Hennicke M, Cheikh-Ali Z, Liebisch T, Maciá-Vincente JG, Bode HB, Piepenbring M (2016) Distinguishing commercially grown *Ganoderma lucidum* from *Ganoderma lingzhi* from Europe and East Asia on the basis of morphology, molecular phylogeny, and triterpenic acid profiles. Phytochemistry 127:29
41. Li TH, Hu HP, Deng WQ (2015) *Ganoderma leucocontextum*, a new member of the *G. lucidum* complex from southwestern China. Mycoscience 56:81

42. Zhou LW, Cao Y, Wu SH, Vlasák J, Li D-W, Li M-J, Di Y-C (2015) Global diversity of the *Ganoderma lucidum* complex (Ganodermataceae, Polyporales), inferred from morphology and multilocus phylogeny. Phytochemistry 114:7
43. Cao Y, Wu SH, Dai YC (2012) Species clarification of the prize medicinal *Ganoderma* mushroom "Lingzhi". Fungal Divers 56:49
44. Jargalmaa S, Eimes JA, Park MS, Park JY, Oh S-Y, Lim YW (2017) Taxonomic evaluation of selected *Ganoderma* species and database sequence validation. PeerJ 5:e3596
45. Bidartondo M (2008) Preserving accuracy in GenBank. Science 319:1616
46. Boyle B, Hopkins N, Lu ZY, Garay JAR, Mozzherin D, Rees T, Matasci N, Narro ML, Piel WH, McKay SJ, Lowry S, Freeland C, Peet RK, Enquist BJ (2013) The taxonomic name resolution service: an online tool for automated standardization of plant names. BMC Bioinformatics 14:16
47. Leray M, Knowlton N, Ho SL, Nguyen BN, Machida RJ (2019) GenBank is a reliable resource for 21st century biodiversity research. Proc Natl Acad Sci USA 116:22651
48. Mei Z, Yang LQ, Khan MA, Yang M, Wei C, Yang W, Peng X, Tania M, Zhang H, Li X, Fu J (2014) Genotyping of *Ganoderma* species by improved random amplified polymorphic DNA (RAPD) and inter-simple sequence repeat (ISSR) analysis. Biochem Syst Ecol 56:40
49. Pawlik A, Janusz G, Debska I, Siwulski M, Frac M, Rogalski J (2015) Genetic and metabolic intraspecific biodiversity of *Ganoderma lucidum*. Biomed Res Int 726149
50. Wasser SP (2005) Reishi or ling zhi (*Ganoderma lucidum*). In: Coates PM, Blackman MR, Cragg GJ, Levine M, Moss J, White J (eds) Encyclopedia of dietary supplements. Marcel Dekker, New York, NY, p 680
51. Chen B, Ke B, Ye L, Jin S, Jie F, Zhao L, Wu X (2017) Isolation and varietal characterization of *Ganoderma resinaceum* from areas of *Ganoderma lucidum* production in China. Sci Hort 224:109
52. Hiesey WM (1953) Comparative growth between and within climatic races of *Achillea* under controlled conditions. Evolution 7:297
53. López-Vinyallonga S, Soriano I, Susanna A, Montserra JM, Roquet C, Garcia-Jacas N (2015) The polyploid series of the *Achillea millefolium* aggregate in the Iberian Peninsula investigated using microsatellites. PLoS One 10:e0129861
54. Nunez-Farfan J, Schlichting CD (2001) Evolution in changing environments: the synthetic work of Clausen, Keck, and Hiesey. Quart Rev Biol 76:433
55. Applequist W, Moerman W (2011) Yarrow (*Achillea millefolium* L.): a neglected panacea? A review of ethnobotany, bioactivity, and biomedicals research. Econ Bot 65:209
56. Murovec J, Erzen JJ, Flajsman M, Vodnik D (2022) Analysis of morphological traits, cannabinoid profiles, *THCAS* gene sequences, and photosynthesis in wide and narrow leaflet high-cannabidiol breeding populations of medical *Cannabis*. Front Plant Sci 13:786161
57. Skinner M, Parker BL, Ghalehgolabbehbahani A (2020) Saffron by the numbers. North American Center for Saffron Research and Development University of Vermont, Burlington, VT
58. Bhandari PR (2015) *Crocus sativus* L. (saffron) for cancer chemoprevention: a mini review. J Trad Complement Medic 5:81
59. Heitmar R, Brown J, Kyrou I (2019) Saffron (*Crocus sativus* L.) in ocular diseases: a narrative review of the existing evidence from clinical studies. Nutrients 11:649
60. Heinrich M, Scotti F, Booker A, Fitzgerald M, Kum KY Löbel K (2019) Unblocking high-value botanical value chains: is there a role for blockchain systems? Front Pharmacol 10:00396
61. Applequist W (2006) The identification of medicinal plants. A handbook of the morphology of botanicals in commerce. American Botanical Council, Austin, TX
62. Cooper JW, Denston TC, Riley M (1931) A textbook of pharmacognosy. Sir Isaac Pitman and Sons, Ltd., London, UK
63. Flueckiger FA, Tschirch A (1887) The principles of pharmacognosy. William Wood and Co., New York, NY
64. Leon C, Lin YL (2018) Chinese medicinal plants, herbal drugs and substitutes: an identification guide. Royal Botanic Gardens, Kew, UK

65. Mansfield W (1926) Mansfield's materia medica and pharmacognosy. William Mansfield, Albany, NY
66. Wichtl M (2004) Herbal drugs and phytopharmaceuticals: a handbook for practice on a scientific basis. Germany Medpharm/CRC Press, Boca Raton, FL, Stuttgart, Germany
67. Zhao Z, Chen H, Guo P, Brand E (2014) Chinese medicinal identification. Paradigm Publications, Taos, NM
68. United States Department of Agriculture (1995) Blueberries: grade standards. Fed Reg 60:11242
69. United States Department of Agriculture (2017) Index of official visual aids, BLC-CP-1. https://www.ams.usda.gov/sites/default/files/media/Official%20Inventory%20of%20FV%20Inspection%20Aids.pdf
70. The State Pharmacopoeia Commission of PR China (2015) Pharmacopoeia of the People's Republic of China, vol 1. China Medical Science Press, Beijing, PRC
71. Bensky D, Gamble A (1993) Chinese herbal medicine: materia medica, 2nd edn. Eastland Press, Seattle, WA, USA, Seattle
72. Bensky D, Clavey S, Stoeger E (2004) Chinese herbal medicine: materia medica, 3rd edn. Eastland Press, Seattle, WA
73. Hu Z, Wang ZT, Xie PS (2015) Monographs for quality evaluation of Chinese crude drugs. SCPG Publishing Corporation, New York, NY
74. Zhao Z (2016) An illustrated Chinese materia nedica in Hong Kong. Hong Kong Baptist University, Hong Kong
75. Fitzgerald M, Heinrich M, Booker A (2020) Medicinal plant analysis: a historical and regional discussion of emergent complex techniques. Front Pharmacol 10:1480
76. Henkin JM, Ren Y, Soejarto DD, Kinghorn AD (2017) The search for anticancer agents from tropical plants. Prog Chem Org Nat Prod 107:1
77. Lozano I (2001) The therapeutic use of *Cannabis sativa* L. in Arabic medicine. J Cannabis Ther 1:63
78. Friedman D, Sirven JI (2017) Historical perspective on the medical use of cannabis for epilepsy: ancient times to the 1980s. Epilepsy Behav 70:298
79. O'Shaughnessy WB (1843) On the preparations of the Indian hemp or gunju (*Cannabis indica*). Their effects on the animal system in health and their validity in the treatment of tetanus and other convulsive diseases. Prior Med Retrosp Med Sci 5:303
80. Sutton A (2022) United States pharmacopeial convention. In: Cannabis testing forum-focus on medicinal use. Rockville, MD, USA, 7–8 Dec 2022 (Unpublished)
81. Herper M (2013) The cost of creating a new drug not $5 billion. Pushing big pharma to change. Forbes, Pharma & Healthcare. https://www.forbes.com/sites/matthewherper/2013/08/11/how-the-staggering-cost-of-inventing-new-drugs-is-shaping-the-future-of-medicine/?sh=194b015913c3
82. Simmler C, Anderson JR, Gauthier L, Lankin DC, McAlpine JB, Chen S-N, Pauli GF (2015) Metabolite profiling and classification of DNA-authenticated licorice botanicals. J Nat Prod 78:2007
83. Hajirahimkhan A, Simmler C, Yang Y, Anderson JR, Chen S-N, Nikolic D, Dietz BM, Pauli GF, van Bremen RB, Bolton JL (2013) Evaluation of estrogenic activity of licorice species in comparison with hops used in botanicals for menopausal symptoms. PLoS One 8:e67947
84. Yang R, Yuan BC, Ma YS, Zhou S, Liu Y (2017) The anti-inflammatory activity of licorice, a widely used Chinese herb. Pharm Biol 55:5
85. Trease GE, Evans WC (1966) A textbook of pharmacognosy 9th ed. Baillere, Tindall and Cassell, London, UK
86. Ministry of Health and Welfare (2019) Taiwan herbal pharmacopeia, 3rd ed. (English). Ministry of Health and Welfare, Taipei, Taiwan
87. ISO Guide 30:1992, Amendment 1 (2008) Terms and definitions used in connection with reference materials-Amendment 1: revision of definitions for reference material and certified reference material. https://www.iso.org/standard/46210.html
88. Eurofins (2022) The role of botanical reference materials in dietary supplement testing. https://www.eurofinsus.com/food-testing/resources/botanical-reference-materials-in-supplement-testing/

Roy Upton RH, DipAyu, has been working and practicing professionally as an herbalist since 1981. He is trained in Ayurvedic, Traditional Chinese, and Western herbal medicine, and also has studied and worked extensively with Native American and Caribbean ethnobotanical traditions. He is the Founder, Executive Director, and Editor of the American Herbal Pharmacopoeia (AHP). In addition, he serves on botanical expert advisory committees of the American Botanical Council (Austin, Texas), and of the Journal of Alternative and Complementary Medicine, Chinese Herbal Medicines, World Journal of Traditional Chinese Medicine, the Botanical Safety Handbook, and Herbs of Commerce.

Roy Upton has received outside recognition, inclusive of the James Lind Scientific Achievement Award (2004), the James Duke Excellence in Botanical Literature Award (Technical) from the American Botanical Council (2011), the Varro E. Tyler Excellence in Botanical Research Award from the American Society of Pharmacognosy (2012), and the Outstanding International Ethnopharmacologist Award from the International Society for Ethnopharmacology (2020).

An essential part of his work is defending the rights of all people to access herbal medicines, and to facilitate herbal medicine integration into the fabric of both community and national health care systems as well as to bridge gaps between traditional herbal and scientific perspectives through the integration of both.

Deoxyribonucleic Acid Barcoding for the Identification of Botanicals

Natascha Techen, Iffat Parveen, and Ikhlas A. Khan

Contents

N. Techen (✉) · I. Parveen · I. A. Khan
School of Pharmacy, National Center for Natural Product Research, The University of Mississippi, P.O. Box 1848, University, MS 38677-1848, USA
e-mail: ntechen@olemiss.edu

I. Parveen
e-mail: iparveen@olemiss.edu

I. A. Khan
e-mail: ikhan@olemiss.edu

A. D. Kinghorn, H. Falk, S. Gibbons, Y. Asakawa, J.-K. Liu, V. M. Dirsch (eds.), *Progress in the Chemistry of Organic Natural Products 122*, Progress in the Chemistry of Organic Natural Products 122, https://doi.org/10.1007/978-3-031-26768-0_4

1 Introduction

A botanical is a plant or plant part valued for its medicinal or therapeutic properties, flavor, or scent. Herbs are a subset of botanicals. Products made from botanicals used to maintain or improve health are called herbal products, botanical products, or phytomedicines. The natural herbal products industry uses botanicals or herbs as raw materials for producing herbal products or dietary supplements. Recently, the demand for natural herbal products has increased tremendously, leading to ingredient adulteration problems and to counterfeit herbal products. The current methods for botanical identification are based mostly on detecting therapeutic secondary metabolites using various chromatographic and spectroscopic methods, such as those involving TLC, HPLC, MS, and NMR procedures (see chapter on "Quality Consistency of Herbal Products: Chemical Evaluation", in this volume). However, sample metabolite variations as a result of plant sourcing from different geographical locations, as well as seasonal variations and changes in storage conditions, and since different species may produce the same major constituents, all tend to make the determination of phytochemical profiles by the above-mentioned methods unsuitable for identifying all botanicals. Hence, more orthogonal methods are required for the botanical identification of natural herbal products. Molecular markers such as deoxyribonucleic acid (DNA) barcodes have been used successfully for identification of biological species. The present contribution deals with the utilization of DNA barcodes and other molecular methods to identify botanicals.

Traditionally, plant specimens can be identified using macro-morphological and microscopic examinations or with chemical profiling methods if their marker compounds are known and established. However, as indicated above, such methods may be deficient and not able to identify botanical material unambiguously. Consequently, there is a strong interest currently in the development of additional identification methods, such as DNA barcoding. The concept behind this process is that DNA is present in almost every cell of the plant body and is not affected by geographical location or seasonal variations, unlike plant secondary metabolites. On the other hand, DNA barcoding is not able to distinguish between the various tissues of a given plant species. Genomic DNA is located in the nucleus, chloroplasts, and mitochondria of a plant cell (Fig. 1).

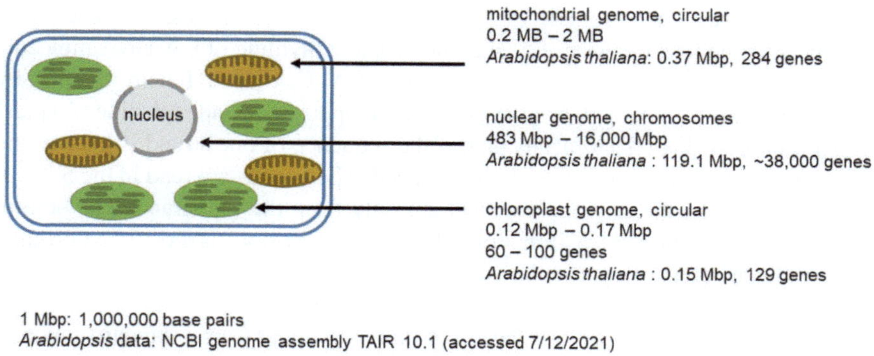

Fig. 1 Genomes located in the plant cell

1.1 Definitions of Gene, Genomic Region, and Loci

Before providing more specific information on deoxyribonucleic acid (DNA) barcoding, some terminology needs to be explained. A "gene" is a segment of DNA (and in some viruses formed by ribonucleic acid, RNA) on a specific location on a chromosome, and it is a functional unit of heredity (Fig. 2). The position of a gene on a chromosome is called the "locus" (plural = loci). Genes have a start (promoter) and a stop region (terminator). They can consist of interchanging sections of exons (coding region) and introns (non-coding regions). Genes are often separated by intergenic (non-coding) sequences, which are also referred to as spacers. Interestingly, many genes do not provide instructions for protein synthesis.

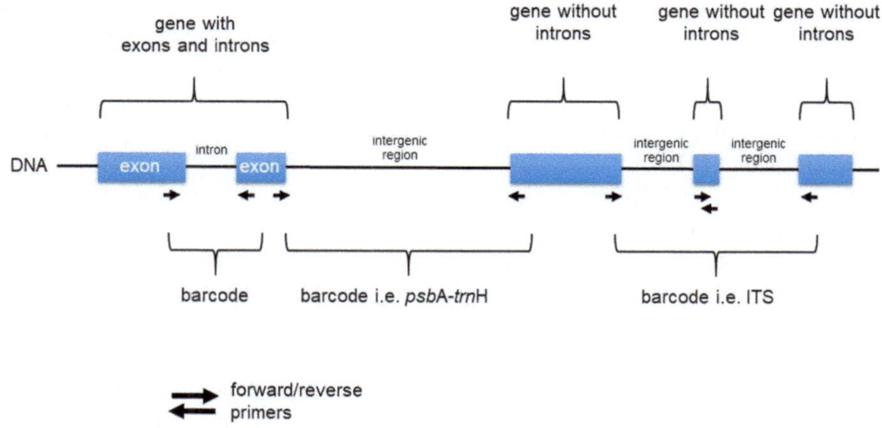

Fig. 2 Genes and genomic regions

The DNA sequence is made of four different units, the bases, or nucleotides, namely, adenine (A), cytosine (C), guanine (G), and thymine (T). A DNA molecule consist of two intertwined strands of a series of A, C, G, and T held together by hydrogen bonds formed between the bases of the opposite strand. Adenine (A) pairs with T, and C with G. That means that, if the sequence of one strand is known, the complementary strand sequence can be deduced. The strands are read in the $5' \rightarrow 3'$ direction. Usually, the sequences shown are only from the top strand. The bottom strand is shown in the $3' \leftarrow 5'$ direction. The "genomic region" is more of a universal term, and can refer to a particular locus or gene, a part of it, or a location in the genome, which is not associated with any locus or gene.

Hebert et al. suggested in 2004 the use of short sequences from a standard genomic region in order to identify species [1]. In the animal kingdom, this is the mitochondrial gene, cytochrome c oxidase I (COI or COX1), although this has been found unsuitable for use in plants. For plants, no single genomic region is found that applies to all species of a genus. The reason is for this is that some species have evolved recently, and hence their genomic information is remarkably similar. A gene region that shows sequence variation to distinguish individual species from a particular family or genus may work nicely, but this may not be helpful in distinguishing species from a different family or genus.

An ideal barcode is a single genomic region that can be used to identify all terrestrial plants. It should consist of an area where the sequence differs between species (diagnostic area). However, the flanking regions are mostly conserved (identical across species) to allow the binding of universal primers to allow the polymerase chain reaction (PCR).

In 2011, at the Fourth International Barcode of Life Conference, three barcodes were approved for plant identification: chloroplast genes ribulose-bisphosphate carboxylase (*rbcL*), maturase K (*matK*), and the non-coding chloroplast region between the genes *psbA* and *trnH* (*psbA-trnH*) [2]. A few years later, the China Plant BOL Group presented the effectiveness of the nuclear ribosomal internal transcribed spacer (ITS) as a universal DNA barcode from the nuclear genome. It has been proposed that ITS2, which is part of ITS, is an even better barcode, because of its short length, high amplification and sequencing efficiency, and high variation between species [3].

Since the use of DNA barcoding commenced, various gene regions have been used for species identification for a genus of interest. Consequently, whatever gene region is deemed appropriate may be applied. This includes, for example, various gene regions from the chloroplast or gene regions from the nuclear genome, as mentioned previously. A large amount of DNA data are available as a result of the development of high-throughput sequencing. It has been suggested that several chloroplast genomic regions should be used, or even the whole plastid genome, as the barcode for species identification [4–8]. Unfortunately, none of the identified markers can be applied universally for the identification of 100% of all plant species. Most studies combine two or more genomic regions for plant species identification and phylogenetic analysis.

There is no guarantee that the chosen genomic region for barcoding can help to distinguish between species if the sequence variation between species is limited or indeed does not exist. If the intraspecific distances (sequence differences for a given species) are below 1%, and the interspecific distances (sequence differences between two or more species) are above 1%, then there is a so-called "barcoding gap". The presence of a "barcoding gap" gives assurance that genetic distance can be used to assign species names to specimens. However, closely related species may have a very variable region, which may be too variable or not consistent enough within a group of specimen to clearly assign species names to them. Nevertheless, no matter what gene region is used for species identification, the entire process depends on a well-curated database with several representative samples of a voucher specimen.

1.2 Nucleotide Polymorphism Can Become Diagnostic Sequences

A single nucleotide variation or polymorphism (SNP) is the most common type of genetic variation. The SNP could be, for example, the replacement of the nucleotide cytosine (C) with the nucleotide thymine (T) in a certain genomic region. The insertion and/or deletion of less than 1 kb nucleotides in each DNA sequence is classified as a genetic variation, and these are referred to as "insertion-deletion mutations (Indels)". Their use may be important in distinguishing between species.

The DNA restriction endonucleases are the enzymes that can be used to "read" a DNA sequence and enzymatically cut it at their recognition sequence. For example, the EcoRI endonuclease recognizes the palindromic sequence GAATTC and will cleave the double-stranded DNA between residues G and A on each strand. To date, over 200 restriction endonucleases are known and commercially available. The genetic variations-SNPs or Indels (Fig. 3) can create or abolish endonuclease recognition sites. Consequently, DNA or polymerase chain reaction products from different specimens can be cleaved or not if exposed to a DNA restriction endonuclease.

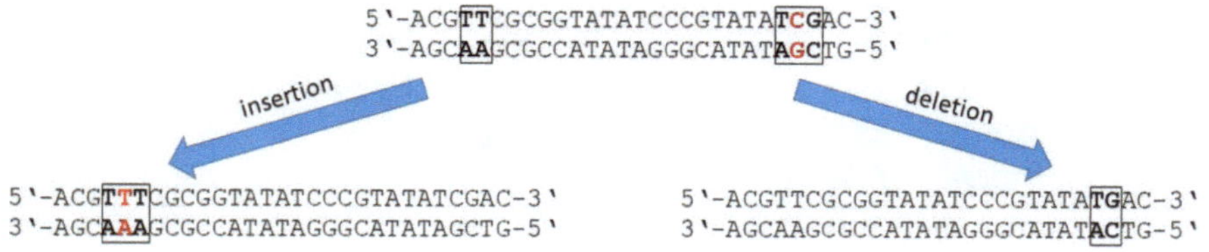

Fig. 3 Indel mutations: nucleotide insertion and deletion

2 Polymerase Chain Reaction

Using the polymerase chain reaction (PCR) procedure (Fig. 4), new DNA strands, exact copies of the template DNA, are produced through a three-step cyclic process of denaturation, annealing of primers (short pieces of DNA matching the target sequence), and extension. The first step is to separate the DNA strands at about 95°C so that in the second step, annealing, the DNA primers can bind to the separated strands. A thermostable DNA polymerase then extends the primers by adding nucleotides using the DNA strand as a template. This process creates a copy of the target DNA between the two primers. These three steps are cycled 20–30 times (with more or fewer cycles). In each cycle, the amount of target sequence is doubled, and, in an ideal case, the reaction continues in an exponential manner until both the primers and nucleotides are used up, and the DNA polymerase loses its activity. Various DNA polymerases are available for this process. The standard amplification length is usually smaller than 3 kb in length, and larger amplicons are also possible.

The duplication process is not flawless, however. Based on the polymerase used, the error rate in generating copies varies from polymerase to polymerase (*Taq* DNA polymerase: $1–20 \times 10^{-5}$ errors/bp/duplication). The classic polymerase chain reaction uses 94°C for denaturation, 50°C for annealing, and 72°C for the extension step. Many variations of these conditions have been applied depending on the polymerase chain reaction method employed. Automatic thermal cyclers (PCR machines) are comprised of metal blocks wherein tubes containing the polymerase chain reactions can be inserted. Most polymerase chain reactions take 2 h on average for completion.

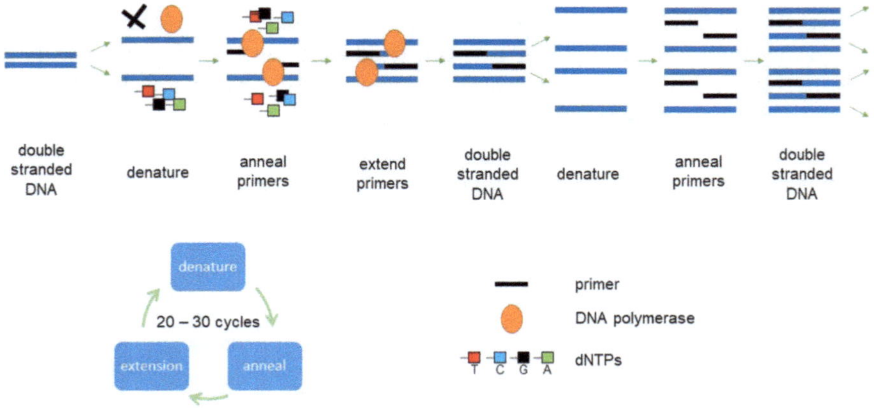

Fig. 4 Polymerase chain reaction (PCR) method

2.1 *Gel Analysis: Agarose and Polyacrylamide*

Polymerase chain reaction products, also referred to as "amplicons", can be analyzed by agarose gel electrophoresis. Agarose is a polymer formed into a gel-like substance with a similar consistency to gelatin, which, when hot, is poured into a casting mold holding a "comb", creating small pouches. Once cooled, the agarose solution takes the shape of the tray, and the thus formed "comb" will leave small indentations where the PCR product(s) can be loaded. The resulting gel is a matrix with small holes between each agarose molecule, representing effectively an obstacle course for the substances to be analyzed. The electrophoresis buffer used to prepare the agarose gel allows the DNA or polymerase chain reaction product(s) to be separated by their overall charge and size. The sample is loaded at the negative pole, and, by applying an electric field, will move towards the positively charged pole. The larger molecules move more slowly, while the smaller molecules move faster through the holes. The detection capability and sizing limits of agarose gel depends on the percentage of the gel. This procedure can separate DNAs up to 20 kb in size. A high agarose percentage of 2–3% allows the separation of fragments that differ by 20 bp in length of an overall size up to 500 bp. Polyacrylamide gel electrophoresis provides a higher resolution and is therefore very useful for separating small DNA fragments. This type of gel is made up of a cross-linked acrylamide polymer and, unlike the use of agarose gels, is cast vertically. Polyacrylamide gels can separate 5–500 bp fragments of DNA (Fig. 5).

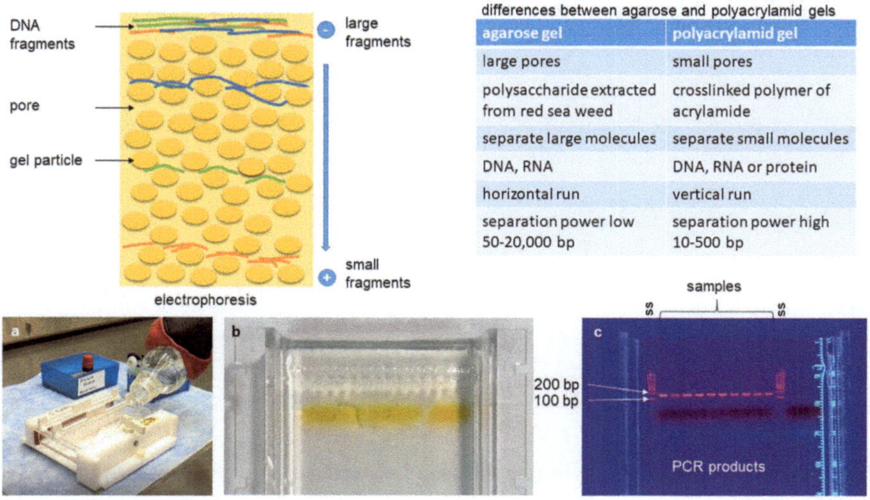

differences between agarose and polyacrylamid gels

agarose gel	polyacrylamid gel
large pores	small pores
polysaccharide extracted from red sea weed	crosslinked polymer of acrylamide
separate large molecules	separate small molecules
DNA, RNA	DNA, RNA or protein
horizontal run	vertical run
separation power low 50-20,000 bp	separation power high 10-500 bp

Fig. 5 Agarose gel versus polyacrylamide electrophoresis. Assembly and pouring of an agarose gel (**a**), orange DNA loading dye at the lower part of the gel helps determine how far samples move in the agarose gel (**b**), DNA, such in as PCR procedures, is visible under UV light (**c**). SS = molecular weight standard for DNA

3 Deoxyribonucleic Acid Dyes

Ethidium bromide is an intercalating agent that can accumulate between two DNA strands. The dye can be added when the gel is cast or be poured into a buffer to stain an agarose or polyacrylamide gel after electrophoresis. The DNA fragments are visualized under UV light and compared to a molecular size standard, which are run in gel pouches or wells next to the samples. Other types of dyes to stain DNA are available commercially (e.g. SYBR® Safe, GelRed™, GelStar™, and others).

4 Primers

Primers are essential for polymerase chain reactions, as they initiate DNA synthesis. Primers are stretches of DNA, and, depending on the type of experiment to be conducted, they are 6–40+ nucleotides in length and complementary to the target sequence (Fig. 6). The DNA polymerase can only begin synthesizing a complementary strand using nucleotides after a primer has bound to the template. Primers can be designed to match the template perfectly or to contain desired mutations (i.e. base changes, deletions, or insertions). Primer sequences and polymerase chain reaction conditions can be optimized to match specific targets. The resulting presence or absence of a polymerase chain reaction product can help to distinguish between specimens. Various software programs (e.g. Primer-BLAST, and Primer3Plus) are available to design primers for polymerase chain reaction applications, such as the use of species-specific primers.

```
DNA          5'-TCAGTGACGCGTTTCGCTAGGTATATCGTAACCGTATATAGGTCCTCCGACGGTTACCGTACCAGTCTGA-3'
             3'-AGTCAGTGCGCAAAGCGATCCATATAGCATTGGCATATATCCAGGAGGCTGCCAATGGCATGGTCAGACT-5'

forward         5'-GACGCGTTTCGCTAG-3'
primer       3'-AGTCACTGCGCAAAGCGATCCATATAGCATTGGCATATATCCAGGAGGCTGCCAATGGCATGGTCAGACT-5'

reverse      5'-TCAGTGACGCGTTTCGCTAGGTATATCGTAACCGTATATAGGTCCTCCGACGGTTACCGTACCAGTCTGA-3'
primer                                                        3'-CTGCCAATGGCATGG-5'

PCR product     5'-GACGCGTTTCGCTAGGTATATCGTAACCGTATATAGGTCCTCCGACGGTTACCGTACC-3'
                3'-CTGCGCAAAGCGATCCATATAGCATTGGCATATATCCAGGAGGCTGCCAATGGCATGG-5'
```

Fig. 6 Primers and resulting polymerase chain reaction product

5 Polymerase Chain Reaction Methods

5.1 *Simple Sequence Repeats, Microsatellites, Short Tandem Repeats, and Random Amplification of Polymorphic Deoxyribonucleic Acid*

Many variations of the classic polymerase chain reaction have now been developed to cater to the various needs of amplifying single targets, such as accepted barcodes or multiple gene regions using the random amplified polymorphic DNA (RAPD) (Fig. 7), simple sequence repeats (SSR), or microsatellite methods (Fig. 8).

Random amplification of polymorphic DNA (RAPD) (Fig. 7) is a polymerase chain reaction method that uses arbitrary primers that bind to random sites of DNA and then amplify it. The advantage of this method is that it can be used for amplification of any target DNA. However, it often lacks reproducibility, making it less reliable for identifying botanical samples.

Simple sequence repeats (SSRs), microsatellites, or short tandem repeats (STRs) are areas in the genome consisting of repetitive sequences of one to six base core sequences (mainly two to four). For example, the nucleotides A and C are repeated like beads on a necklace several times (8–50) as in, e.g. ACACACAC.

5.2 *Sequence Characterized Amplified Region Marker*

The reproducibility of polymerase chain reaction methods can be improved by converting fingerprint-like band patterns of random amplification of polymorphic

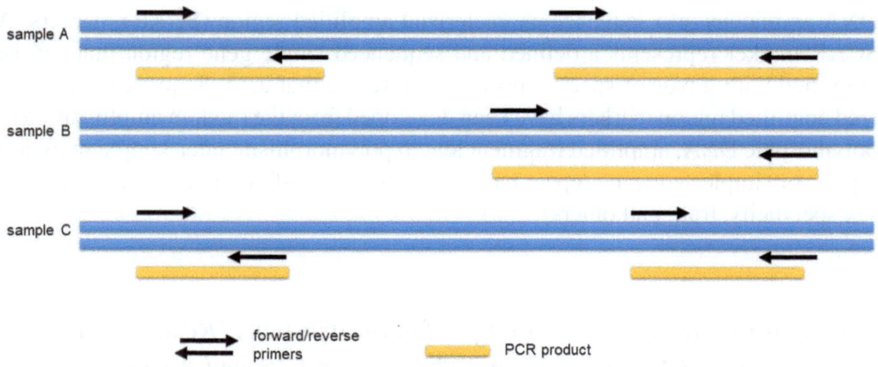

Fig. 7 Random amplification of polymorphic DNA (RAPD). The primers can bind to random locations in the genome and result in multiple PCR products. Analyzing the products on an agarose or polyacrylamide gel will lead to unique patterns of DNA pieces that help distinguish between samples

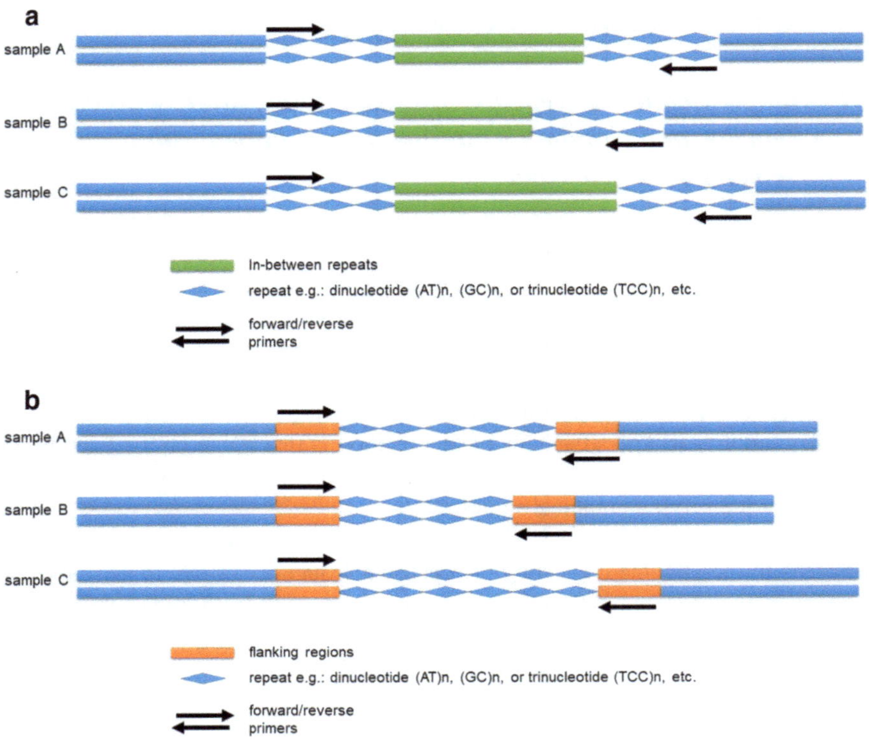

Fig. 8 Simple sequence repeats/microsatellite methods. Generation of multiple PCR products using primers binding to **a** repetitive sequences or their **b** flanking areas. Analyzing the products on an agarose or polyacrylamide gel will result in unique patterns of DNA pieces that help distinguish between samples

DNA amplicons into sequence characterized amplified region (SCAR) markers. A SCAR marker represents a defined and sequenced single gene region that can be amplified with a pair of specific primers tailored to that area. Sequence characterized amplified region markers have been developed from the random amplification of polymorphic DNA, amplified fragment length polymorphism, inter-simple sequence repeat, or simple sequence repeat band patterns to identify plant material (varieties, sex specificity, trait, and others).

5.3 Amplified Fragment Length Polymorphisms, Restriction Fragment Length Polymorphism, and Cleaved Amplified Polymorphic Sequence

The following methods generate unique patterns of DNA pieces, like barcodes helpful for keeping track of inventory or shipping packages. This allows the identification of

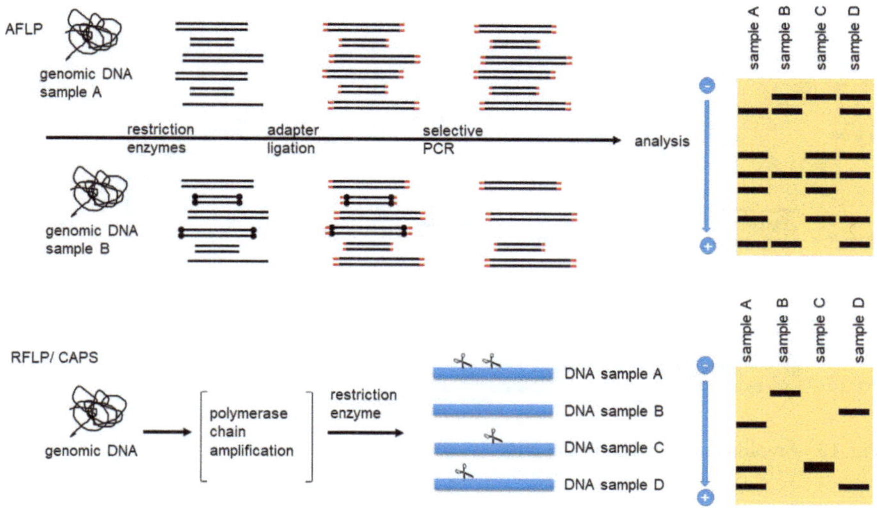

Fig. 9 Amplified fragment length polymorphism (AFLP), restriction fragment length polymorphism (RFLP), and cleaved amplified polymorphic sequence (CAPS)

individuals according to their patterns. The number of bands can range from several tens to several hundred, as generated from multiple loci.

Amplified fragment length polymorphism (AFLP) is a PCR-based fingerprinting technique where enzymes cleave genomic DNA, followed by ligation of adapters and selective PCR using specific primers. These primers recognize differences in a sequence caused by SNPs or Indels (Fig. 3) and result in the absence or presence of polymerase chain reaction products generating polymorphic fragments from DNA of different individuals [9, 10]. Amplified fragment length polymorphism is highly reproducible and can help distinguish between different species populations.

The method of restriction fragment length polymorphism (RFLP) is a technique in which specimens can be distinguished by the pattern of cleaved DNA, showing differences in the size of DNA fragments. It is based on the principle that a change in a DNA sequence can alter restriction enzyme recognition sequences, resulting in the absence or presence of a cutting site. This method can be applied to purify whole genomic DNA. The cleaved amplified polymorphic sequence (CAPS) method is an extension of RFLP where a PCR product is subjected to a restriction nuclease, resulting in a cleaved or non-cleaved product (Fig. 9).

5.4 Amplification Refractory Mutation System

The amplification refractory mutation system (ARMS) uses sequence-specific primers that can discriminate between template DNA samples that differ by one to five nucleotides. For example, ARMS can detect single base changes or small

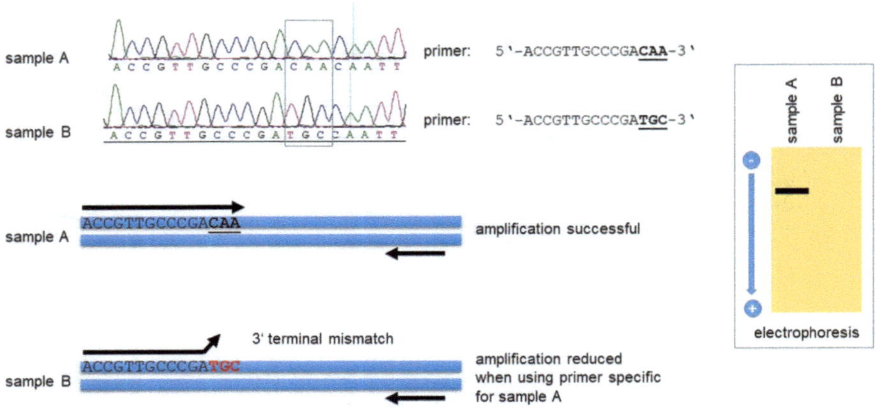

Fig. 10 Amplification refractory mutation system (ARMS)

deletions (see Fig. 3). The primers are designed so that the last nucleotide closest to the end (3′ end) that will be extended by the DNA polymerase is specific to the polymorphism to be detected. Only if the primers match the target DNA will a PCR product result. The absence or presence of a product can then be used to diagnose for the absence or presence of the target (Fig. 10).

5.5 High-Resolution Melt Analysis, Real-Time Polymerase Chain Reaction, and Quantitative Polymerase Chain Reaction

High-resolution melt (HRM) analysis (Fig. 11) is a special PCR method to study the thermal stability of an amplicon of < 200 bp. The first step is the polymerase chain reaction amplification in the presence of a fluorescent dye (intercalating dye) that appears brightly active when it occurs in between the two generated DNA strands. In the absence of double-stranded DNA, the fluorescence is only evident at a low level. The HRM analysis starts after the PCR process. The reaction temperature is increased from 50°C to about 95°C, causing the two DNA strands to separate or "melt". The amount of double-stranded DNA decreases due to "melting" and the fluorescence is reduced. A plot made of the level of fluorescence vs. the temperature results in a "melting curve". Identical DNA sequences will result in the same melting curve. Only DNA sequences with sequence differences, for example when due to a single nucleotide polymorphism, result in different melting curves and can distinguish between specimens. The most commonly used intercalating dye is SYBR® Green I.

Real-time detection of PCR products (RT-PCR) requires a fluorescent reporter molecule to be present when performing polymerase chain reactions. With an increasing amount of an amplicon, the fluorescence also increases and is monitored during amplification cycles. The amount of amplified polymerase chain reaction product is proportional to the measured fluorescence. Real-time PCR (75–150 bp) therefore can be used to either measure the presence or absence of a sequence (qualitative PCR) or the copy number (quantitative PCR) of the target DNA sequence (Figs. 12 and 13).

Both the SYBR Green and TaqMan assays are being used to monitor the amplification process of real time PCR by generating a fluorescence that can be measured. The difference between the two methods is that the SYBR green dye accumulates between the nucleic acid strands. The TaqMan method uses a fluorogenic probe, where the fluorophore is released when the probe binds to the target that accumulates during PCR.

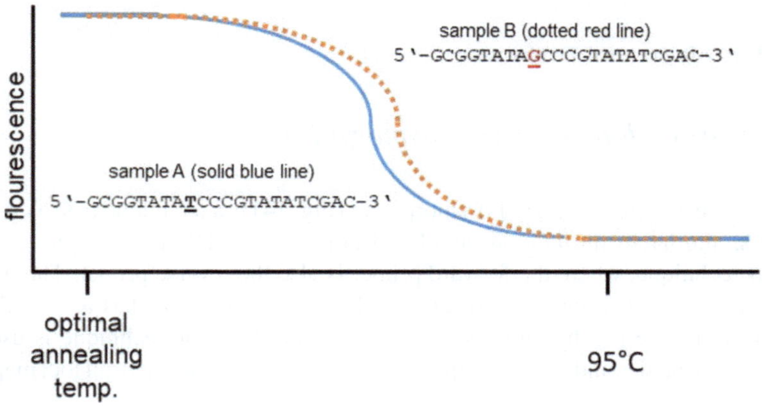

Fig. 11 High-resolution melt (HMR) analysis

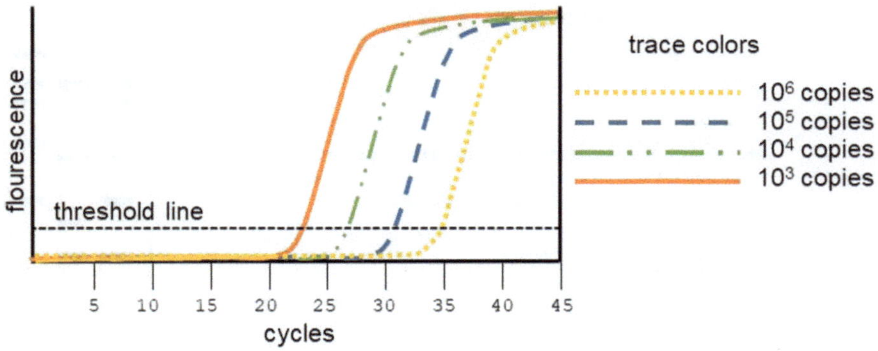

Fig. 12 Real-time polymerase chain reaction (RT-PCR), quantitative polymerase chain reaction (qPCR)

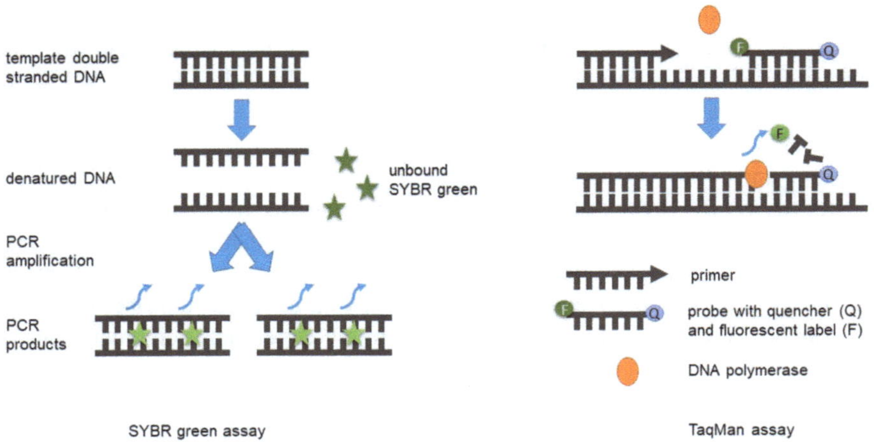

Fig. 13 SYBR green and TaqMan assay

5.6 *Start Codon Targeted Polymorphism*

Start Codon Targeted (SCoT) Polymorphism (Fig. 14) is a technique based on single primers, similar to the random amplified polymorphic DNA or simple sequence repeats technique, where the forward primer is also the reverse primer. The primers target the conserved regions surrounding the ATG translation start and work well on genes that are nearby on opposite strands. Currently, the technique is used for population genetics and cultivar improvement, such as quantitative trait loci mapping.

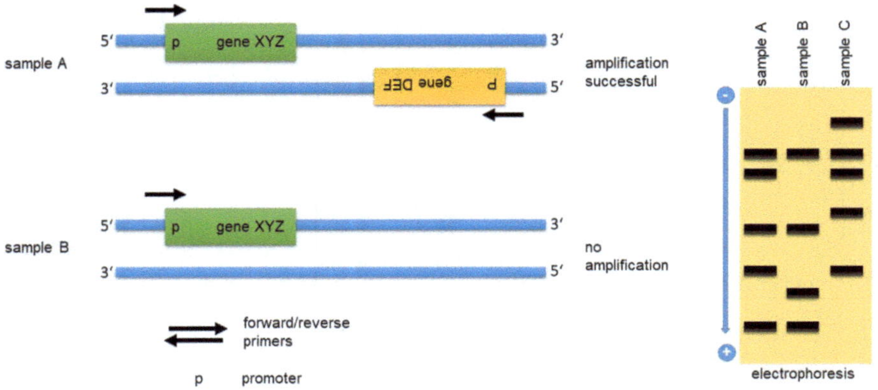

Fig. 14 Start codon targeted (SCoT) polymorphism

6 Genomic Regions for Botanical Identification

Various chloroplast genes have been used for species identification and authentication, such as *rbcL* (ribulose-bis-phosphate carboxylase), *matK* (maturase K), *ycf5*, *rpoC1*, *ycf1b*, *rpoB*, and the non-coding chloroplast region between genes such as *psbA-trnH*, *atpF-atpH*, and *trnL-trnF*. The nuclear genomic regions ITS (or part of it: ITS1 and ITS2), low copy nuclear gene *WAXY* coding for granule-bound starch synthase (*GBSS1*), and the transcription factor *LEAFY* (*LFY*) have also been used for species identification.

For example, Thakur et al. analyzed the *matK*, *psbA-trnH, and trnL* of some medicinally important plant species of the family Lamiaceae in India [11]. In turn, Nithaniyal et al. analyzed *rbcL* and *matK* barcodes from 521 medicinal plant species and found that the species resolution was 74.4%, 90.2%, and 93.0% for *rbcL, matK,* and *rbcL* + *matK*, respectively [12]. The nuclear ribosomal internal transcribed spacers and chloroplasts *rbcL, matK*, *psbA-trnH,* and *trnL-trnF* were screened for their potential as DNA barcodes for the identification of 48 *Phyllanthus* taxa used medicinally in Brazil [13]. Four genomic regions, *rbcL, matK, trnH-psbA,* and ITS2, were analyzed to help identify market samples of Ashwagandha (*Withania somnifera*), also known as Indian Ginseng. Authentication was successful using ITS2 and *trnH*-psbA [14].

DNA authentication of St John's wort (*Hypericum perforatum*) was performed using a developed qPCR assay based on a species-specific primer pair spanning the ITS1 and ITS2 regions and a next-generation sequencing (NGS) assay [15]. Differentiation of *Mitragyna speciosa*, a narcotic plant, from allied *Mitragyna* species using DNA barcoding-high-resolution melting (Bar-HRM) analysis was performed [16]. Five Licorice (*Glycyrrhiza*) species were analyzed using high-resolution melting (HRM) analysis of nine simple sequence repeats (SSR) [17]. The complete chloroplast genome was analyzed to find barcodes for the three Apiaceae species *Ledebouriella seseloides, Peucedanum japonicum*, and *Glehnia littoralis* [18]. Sequence characterized amplified region markers were identified to authenticate and distinguish between the herbal medicinal plants *Dipsacus asper*, *D. japonicus* (Caprifoliaceae), and *Phlomoides umbrosa* (Lamiaceae) [19]. Kreuzer et al. analyzed $\sim$ 160 kbp of chloroplast DNA for 57 species and identified sequences for barcoding that are clade-specific rather than species-specific [20].

When cultivars of a species need to be distinguished, the sequence differences in the genome may be rare. Zhang analyzed three *Pogostemon cablin* (Blanco) Benth. (Patchouli) cultivars by sequencing the plastid genomes [21]. One single variable locus, cpSSR, was found to be able to help distinguish between the cultivars.

A combination of DNA barcoding and chemical profiling can be helpful in characterizing either a species or a variety of a species. Deoxyribonucleic acid barcoding and ultra-high-performance liquid chromatography/triple quadrupole mass spectrometry was used to analyze 33 samples of valerian, linden, tea, and chamomile [22]. Six genomic regions (*matK*, *rbcL*, *rpoC1*, *psbA-trnH*, ITS1, and ITS2) were analyzed in addition to using thin-layer chromatography of four different species of the genus

Galphimia that produced galphimines [23]. A method combining DNA barcoding (ITS) and high-performance liquid chromatography was applied to help distinguish Fritillariae Cirrhosae Bulbus (known as "chuanbeimbu" in Chinese traditional medicine) from other types of dried Fritillariae Bulbus [24].

7 Sequencing Methods

Several methods are available to derive the sequence of DNA and DNA fragments, and these are described briefly in the following subsections.

7.1 *Sanger Sequencing*

The initial method of sequencing was the Sanger dideoxy synthesis method, developed in 1977, and is still in use for many current applications where high throughput is not required [25, 26]. It generates about 600–1,000 bases of sequences. This method uses specific (dideoxy) nucleotides that terminate chain extension. These nucleotides are labeled radioactively or fluorescently for detection. The template DNA is copied in a similar manner to a PCR, with the difference being that the elongation stops at random locations due to the incorporation of the dideoxy nucleotides. This method generates a pool of DNA fragments with a chain-terminating labeled nucleotide at every position of the target DNA. These fragments are analyzed using slab or capillary gel electrophoresis, in which shorter fragments run faster than longer fragments. At the end of the capillary is a sensor detecting the attached label. The identified label is read and the DNA sequence deducted (Fig. 15).

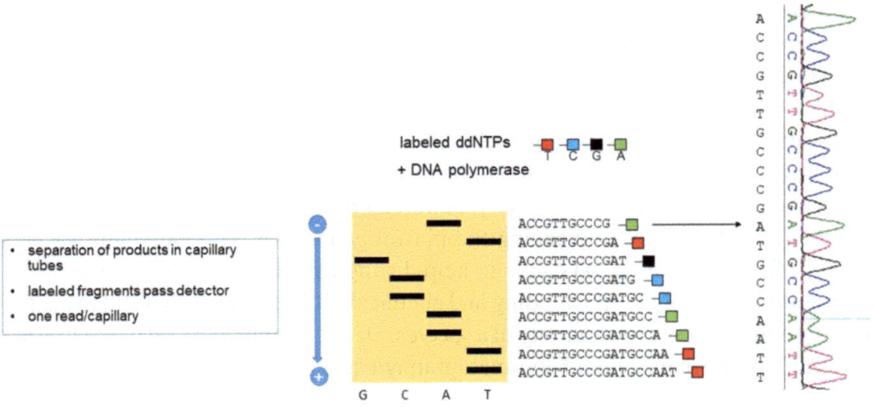

Fig. 15 Sanger sequencing and a resulting chromatogram

7.2 *Second- or Next-Generation Sequencing*

With next-generation sequencing, as described in the following subsections, large quantities of DNA, such as whole genomes or transcriptomes, can be sequenced at the same time (in parallel) more quickly and cheaply. Second-generation sequencing can be classified into two groups: (a) sequencing by hybridization and (b) sequencing by synthesis. Reactions are run in parallel, often in tiny amounts, such as in nanoliter or picoliter volumes.

7.2.1 Sequencing by Hybridization

Sequencing by hybridization works by repeatedly hybridizing and washing away unwanted non-matching DNA to an array of oligonucleotides arranged on filters. This method has been utilized successfully for diagnostic approaches, such as identifying single nucleotide polymorphisms in specific gene regions that are linked to diseases or identifying chromosome abnormalities.

7.2.2 Sequencing by Synthesis

Sequencing by synthesis (Fig. 16) works by either securing DNA to a substrate or by distributing millions of individual DNA molecules into separate wells or chambers. This approach has a much higher error rate than the original Sanger sequencing method, but due to the large number of sequences covered (i.e. referring to how many times a location has been sequenced repeatedly, as in "parallel sequencing"), this helps identify sequencing errors. These errors can often be corrected by aligning the sequences into a consensus sequence, with the reads being shorter (50–300 bases) than the original Sanger sequences (600–1,000 bases). A further difference from Sanger sequencing is that the incorporation of labeled nucleotides is ongoing during DNA synthesis and is not terminated. Each incorporation is detected and recorded.

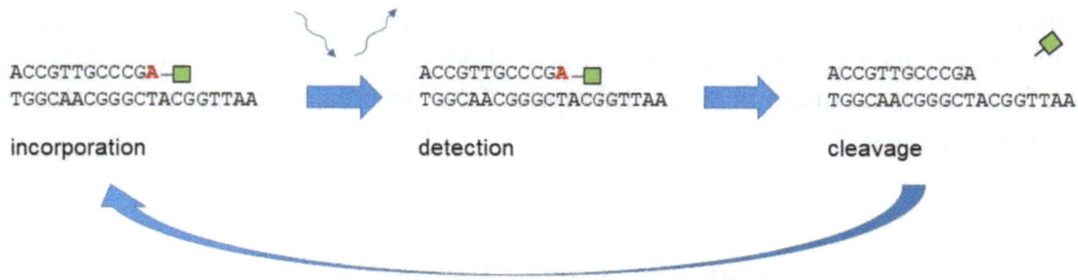

Fig. 16 Sequencing by synthesis

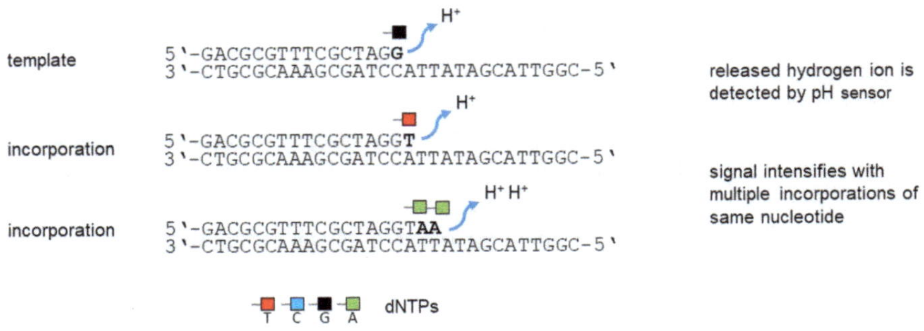

Fig. 17 Ion torrent sequencing

7.2.3 Ion Torrent Sequencing

Ion torrent sequencing reactions (Fig. 17) occur in millions of wells that occur on a semiconductor chip and enable large numbers of parallel sequencing steps. This type of sequencing is based on incorporating nucleotides, which release hydrogen ions, thereby changing the pH of the solution, which then can be measured. If no pH change occurs, then a nucleotide is not incorporated. The DNA sequence can be determined by constantly washing and flooding the sequencing chamber with one of the four nucleotides simultaneously and measuring the resultant pH.

7.2.4 Illumina Technology

Illumina technology requires the use of randomly fragmented DNA of a determined size of about 500 bp with ligated adapters at each end. From each DNA fragment, copies are made in clonal "clusters" using fluorescently labeled nucleotides that terminate the DNA synthesis for each reaction. Illumina bridge amplification is illustrated in Fig. 18. Contrary to Sanger sequencing, the nucleotides are unblocked for the next round so that a newly labeled nucleotide can be incorporated. This process is repeated for approximately 50–300 rounds. Many applications make use of Illumina technology, such as whole genome sequencing, RNA sequencing, targeted sequencing, and exome sequencing. This method requires special care to be taken to perform the exact quantification of template DNA to create the optimal number of sequencing clusters.

7.3 Third- and Fourth-Generation Sequencing

Third-generation sequencing moves away from sequencing short fragments in favor of very long molecules, and can use both DNA and RNA (ribonucleic acid) as templates while maintaining "parallel sequencing".

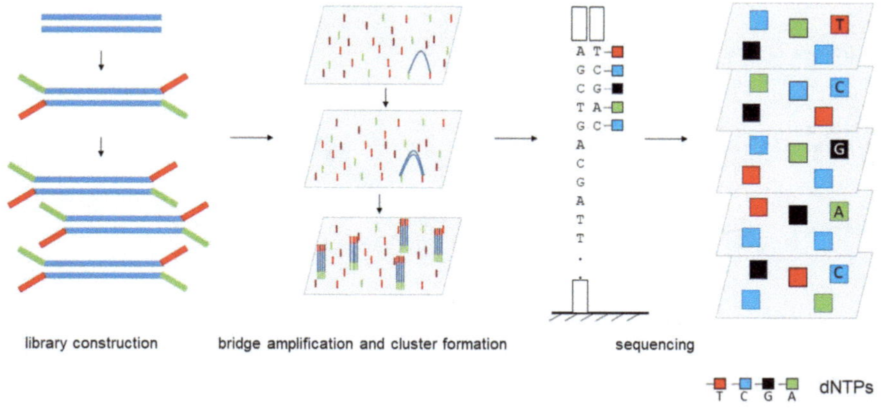

Fig. 18 Illumina bridge amplification. The DNA attaches to the flow cell via complementary sequences. The strand bends over and attaches to a second oligo forming a bridge. A polymerase synthesizes the reverse strand. The two strands release and straighten. Each forms a new bridge (bridge amplification). The result is a cluster of DNA forward and reverses strand clones

Single-molecule real-time (SMRT) sequencing utilizes an engineered DNA polymerase and binds it with DNA to the bottom of a well. The DNA polymerase then incorporates fluorescently labeled nucleotides, each with a different phosphor-linked fluorophore, and when a nucleotide is incorporated, an image is taken. After incorporation, the fluorophore is released and then is no longer detectable. With this method, as developed by PacBio (Pacific Biosciences of California, Inc.), very long circularized DNA fragments can be sequenced repeatedly, with an average read length of 10–15 kb and up to 30–50 kb or more in parallel (Fig. 19). The minimum sequence length is 250 bp. Moreover, DNA sequences that include modified bases, such as methylated adenine and cytosine, can also be identified. The high error rate of this method can be overcome by sequencing the template multiple times, aligning the reads, and deducing its consensus sequence. This single molecule real-time technique has been shown to be useful for genome assemblies, such as complete bacterial genomes.

Fourth-generation sequencing utilizes the change of current when a nucleotide passes through a small hole, i.e. a pore. Each nucleotide that passes through the pore results in a characteristic current disruption, which can be measured, and the responsible nucleotide identified. This method can recognize some base modifications. Oxford Nanopore Technologies has commercialized this nanopore-based sequencing procedure (Fig. 20) and it has been used to sequence both RNA and DNA. Several enzymes are needed for this process, such as a highly processive DNA polymerase (phi29), as well as an accessory protein DNA helicase, exonuclease 1, to help unwind and move the then single-stranded template through the nanopore. The recording occurs in real-time with an average of 10 kb read length and can be performed on a small, cigarette lighter-sized device. Once a template fragment has passed through the pore, it can be reused for another molecule. The optimal DNA

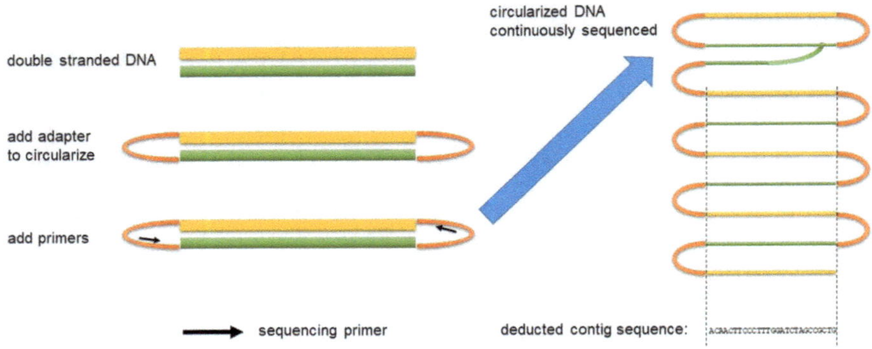

Fig. 19 PacBio single molecule real-time (SMRT) amplification

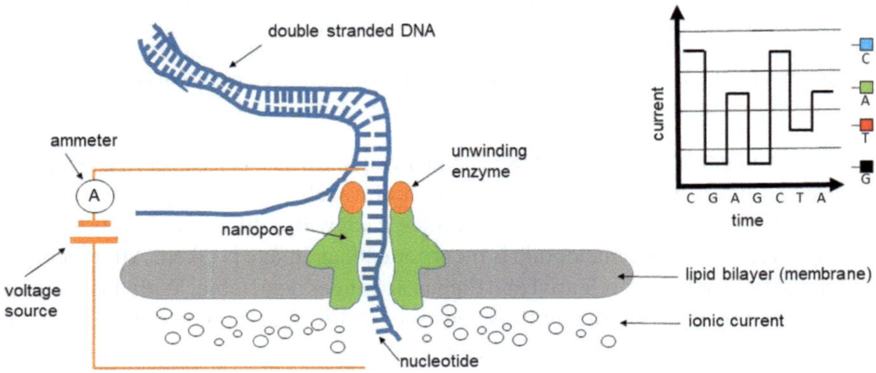

Fig. 20 Nanopore sequencing. Single-stranded DNA is passed through the pore with the help of an unwinding and motor enzyme. Changes in the current while the DNA passes through the pore help identify the nucleotides. This design allows long DNA molecules to pass through and the sequencing of both strands

size for sequencing is 200 bp to 8 kb. Shorter sequences than 200 bp cannot be identified, since their detection and base-calling are not possible. Nanopore sequencing technology has been used for metagenomic and environmental samples.

7.4 Amplicon Metabarcoding and Whole Genome Sequencing

Amplicon metabarcoding uses targeted sequencing, as for example, to sequence the ITS2 region of all the DNA in a mixed plant material sample using next-generation/high-throughput sequencing. However, this can fail due to fragmented DNA, e.g. as found in processed plant material. Whole genome sequencing, including low-coverage genome skimming, has the advantage of having no need for a specific marker selection, and reduces negative bias associated with amplification by the polymerase chain reaction. Whole genome sequencing data may need to be reduced to more limited data, such as chloroplast reads, due to the lack of annotated genomes of a species of interest or from using non-universal identifying markers for model clades of investigated botanicals.

8 Deoxyribonucleic Acid Quality and Quantity

For some DNA-based detection or sequencing methods, it is imperative to have good quality DNA and in sufficient quantity (Fig. 21). Unfortunately, there are large differences in the outcome of DNA extraction processes, depending on the starting material. Raw, freshly harvested plant material is usually the best source. It is important to avoid degradation due to unfavorable storage conditions (UV, light, temperature, bacteria, fungi) for experiments requiring high-quality DNA and under good laboratory practice to avoid cross-contamination with other samples. Plant material that has been dried, ground into a powder, or further processed to create botanical extracts then shows both a lower quality and quantity of DNA. Storage time, storage conditions, and processing methods affect the quality and quantity of DNA in the plant material. Ragupathy et al. documented an investigation of DNA from raw botanicals to botanical extracts and found that the latter had the lowest quality and highest degree of degradation present [27]. Plant metabolites or excipients mixed with plant material (raw or extracts) can have an influence on DNA extraction and downstream applications. Numerous optimized protocols have been published to remove potentially hindering metabolites such as polysaccharides, flavonoids, other phenols, and

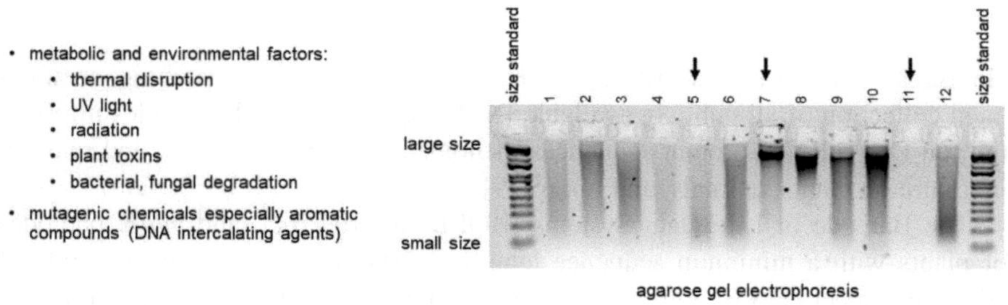

Fig. 21 DNA quality and quantity. High-quality DNA is visible as a thick band at the top of the agarose gel (sample #7), low-quality DNA is fragmented and shows as a smear (sample #5). Some samples, such as heavily processed plant material found in tablets or capsules, may not result in detectable DNA (sample #11)

terpene lactones. There are also several commercially available kits for various needs, such as DNA extraction from blood, food, and plants.

9 Barcoding of Fragmented DNA

Finding useful gene regions to help with species identification is even a bigger challenge when it is necessary to be applied to damaged DNA that is found often in processed plant material. Routinely, these DNA specimens are of low quality and quantity, and therefore not suitable to target for amplification of full-length barcodes that are often longer than 300 bp. The damaged DNA is present in fragments, often only allowing the amplification of polymerase chain reaction products of $\leq$ 250 bp. Developing DNA mini-barcodes helps to overcome this problem. Here, primers are designed to amplify small PCR products of roughly 70–250 bp that can use fragmented DNA as templates. The development of these mini-barcodes and their corresponding primers is often difficult, as several requirements need to be fulfilled, involving primer length, primer composition, and primer melting temperature (T_M). Also, there is a need to avoid dimer formation that might result in a small amplicon with a characteristic sequence variation. If the designed primers are not species-specific, the resulting amplicons may need to be sequenced to show the differences. Direct Sanger sequencing of the $\leq$ 250 bp amplicons is not recommended, as this method requires at least 400 bp for high-quality results. Cloning the products and then sequencing the transformants may be an option, but this adds time and costs in identifying a single species sample.

10 Deoxyribonucleic Acid Databases

GenBank® is the National Institutes of Health genetic sequence database, an annotated collection of all publicly available DNA sequences [28]. It provides information on DNA, RNA, protein sequences, genomes, available software, and relevant literature on plants, animals, and fungi [29].

The Barcode of Life Data Systems (BOLD) affords *rbcL* and *mat*K barcodes of plants with a minimum sequence length of 500 bp [30, 31]. The Barcode of Wildlife Project (BWP) aims to provide a public, free-for-use reference library of DNA barcodes for 2,000 endangered species, both plants and animals, as protected under the Convention on International Trade in Endangered Species of Wild Fauna and Flora (CITES) [32]. The partner countries Kenya, Mexico, Nigeria, and South Africa have selected 774 priority species for this project.

The International Barcode of Life (IBOL) has overseen the completion of one major program of barcoding 500,000 species, by investing $150 million from research organizations in 25 nations [33]. Plant barcoding studies are based on *rbcL*, *matK*, *trnH-psbA*, and the internal transcribed spacer. Animal barcoding studies focus on *COXI*, and fungal studies are based on the internal transcribed spacer (ITS).

The Medicinal Materials DNA Barcode Database (MMDBD) provides 65,520 sequences of 2,225 species (accessed 6/28/2022) of nuclear, mitochondrial, and chloroplast genomic regions to help identify and authenticate plant and animal species [34, 35].

Vassou et al. analyzed the *rbcL* barcodes from 347 medicinal plants used in Ayurvedic medicine [36]. In addition, the *rbcL* barcodes of 27 medicinal plants from public databases were used to generate a Reference DNA Barcode Library (API-RDBL) to authenticate 100 raw plant materials in the form of powders and seeds.

Gong et al. constructed a DNA barcode reference library for herbaceous plants occurring in southern China [37]. The database contains the internal transcribed spacer 2 sequences of 359 species, and offers a 67.56% success rate in terms of species identification, and a 96.57% success rate at the genus level.

In Fig. 22, an example of a plant species identification process using DNA barcoding is illustrated.

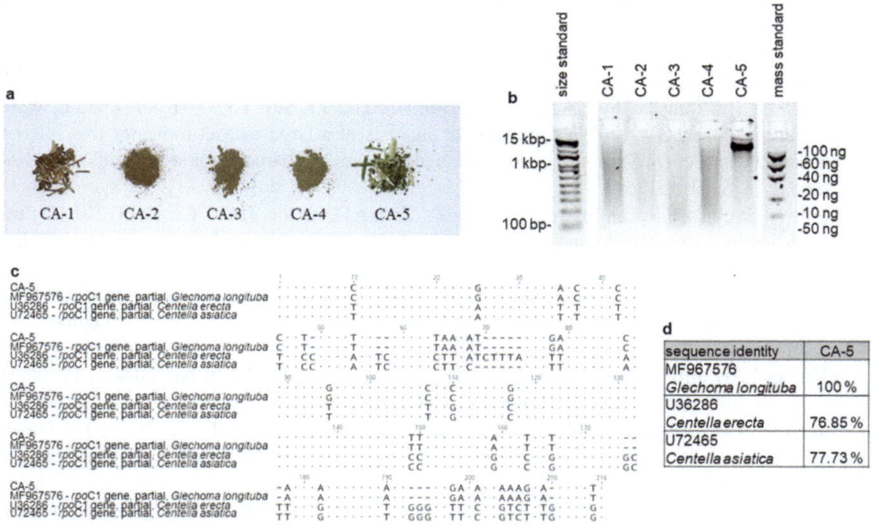

Fig. 22 Analysis for species identification of purchased coarse and ground sample material claimed to be a *Centella* sp. (**a**). Whole genomic DNA was isolated and analyzed on an agarose gel for quality and quantity. The smear obtained indicated low-quality DNA down the lane. Only sample CA-5 resulted in high quality and quantity DNA (**b**). PCR amplified a part of the gene region *rpo*C1A, and the resulting products as sequenced using the Sanger method (**c**). The analysis and comparison with available sequences showed that the derived *rpo*C1 sequence of sample CA-5 had only a 77% identity to *Centella* species, but rather a 100% identity to *Glechoma longituba* (**d**). It was concluded that CA-5 is not a *Centella* sp. The samples CA-1 to CA-4 showed a 100% sequence identity to *Centella* species (not shown). Dashes in the sequence alignments are deletions in the sequence, and the dots indicate identical bases found in all aligned sequences

11 Conclusions

The DNA extraction or PCR method to be applied to a plant species sample of interest must be chosen carefully to avoid either a false-negative or a false-positive identification. Despite the large number of amplification and sequencing methods available today, no single genomic region or method works for every plant for its identification. While a particular genomic region may work for a certain plant genus to help distinguish a species, the same region may not show diagnostic sequence differences of a genus from another family. With the increase in high-throughput sequencing of whole genomes or transcriptomes, more data on species of interest will become available at Genbank and at the 1000 Plants initiative in the future [28, 38]. This enhanced data pool shows promise for mining with state-of-the-art software to find genomic region(s) that can help identify botanicals more effectively.

References

1. Hebert PD, Stoeckle MY, Zemlak TS, Francis CM (2004) Identification of birds through DNA barcodes. PLoS Biol 2:e312
2. International Barcode of Life Conference (2011) http://www.dnabarcodes2011.org/
3. China Plant BOL Group, Li DZ, Gao LM, Li HT, Wang H, Ge XJ, Liu JQ, Chen ZD, Zhou SL, Chen SL, Yang JB, Fu CX, Zeng CX, Yan HF, Zhu YJ, Sun YS, Chen SY, Zhao L, Wang K, Yang T, Duan GW (2011) Comparative analysis of a large dataset indicates that internal transcribed spacer (ITS) should be incorporated into the core barcode for seed plants. Proc Natl Acad Sci USA 108:19641
4. Kane N, Sveinsson S, Dempewolf H, Yang JY, Zhang D, Engels JMM, Cronk Q (2012) Ultra-barcoding in cacao (*Theobroma* spp.; Malvaceae) using whole chloroplast genomes and nuclear ribosomal DNA. Am J Bot 99:320
5. Vaughn JN, Chaluvadi SR, Rangan LT, Bennetzen JL (2014) Whole plastome sequences from five ginger species facilitate marker development and define limits to barcode methodology. PLoS One 9:e108581
6. Coissac E, Hollingsworth PM, Lavergne S, Taberlet P (2016) From barcodes to genomes: extending the concept of DNA barcoding. Mol Ecol 25:1423
7. Zhang N, Erickson DL, Ramachandran P, Ottesen AR, Timme RE, Funk VA, Luo Y, Handy SM (2017) An analysis of *Echinacea* chloroplast genomes: implications for future botanical identification. Sci Rep 7:216
8. Manzanilla V, Kool A, Nhat LN, Van HN, Thu HLT, de Boer HJ (2018) Phylogenomics and barcoding of *Panax*: toward the identification of ginseng species. BMC Evol Biol 18:44
9. Vos P, Hogers R, Bleeker M, Reijans M, van de Lee T, Hornes M, Frijters A, Pot J, Peleman J, Kuiper M (1995) AFLP: a new technique for DNA fingerprinting. Nucleic Acids Res 23:4407
10. Zabeau M, Vos P (1993) Selective restriction fragment amplification: a general method for DNA fingerprinting. Eur Pat Office Publ 0 534 858 Al
11. Thakur VV, Tripathi N, Tiwari S (2021) DNA barcoding of some medicinally important plant species of Lamiaceae family in India. Mol Biol Rep 48:3097
12. Nithaniyal S, Vassou SL, Poovitha S, Raju B, Parani M (2017) Identification of species adulteration in traded medicinal plant raw drugs using DNA barcoding. Genome 60:139
13. Inglis PW, Mata LR, da Silva MJ, Vieira RF, Alves R de BN, Silva DB, Azevedo VCR (2018) DNA barcoding for the identification of *Phyllanthus* taxa used medicinally in Brazil. Planta Med 84:1300

14. Amritha N, Bhooma V, Parani M (2020) Authentication of the market samples of Ashwagandha by DNA barcoding reveals that powders are significantly more adulterated than roots. J Ethnopharmacol 256:112725
15. Howard C, Hill E, Kreuzer M, Mali P, Masiero E, Slater A, Sgamma T (2019) DNA authentication of St John's wort (*Hypericum perforatum* L.) commercial products targeting the ITS region. Genes 10:286
16. Tungphatthong C, Urumarudappa SKJ, Awachai S, Sooksawate T, Sukrong S (2021) Differentiation of *Mitragyna speciosa*, a narcotic plant, from allied *Mitragyna* species using DNA barcoding-high-resolution melting (Bar-HRM) analysis. Sci Rep 11:6738
17. Li J, Xiong C, He X, Lu Z, Zhang X, Chen X, Sun W (2018) Using SSR-HRM to identify closely related species in herbal medicine products: a case study on Licorice. Front Pharmacol 9:407
18. Lee HO, Joh HJ, Kim K, Lee SC, Kim NH, Park JY, Park HS, Park MS, Kim S, Kwak M, Kim K-Y, Lee WK, Yang T-J (2019) Dynamic chloroplast genome rearrangement and DNA barcoding for three Apiaceae species known as the medicinal herb "Bang-poong". Int J Mol Sci 20:2196
19. Park I, Yang S, Kim WJ, Noh P, Lee HO, Moon BC (2018) Authentication of herbal medicines *Dipsacus asper* and *Phlomoides umbrosa* using DNA barcodes, chloroplast genome, and sequence characterized amplified region (SCAR) marker. Molecules 23:1748
20. Kreuzer M, Hawkins JA, Howard C, Adhikari B, Pendry CA (2019) Phylogenomic approaches to DNA barcoding of herbal medicines: developing clade-specific diagnostic characters for *Berberis*. Front Plant Sci 10:586
21. Zhang C, Liu T, Yuan X, Huang H, Yao G, Mo X, Xue X, Yan H (2019) The plastid genome and its implications in barcoding specific-chemotypes of the medicinal herb *Pogostemon cablin* in China. PLoS One 14:e0215512
22. Sanchez M, Gonzalez-Burgos E, Divakar PK, Gomez-Serranillos MP (2020) DNA-based authentication and metabolomics analysis of medicinal plants samples by DNA barcoding and ultra-high-performance liquid chromatography/triple quadrupole mass spectrometry (UHPLC-MS). Plants 9:1601
23. Gesto-Borroto R, Cardoso-Taketa A, Yactayo-Chang JP, Medina-Jimenez K, Hornung-Leoni C, Lorence A, Villarreal ML (2019) DNA barcoding and TLC as tools to properly identify natural populations of the Mexican medicinal species *Galphimia glauca* Cav. PLoS One 14:e0217313
24. Zhong Y, Wang H, Wei Q, Cao RZ, Hailong HY, Wang L (2019) Combining DNA barcoding and HPLC fingerprints to trace species of an important traditional Chinese medicine Fritillariae Bulbus. Molecules 24:3269
25. Sanger F, Coulson AR (1979) A rapid method for determining sequences in DNA by primed synthesis with DNA polymerase. J Mol Biol 94:441
26. Sanger F, Nicklen S, Coulson AR (1977) DNA sequencing with chain-terminating inhibitors. Proc Natl Acad Sci USA 74:5463
27. Ragupathy S, Faller AC, Shanmughanandhan D, Kesanakurti P, Shaanker RU, Ravikanth G, Sathishkumar R, Mathivanan N, Song J, Han J, Newmaster S (2019) Exploring DNA quantity and quality from raw materials to botanical extracts. Heliyon 5:e01935
28. Genbank. www.ncbi.nlm.nih.gov
29. Benson DA, Cavanaugh M, Clark K, Karsch-Mizrachi I, Lipman DJ, Ostell J, Sayers EW (2013) GenBank. Nucl Acids Res 41:D36
30. The Barcode of Life Data Systems (BOLD). https://boldsystems.org
31. Ratnasingham S, Hebert PDN (2007) BOLD: the barcode of life data system. Mol Ecol Notes 7:355. www.barcodinglife.org
32. The Barcode of Wildlife Project (BWP). www.barcodeofwildlife.org/cbol.html
33. The International Barcode of Life (IBOL). https://ibol.org
34. Medicinal Materials DNA Barcode Database (MMDBD). https://rdccm.cuhk.edu.hk/mherbsdb
35. Wong TH, But GW, Wu HY, Tsang SS, Lau DT, Shaw PC (2018) Medicinal Materials DNA Barcode Database (MMDBD) version 1.5-one-stop solution for storage, BLAST, alignment and primer design. Database (Oxford):bay112

36. Vassou SL, Nithaniyal S, Raju B, Parani M (2016) Creation of reference DNA barcode library and authentication of medicinal plant raw drugs used in Ayurvedic medicine. BMC Complement Altern Med 16(Suppl 1):186
37. Gong L, Qiu XH, Huang JXW, Bai JQ, Zhang J, Su H, Xu CM, Huang ZH (2018) Constructing a DNA barcode reference library for southern herbs in China: a resource for authentication of southern Chinese medicine. PLoS One 13:e0201240
38. The 1000 plants (OneKP or 1KP) initiative. www.onekp.com

Natascha Techen received her B.S., M.S., and Ph.D. degrees in Plant Molecular Biology at the University of Hamburg in Germany. She joined the National Center for Natural Products Research at the University of Mississippi, USA in 2002. She is working as a member of the group of Prof. Ikhlas A. Khan on the identification of genomic markers, also known as "DNA-barcoding", which can help with the authentication of plant material and dietary supplements and their adulteration using various molecular methods. In addition, her research has helped with the genetic improvement of crops, medicinal, and ornamental plants for higher yield and nutrient value, disease and pest tolerance, non-invasiveness, and ornamental traits. She is currently working on the development of DNA mini-barcodes (< 250 bp) to assist with identification using fragmented DNA present in processed plant material.

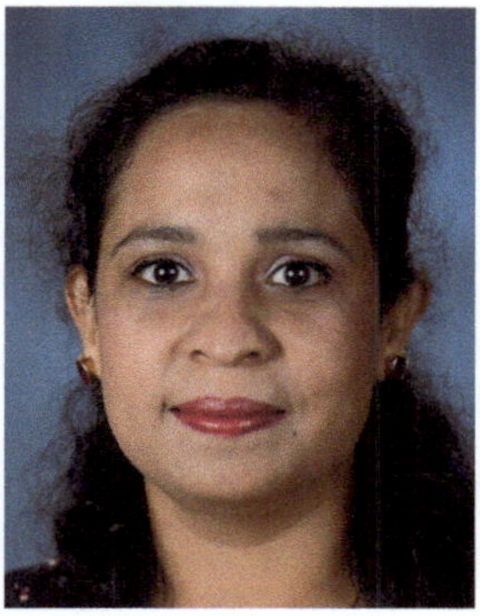

Iffat Parveen is a Research Scientist at the National Center for Natural Products Research (NCNPR), University of Mississippi, USA. Her passion for the plants began as a student exploring the botanical gardens of Northern India for botanical field trips. She graduated with a Ph.D. degree in Botany from the University of Delhi in India. Her doctoral research was focused on developing molecular methods such as DNA barcodes for the identification of different orchids, an economically valuable group of ornamental plants. Her main research areas focus on plant biodiversity, taxonomy, and molecular systematics. Due to her expertise in plant molecular biology she joined NCNPR as a postdoctoral fellow. At NCNPR, her research work is focused mainly on the identification and authentication of botanicals and medicinal plants using different molecular markers and a DNA barcoding approach.

Ikhlas A. Khan is Research Professor and Director of the both the National Center for Natural Products Research (NCNPR) and the FDA Center of Excellence, as well as Distinguished Professor of Pharmacognosy in the Department of BioMolecular Sciences, all at the University of Mississippi. He also serves as Coordinator for Natural Products Research in the Center for Water and Wetland Diseases. Prof. Khan received B.S. (Chemistry, 1980) and M.S. (Organic Chemistry, 1982) degrees from the Aligarh Muslim University, Aligarh, India and a Ph.D. degree in Pharmacy from the Institute für Pharmaceutische Biology in Munich, West Germany in 1987. He then performed postdoctoral studies at the Swiss Federal Institute of Technology (ETH) in Zurich, and was appointed initially as a Research Scientist at NCNPR in 1992. Prof. Khan's primary research interests include analytical fingerprinting for the standardization of herbal products, bioanalytical approaches to the improvement of product quality and safety, microbial biotransformation, and the isolation of natural products of potential commercial value. Prof. Khan has authored or co-authored over 840 original research and review articles. He has presented invited lectures in many countries, and has served on numerous committees, scientific boards, and task forces.

Prof. Khan has received many awards and honors throughout his career, including fellowships of the American Institute of Chemists and of the Royal Society of Chemistry (London), and is an Honorary Member of the American Society of Pharmacognosy (ASP). He was awarded the Varro E. Tyler Award of ASP (2011) and the U.S. FDA Center for Food Safety and Applied Nutrition (CFSAN) Directors' Special Citation Award (2012) as well as the Wiley Award of the Association of Official Agricultural Chemists International (2018). He holds honorary or adjunct professor titles at several universities, including the Chinese University of Hong Kong, Heilongjiang University of Chinese Medicine, Hunan University of Chinese Medicine, and Soochow University in the People's Republic of China, and King Saud University in Saudi Arabia.

MIX
Papier aus verantwortungsvollen Quellen
Paper from responsible sources
FSC® C105338

If you have any concerns about our products,
you can contact us on
ProductSafety@springernature.com

In case Publisher is established outside the EU,
the EU authorized representative is:
Springer Nature Customer Service Center GmbH
Europaplatz 3, 69115 Heidelberg, Germany

Printed by Libri Plureos GmbH
in Hamburg, Germany